Research Notes in Mathematics

Submission of proposals for consideration
Suggestions for publication, in the form of outlines and representative samples, are invited by the editorial board for assessment. Intending authors should contact either the main editor or another member of the editorial board, citing the relevant AMS subject classifications. Refereeing is by members of the board and other mathematical authorities in the topic concerned, located throughout the world.

Preparation of accepted manuscripts
On acceptance of a proposal, the publisher will supply full instructions for the preparation of manuscripts in a form suitable for direct photolithographic reproduction. Specially printed grid sheets are provided and a contribution is offered by the publisher towards the cost of typing.

Illustrations should be prepared by the authors, ready for direct reproduction without further improvement. The use of hand-drawn symbols should be avoided wherever possible, in order to maintain maximum clarity of the text.

The publisher will be pleased to give any guidance necessary during the preparation of a typescript, and will be happy to answer any queries.

Important note
In order to avoid later retyping, intending authors are strongly urged not to begin final preparation of a typescript before receiving the publisher's guidelines and special paper. In this way it is hoped to preserve the uniform appearance of the series.

Titles in this series

1 Improperly posed boundary value problems
A Carasso and A P Stone
2 Lie algebras generated by finite dimensional ideals
I N Stewart
3 Bifurcation problems in nonlinear elasticity
R W Dickey
4 Partial differential equations in the complex domain
D L Colton
5 Quasilinear hyperbolic systems and waves
A Jeffrey
6 Solution of boundary value problems by the method of integral operators
D L Colton
7 Taylor expansions and catastrophes
T Poston and I N Stewart
8 Function theoretic methods in differential equations
R P Gilbert and R J Weinacht
9 Differential topology with a view to applications
D R J Chillingworth
10 Characteristic classes of foliations
H V Pittie
11 Stochastic integration and generalized martingales
A U Kussmaul
12 Zeta-functions: An introduction to algebraic geometry
A D Thomas
13 Explicit *a priori* inequalities with applications to boundary value problems
V G Sigillito
14 Nonlinear diffusion
W E Fitzgibbon III and H F Walker
15 Unsolved problems concerning lattice points
J Hammer
16 Edge-colourings of graphs
S Fiorini and R J Wilson
17 Nonlinear analysis and mechanics: Heriot-Watt Symposium Volume I
R J Knops
18 Actions of fine abelian groups
C Kosniowski
19 Closed graph theorems and webbed spaces
M De Wilde
20 Singular perturbation techniques applied to integro-differential equations
H Grabmüller
21 Retarded functional differential equations: A global point of view
S E A Mohammed
22 Multiparameter spectral theory in Hilbert space
B D Sleeman
24 Mathematical modelling techniques
R Aris
25 Singular points of smooth mappings
C G Gibson
26 Nonlinear evolution equations solvable by the spectral transform
F Calogero
27 Nonlinear analysis and mechanics: Heriot-Watt Symposium Volume II
R J Knops
28 Constructive functional analysis
D S Bridges
29 Elongational flows: Aspects of the behaviour of model elasticoviscous fluids
C J S Petrie
30 Nonlinear analysis and mechanics: Heriot-Watt Symposium Volume III
R J Knops
31 Fractional calculus and integral transforms of generalized functions
A C McBride
32 Complex manifold techniques in theoretical physics
D E Lerner and P D Sommers
33 Hilbert's third problem: scissors congruence
C-H Sah
34 Graph theory and combinatorics
R J Wilson
35 The Tricomi equation with applications to the theory of plane transonic flow
A R Manwell
36 Abstract differential equations
S D Zaidman
37 Advances in twistor theory
L P Hughston and R S Ward
38 Operator theory and functional analysis
I Erdelyi
39 Nonlinear analysis and mechanics: Heriot-Watt Symposium Volume IV
R J Knops
40 Singular systems of differential equations
S L Campbell
41 N-dimensional crystallography
R L E Schwarzenberger
42 Nonlinear partial differential equations in physical problems
D Graffi
43 Shifts and periodicity for right invertible operators
D Przeworska-Rolewicz
44 Rings with chain conditions
A W Chatters and C R Hajarnavis
45 Moduli, deformations and classifications of compact complex manifolds
D Sundararaman
46 Nonlinear problems of analysis in geometry and mechanics
M Atteia, D Bancel and I Gumowski
47 Algorithmic methods in optimal control
W A Gruver and E Sachs
48 Abstract Cauchy problems and functional differential equations
F Kappel and W Schappacher
49 Sequence spaces
W H Ruckle
50 Recent contributions to nonlinear partial differential equations
H Berestycki and H Brezis
51 Subnormal operators
J B Conway
52 Wave propagation in viscoelastic media
F Mainardi
53 Nonlinear partial differential equations and their applications: Collège de France Seminar. Volume I
H Brezis and J L Lions
54 Geometry of Coxeter groups
H Hiller
55 Cusps of Gauss mappings
T Banchoff, T Gaffney and C McCrory

56 An approach to algebraic K-theory
A J Berrick
57 Convex analysis and optimization
J-P Aubin and R B Vintner
58 Convex analysis with applications in the differentiation of convex functions
J R Giles
59 Weak and variational methods for moving boundary problems
C M Elliott and J R Ockendon
60 Nonlinear partial differential equations and their applications: Collège de France Seminar. Volume II
H Brezis and J L Lions
61 Singular systems of differential equations II
S L Campbell
62 Rates of convergence in the central limit theorem
Peter Hall
63 Solution of differential equations by means of one-parameter groups
J M Hill
64 Hankel operators on Hilbert space
S C Power
65 Schrödinger-type operators with continuous spectra
M S P Eastham and H Kalf
66 Recent applications of generalized inverses
S L Campbell
67 Riesz and Fredholm theory in Banach algebra
B A Barnes, G J Murphy, M R F Smyth and T T West
68 Evolution equations and their applications
F Kappel and W Schappacher
69 Generalized solutions of Hamilton-Jacobi equations
P L Lions
70 Nonlinear partial differential equations and their applications: Collège de France Seminar. Volume III
H Brezis and J L Lions
71 Spectral theory and wave operators for the Schrödinger equation
A M Berthier
72 Approximation of Hilbert space operators I
D A Herrero
73 Vector valued Nevanlinna Theory
H J W Ziegler
74 Instability, nonexistence and weighted energy methods in fluid dynamics and related theories
B Straughan
75 Local bifurcation and symmetry
A Vanderbauwhede
76 Clifford analysis
F Brackx, R Delanghe and F Sommen
77 Nonlinear equivalence, reduction of PDEs to ODEs and fast convergent numerical methods
E E Rosinger
78 Free boundary problems, theory and applications. Volume I
A Fasano and M Primicerio
79 Free boundary problems, theory and applications. Volume II
A Fasano and M Primicerio
80 Symplectic geometry
A Crumeyrolle and J Grifone
81 An algorithmic analysis of a communication model with retransmission of flawed messages
D M Lucantoni
82 Geometric games and their applications
W H Ruckle
83 Additive groups of rings
S Feigelstock
84 Nonlinear partial differential equations and their applications: Collège de France Seminar. Volume IV
H Brezis and J L Lions
85 Multiplicative functionals on topological algebras
T Husain
86 Hamilton-Jacobi equations in Hilbert spaces
V Barbu and G Da Prato
87 Harmonic maps with symmetry, harmonic morphisms and deformations of metrics
P Baird
88 Similarity solutions of nonlinear partial differential equations
L Dresner
89 Contributions to nonlinear partial differential equations
C Bardos, A Damlamian, J I Díaz and J Hernández
90 Banach and Hilbert spaces of vector-valued functions
J Burbea and P Masani
91 Control and observation of neutral systems
D Salamon
92 Banach bundles, Banach modules and automorphisms of C*-algebras
M J Dupré and R M Gillette
93 Nonlinear partial differential equations and their applications: Collège de France Seminar. Volume V
H Brezis and J L Lions
94 Computer algebra in applied mathematics: an introduction to MACSYMA
R H Rand
95 Advances in nonlinear waves. Volume I
L Debnath
96 FC-groups
M J Tomkinson
97 Topics in relaxation and ellipsoidal methods
M Akgül
98 Analogue of the group algebra for topological semigroups
H Dzinotyiweyi
99 Stochastic functional differential equations
S E A Mohammed
100 Optimal control of variational inequalities
V Barbu
101 Partial differential equations and dynamical systems
W E Fitzgibbon III
102 Approximation of Hilbert space operators. Volume II
C Apostol, L A Fialkow, D A Herrero and D Voiculescu
103 Nondiscrete induction and iterative processes
V Ptak and F-A Potra
104 Analytic functions – growth aspects
O P Juneja and G P Kapoor
105 Theory of Tikhonov regularization for Fredholm equations of the first kind
C W Groetsch

106 Nonlinear partial differential equations and free boundaries
J I Díaz

107 Tight and taut immersions of manifolds
T E Cecil and P J Ryan

108 A layering method for viscous, incompressible L_p flows occupying R^n
A Douglis and E B Fabes

109 Nonlinear partial differential equations and their applications: Collège de France Seminar. Volume VI
H Brezis and J L Lions

110 Finite generalized quadrangles
S E Payne and J A Thas

111 Advances in nonlinear waves. Volume II
L Debnath

112 Topics in several complex variables
E Ramírez de Arellano and D Sundararaman

113 Differential equations, flow invariance and applications
N H Pavel

114 Geometrical combinatorics
F C Holroyd and R J Wilson

Finite generalized quadrangles

S E Payne & J A Thas

Miami University/State University of Gent

Finite generalized quadrangles

Pitman Advanced Publishing Program

BOSTON · LONDON · MELBOURNE

PITMAN PUBLISHING LIMITED
128 Long Acre, London WC2E 9AN

PITMAN PUBLISHING INC
1020 Plain Street, Marshfield, Massachusetts 02050

Associated Companies
Pitman Publishing Pty Ltd, Melbourne
Pitman Publishing New Zealand Ltd, Wellington
Copp Clark Pitman, Toronto

First published 1984

AMS Subject Classifications: (main) 51-02, 05B25, 51E20
(subsidiary) 05-02, 51A25, 51BXX

ISSN 0743-0337

Library of Congress Cataloging in Publication Data

Payne, S. E. (Stanley E.)
Finite generalized quadrangles.

Bibliography: p.
Includes index.
1. Finite generalized quadrangles. I. Thas, J. A.
(Joseph Adolf) II. Title.
QA167.2.P39 1984 516 84-9581
ISBN 0-273-08655-3

British Library Cataloguing in Publication Data

Payne, S. E.
Finite generalized quadrangles.—(Research
notes in mathematics; 110)
1. Finite generalized quadrangles
I. Title II. Thas, J. A. III. Series
516′.15 QA167.2

ISBN 0-273-08655-3

Reproduced and printed by photolithography
in Great Britain by Biddles Ltd, Guildford

Preface

When J. Tits [216] introduced the abstract notion of generalized polygon in 1959, the study of finite geometries as objects of interest in their own right was hardly fashionable. But the favorable climate in which the investigation of all kinds of discrete structures thrives today was already being created by pioneers whose names are now familiar to all who work in this broad area. We mention especially B. Segre, whose work on Galois geometries (i.e. projective geometries coordinatized by the Galois fields) has exerted a tremendous positive influence for more than two decades.

The proper subject of this book - finite generalized quadrangles - originally was conceived as an adjunct to the study of finite groups, and this connection remains a healthy one today. However, the present volume may be viewed as an attempt to treat our subject as thoroughly as possible from a combinatorial and geometric point of view. Our goal has been to keep to a minimum the prerequisites needed for reading this work, while at the same time permitting the reader to develop an understanding of the many connections between generalized quadrangles and diverse other structures. In the normal course of education a graduate student would have acquired the necessary linear and abstract algebra early in his studies, and the only additional background needed is a moderate introduction to the finite affine and projective geometries. For this the recent treatise by J.W.P. Hirschfeld [80] is a convenient reference along with the text by D.R. Hughes and F. Piper [86].

In Chapter 1 nearly all the known general results of a combinatorial nature are proved using standard counting tricks, eigenvalue techniques, etc. Most the concepts and terms that play a role throughout the book are introduced in this introductory chapter. Chapter 2 is a continuation of the general combinatorial theme as applied specifically to subquadrangles and certain other substructures such as ovoids and spreads. Then in Chapter 3 the known models are given together with a fairly detailed discussion of their properties as related to the abstract notions introduced previously. Chapter 4 is devoted to a proof of the celebrated theorem of F. Buekenhout and C. Lefèvre

[29] characterizing the finite classical generalized quadrangles as those which can be embedded in projective space. The longest (and perhaps the central) chapter in the book is Chapter 5, in which are given many combinatorial characterizations of the known generalized quadrangles . Here in one unified treatment are results whose proofs (most of which are due to the younger author) were scattered in a variety of overlapping papers. The determination of all generalized quadrangles with sufficiently small parameters has been the work of several authors. This is treated in considerable detail in Chapter 6. In the next chapter there are determined all generalized quadrangles embedded in finite affine spaces, a problem first completely solved by J.A. Thas [196]. At this point the treatment becomes more algebraic. Chapters 8 and 9 study the generalized quadrangles in terms of their collineation groups, but with a minimum of abstract group theory. These two chapters present a nearly complete elementary proof of the celebrated theorem of J. Tits determining all finite Moufang quadrangles. The authors are still hopeful that this program can be completed. Chapter 10 studies group coset geometries and gives an account of the most recently discovered family of examples, due to W.M. Kantor [88]. In the next chapter a fairly general theory of coordinates for generalized quadrangles of order s is developed. The final chapter studies generalized quadrangles as amalgamations of desarguesian planes.

We make no claim to completeness. Important work of M. Walker [229], for example, is completely ignored. But it is our hope that the present volume presents a unified, accessible treatment of a major part of the work done on finite generalized quadrangles, sufficient to obtain a firm grasp on the subject, to find inspiration to add to the current body of knowledge, and to appreciate the many interconnections with other branches of finite geometry.

The manuscript was typed by Mrs. Helen Bogan and Mrs. Zita Oost, the typewriting of the camera-ready copy was done by Mrs. Zita Oost, and the figures have been made by Mrs. Annie Clement. We are most grateful to them for their care and patience.

Contents

1 Combinatorics of finite generalized quadrangles

1.1. AXIOMS AND DEFINITIONS

A (finite) generalized quadrangle (GQ) is an incidence structure $S = (P,B,I)$ in which P and B are disjoint (nonempty) sets of objects called points and lines (respectively), and for which I is a symmetric point-line incidence relation satisfying the following axioms :

(i) Each point is incident with 1+t lines ($t \geqslant 1$) and two distinct points are incident with at most one line.

(ii) Each line is incident with 1+s points ($s \geqslant 1$) and two distinct lines are incident with at most one point.

(iii) If x is a point and L is a line not incident with x, then there is a unique pair $(y,M) \in P \times B$ for which x I M I y I L.

Generalized quadrangles were introduced by J. Tits [216].

The integers s and t are the *parameters* of the GQ and S is said to have *order* (s,t); if s = t, S is said to have *order* s . There is a point-line duality for GQ (of order (s,t)) for which in any definition or theorem the words "point" and "line" are interchanged and the parameters s and t are interchanged. Normally, we assume without further notice that the dual of a given theorem or definition has also been given.

A *grid* (resp., *dual grid*) is an incidence structure $S = (P,B,I)$ with $P = \{x_{ij} \parallel i = 0,\dots,s_1 \text{ and } j = 0,\dots,s_2\}$, $s_1 > 0$ and $s_2 > 0$ (resp., $B = \{L_{ij} \parallel i = 0,\dots,t_1 \text{ and } j = 0,\dots,t_2\}$, $t_1 > 0$ and $t_2 > 0$), $B = \{L_0,\dots,L_{s_1},M_0,\dots,M_{s_2}\}$ (resp., $P = \{x_0,\dots,x_{t_1},y_0,\dots,y_{t_2}\}$), x_{ij} I L_k iff i = k (resp., L_{ij} I x_k iff i = k), and x_{ij} I M_k iff j = k (resp., L_{ij} I y_k iff j = k). A grid (resp., dual grid) with parameters s_1,s_2 (resp., t_1,t_2) is a GQ iff $s_1 = s_2$ (resp., $t_1 = t_2$). Evidently the grids (resp., dual grids) with $s_1 = s_2$ (resp., $t_1 = t_2$) are the GQ with t = 1 (resp., s = 1).

Let S be a GQ, a grid, or a dual grid. Given two (not necessarily distinct) points x,y of S, we write $x \sim y$ and say that x and y are *collinear*, provided that there is some line L for which x I L I y. And $x \not\sim y$ means that x and y are not collinear. Dually, for $L, M \in B$, we write $L \sim M$ or $L \not\sim M$ according as L and M are *concurrent* or nonconcurrent, respectively. If $x \sim y$ (resp., $L \sim M$) we may also say that x (resp., L) is *orthogonal* or *perpendicular* to y (resp., M). The line (resp., point) which is incident with distinct collinear points x,y (resp., distinct concurrent lines L,M) is denoted by xy (resp., LM or $L \cap M$).

For $x \in P$ put $x^{\perp} = \{y \in P \parallel y \sim x\}$, and note that $x \in x^{\perp}$. The *trace* of a pair (x,y) of distinct points is defined to be the set $x^{\perp} \cap y^{\perp}$ and is denoted tr(x,y) or $\{x,y\}^{\perp}$. We have $|\{x,y\}^{\perp}| = s+1$ or $t+1$ according as $x \sim y$ or $x \not\sim y$. More generally, if $A \subset P$, A "perp" is defined by $A^{\perp} = \cap\{x^{\perp} \parallel x \in A\}$. For $x \neq y$, the *span* of the pair (x,y) is $sp(x,y) = \{x,y\}^{\perp\perp} = \{u \in P \parallel u \in z^{\perp}\ \forall z \in x^{\perp} \cap y^{\perp}\}$. If $x \not\sim y$, then $\{x,y\}^{\perp\perp}$ is also called the *hyperbolic line* defined by x and y. For $x \neq y$, the *closure* of the pair (x,y) is $cl(x,y) = \{z \in P \parallel z^{\perp} \cap \{x,y\}^{\perp\perp} \neq \phi\}$.

A *triad* (of points) is a triple of pairwise noncollinear points. Given a triad $T = (x,y,z)$, a *center* of T is just a point of $T^{\perp}$. We say T is *acentric, centric,* or *unicentric* according as $|T^{\perp}|$ is zero, positive, or equal to 1.

Isomorphisms (or collineations), anti-isomorphisms (or correlations), automorphisms, anti-automorphisms, involutions and polarities of generalized quadrangles, grids, and dual grids are defined in the usual way.

1.2. RESTRICTIONS ON THE PARAMETERS

Let $S = (P,B,I)$ be a GQ of order (s,t), and put $|P| = v$, $|B| = b$.

1.2.1. $v = (s+1)(st+1)$ *and* $b = (t+1)(st+1)$.

Proof. Let L be a fixed line of S and count in different ways the number of ordered pairs $(x,M) \in P \times B$ with $x \not\mathrel{I} L$, $x \mathrel{I} M$, and $L \sim M$. There arises $v-s-1 = (s+1)ts$ or $v = (s+1)(st+1)$. Dually $b = (t+1)(st+1)$. □

1.2.2. $s+t$ *divides* $st(s+1)(t+1)$.

Proof. If $E = \{\{x,y\} \parallel x,y \in P \text{ and } x \sim y\}$, then it is evident that (P,E) is a strongly regular graph [17,77] with parameters $v = (s+1)(st+1)$, k (or n_1) $= st+s$, λ (or p^1_{11}) $= s-1$, μ (or p^2_{11}) $= t+1$. The graph (P,E) is called the point graph of the GQ. Let $P = \{x_1,\ldots,x_v\}$ and let $A = (a_{ij})$ be the $v \times v$ matrix over R for which $a_{ij} = 0$ if $i = j$ or $x_i \not\sim x_j$, and $a_{ij} = 1$ if $i \neq j$ and $x_i \sim x_j$, i.e. A is an adjacency matrix of the graph (P,E) (cf.[17]).

If $A^2 = (c_{ij})$, then we have : (a) $c_{ii} = (t+1)s$; (b) if $i \neq j$ and $x_i \not\sim x_j$, then $c_{ij} = t+1$; (c) if $i \neq j$ and $x_i \sim x_j$, then $c_{ij} = s-1$. Consequently $A^2-(s-t-2)A-(t+1)(s-1)I = (t+1)J$. (Here I is the $v \times v$ identity matrix and J is the $v \times v$ matrix with each entry equal to 1.) Evidently $(t+1)s$ is an eigenvalue of A, and J has eigenvalues 0, v with multiplicities v-1, 1, respectively. Since $((t+1)s)^2-(s-t-2)(t+1)s-(t+1)(s-1) = (t+1)(st+1)(s+1) = (t+1)v$, the eigenvalue $(t+1)s$ of A corresponds to the eigenvalue v of J, and so $(t+1)s$ has multiplicity 1. The other eigenvalues of A are roots of the equation $x^2-(s-t-2)x-(t+1)(s-1) = 0$. Denote the multiplicities of these eigenvalues θ_1,θ_2 by m_1,m_2, respectively. Then we have $\theta_1 = -t-1$, $\theta_2 = s-1$, $v = 1+m_1+m_2$, and $s(t+1)-m_1(t+1)+m_2(s-1) = \text{tr } A = 0$. Hence $m_1 = (st+1)s^2/(s+t)$ and $m_2 = st(s+1)(t+1)/(s+t)$. Since $m_1, m_2 \in N$, the proof is complete. □

1.2.3. (*The inequality of D.G. Higman* [77,78]). *If* $s > 1$ *and* $t > 1$, *then* $t \leqslant s^2$, *and dually* $s \leqslant t^2$.

Proof (P.J. Cameron [31]). Let x,y be two noncollinear points of S. Put $V = \{z \in P \parallel z \not\sim x \text{ and } z \not\sim y\}$, so $|V| = d = (s+1)(st+1)-2-2(t+1)s+(t+1)$. Denote the elements of V by $z_1,\ldots,z_d$ and let $t_i = |\{u \in \{x,y\}^\perp \parallel u \sim z_i\}|$. Count in different ways the number of ordered pairs $(z_i,u) \in V \times \{x,y\}^\perp$ with $u \sim z_i$ to obtain

$$\sum_i t_i = (t+1)(t-1)s. \qquad (1)$$

Next count in different ways the number of ordered triples $(z_i,u,u') \in V \times \{x,y\}^\perp \times \{x,y\}^\perp$ with $u \neq u'$, $u \sim z_i$, $u' \sim z_i$, to obtain

$$\sum_i t_i(t_i-1) = (t+1)t(t-1). \qquad (2)$$

From (1) and (2) it follows that $\sum_i t_i^2 = (t+1)(t-1)(s+t)$.
With $d\bar{t} = \sum_i t_i$, $0 \leqslant \sum_i (\bar{t}-t_i)^2$ simplifies to $d \sum_i t_i^2-(\sum_i t_i)^2 \geqslant 0$, which implies $d(t+1)(t-1)(s+t) \geqslant (t+1)^2(t-1)^2s^2$, or $t(s-1)(s^2-t) \geqslant 0$, completing the proof. □

There is an immediate corollary of the proof.

1.2.4. (*R.C. Bose and S.S. Shrikhande* [19]). *If* $s > 1$ *and* $t > 1$, *then* $s^2 = t$ *iff* $d \sum t_i^2 - (\sum t_i)^2 = 0$ *for any pair* (x,y) *of noncollinear points iff* $t_i = \bar{t}$ *for all* $i = 1,\dots,d$ *and for any pair* (x,y) *of noncollinear points iff each triad (of points) has a constant number of centers, in which case this constant number of centers is* $s+1$.

Remark : D.G. Higman first obtained the inequality $t \leq s^2$ by a complicated matrix-theoretic method [77,78]. R.C. Bose and S.S. Shrikhande [19] used the above argument to show that in case $t = s^2$ each triad has 1+s centers, and P.J. Cameron [31] apparently first observed that the above technique also provides the inequality. (See Paragraph 1.4 below for a simplified proof in the same spirit as that of D.G. Higman's original proof.)

1.2.5. *If* $s \neq 1$, $t \neq 1$, $s \neq t^2$, *and* $t \neq s^2$, *then* $t \leq s^2 - s$ *and dually* $s \leq t^2 - t$.

Proof. Suppose $s \neq 1$ and $t \neq s^2$. By 1.2.3 we have $t = s^2 - x$ with $x > 0$. By 1.2.2 $(s+s^2-x) \mid s(s^2-x)(s+1)(s^2-x+1)$. Hence modulo $s+s^2-x$ we have $0 \equiv x(-s)(-s+1) \equiv x(x-2s)$. If $x < 2s$, then $s+s^2-x \leq x(2s-x)$ forces $x \in \{s,s+1\}$. Consequently $x = s$, $x = s+1$, or $x \geq 2s$, from which it follows that $t \leq s^2 - s$. □

1.3. REGULARITY, ANTIREGULARITY, SEMIREGULARITY, AND PROPERTY (H)

Continuing with the same notation as in 1.2 , if $x \sim y$, $x \neq y$, or if $x \not\sim y$ and $|\{x,y\}^{\perp\perp}| = t+1$, we say that the pair (x,y) is *regular* . The point x is *regular* provided (x,y) is regular for all $y \in P$, $y \neq x$. A point x is *coregular* provided each line incident with x is regular. The pair (x,y), $x \not\sim y$, is *antiregular* provided $|z^\perp \cap \{x,y\}^\perp| \leq 2$ for all $z \in P-\{x,y\}$. A point x is *antiregular* provided (x,y) is antiregular for all $y \in P-x^\perp$.

A point u is called *semiregular* provided that $z \in cl(x,y)$ whenever u is the unique center of the triad (x,y,z). And a point u has *property* (H) provided $z \in cl(x,y)$ iff $x \in cl(y,z)$, whenever (x,y,z) is a triad consisting of points in $u^\perp$. It follows easily that any semiregular point has property (H).

1.3.1. *Let* x *be a regular point of the* GQ $S = (P,B,I)$ *of order* (s,t). *Then the incidence structure with pointset* $x^\perp - \{x\}$, *with lineset the set of spans* $\{y,z\}^{\perp\perp}$, *where* $y,z \in x^\perp - \{x\}$, $y \not\sim z$, *and with the natural incidence, is the dual of a net* (*cf.*[17]) *of order* s *and degree* $t+1$. *If in particular* $s = t > 1$, *there arises a dual affine plane of order* s . *Moreover, in this case the incidence structure* π_x *with pointset* $x^\perp$, *with lineset the set of spans* $\{y,z\}^{\perp\perp}$, *where* $y,z \in x^\perp$, $y \neq z$, *and with the natural incidence, is a projective plane of order* s.

Proof. Easy exercise. □

1.3.2. *Let* x *be an antiregular point of the* GQ $S = (P,B,I)$ *of order* s, $s \neq 1$, *and let* $y \in x^\perp - \{x\}$ *with* L *being the line* xy. *An affine plane* $\pi(x,y)$ *of order* s *may be constructed as follows. Points of* $\pi(x,y)$ *are just the points of* $x^\perp$ *that are not on* L . *Lines are the pointsets* $\{x,z\}^{\perp\perp} - \{x\}$, *with* $x \sim z \not\sim y$, *and* $\{x,u\}^\perp - \{y\}$, *with* $y \sim u \not\sim x$.

Proof. Easy exercise. □

Now let $s^2 = t > 1$, so that by 1.2.4 for any triad (x,y,z) we have $|\{x,y,z\}^\perp| = s+1$. Evidently $|\{x,y,z\}^{\perp\perp}| \leq s+1$. We say (x,y,z) is *3-regular* provided $|\{x,y,z\}^{\perp\perp}| = s+1$. Finally, the point x is called *3-regular* iff each triad containing x is 3-regular.

1.3.3. *Let* S *be a* GQ *of order* (s,s^2), $s \neq 1$, *and suppose that any triad contained in* $\{x,y\}^\perp$, $x \not\sim y$, *is 3-regular. Then the incidence structure with pointset* $\{x,y\}^\perp$, *with lineset the set of elements* $\{z,z',z''\}^{\perp\perp}$, *where* z,z',z'' *are distinct points in* $\{x,y\}^\perp$, *and with the natural incidence, is an inversive plane of order* s (*cf.* [49]).

Proof. Immediate. □

For the remainder of this section let x and y be fixed, noncollinear points of the GQ $S = (P,B,I)$ of order (s,t), and put $\{x,y\}^\perp = \{z_0,\dots,z_t\}$. For $A \subset \{0,\dots,t\}$ let $n(A)$ be the number of points that are not collinear with x or y and are collinear with z_i iff $i \in A$.

1.3.4. (i) $n(\phi) = 0$ *iff each triad* (x,y,z) *is centric*.

(ii) $n(A) = 0$ *for each* A *with* $2 \leq |A| \leq t$ *iff* (x,y) *is regular*.

(iii) $n(A) = 0$ *for all* A *with* $3 \leq |A|$ *iff* (x,y) *is antiregular*.

(iv) $n(A) = 0$ *if* $|A| = t$.

Proof. (i) , (ii) and (iii) are immediate. To prove (iv), suppose that $x \not\sim u \not\sim y$, $u \sim z_i$ for $i = 0,\dots,t-1$, and $u \not\sim z_t$. Let L_i be the line incident with z_t and concurrent with uz_i, $i = 0,\dots,t-1$. Then $L_0,\dots,L_{t-1}$, xz_t, yz_t must be t+2 distinct lines incident with z_t, a contradiction. Hence $n(A) = 0$ if $|A| = t$. □

1.3.5. *The following three equalities hold :*

$$\sum_A n(A) = s^2t-st-s+t, \tag{1}$$

$$\sum_A |A|n(A) = t^2s-s, \tag{2}$$

$$\sum_A |A|(|A|-1)n(A) = t^3-t. \tag{3}$$

Proof. Note first that $\sum_A n(A)$ is just the number of points not collinear with x or y. Then count in different ways the number of ordered pairs (u,z_i) with $u \sim z_i$ and u not collinear with x or y, to obtain $\sum_A |A|n(A) = (t+1)(t-1)s$, which is (2). Finally, count in different ways the number of ordered triples (u,z_i,z_j) with $u \in \{z_i,z_j\}^{\perp}$, $z_i \neq z_j$, and u not collinear with x or y. It follows readily that $\sum_A |A|(|A|-1)n(A) = (t+1)t(t-1)$. □

These three basic equations may be manipulated to obtain the following :

$$n(\phi) = s^2t-t^2s-st+\frac{t^3+t}{2}-\frac{1}{2}\sum_{|A|>2}(|A|-1)(|A|-2)n(A), \tag{4}$$

$$\sum_{|A|=1} n(A) = (t^2-1)(s-t) + \sum_{|A|>2}(|A|^2-2|A|)n(A), \tag{5}$$

$$\sum_{|A|=2} n(A) = \frac{1}{2}(t^3-t) - \frac{1}{2}\sum_{|A|>2}(|A|^2-|A|)n(A), \tag{6}$$

$$(t+1)n(\phi) = (s-t)t(s-1)(t+1)+(t-1)\sum_{|A|=2} n(A) + \sum_{2<|A|<t}(|A|-1)(t+1-|A|)n(A). \tag{7}$$

For each integer $\alpha = 0,1,\ldots,t+1$, let $N_\alpha = \sum_{|A|=\alpha} n(A)$. Suppose there are three distinct integers α,β,γ, $0 \leqslant \alpha,\beta,\gamma \leqslant t+1$, for which $\theta \notin \{\alpha,\beta,\gamma\}$ implies that $N_\theta = 0$. Note that we allow $N_\alpha = 0$ also for example. Then equations (1), (2), (3) can be written in matrix form as

$$\begin{pmatrix} 1 & 1 & 1 \\ \alpha & \beta & \gamma \\ \alpha(\alpha-1) & \beta(\beta-1) & \gamma(\gamma-1) \end{pmatrix} \begin{pmatrix} N_\alpha \\ N_\beta \\ N_\gamma \end{pmatrix} = \begin{pmatrix} s^2t-st-s+t \\ t^2s-s \\ t^3-t \end{pmatrix}. \tag{8}$$

The determinant of this linear system is $\Delta = (\alpha-\beta)(\beta-\gamma)(\gamma-\alpha)$, and we may use Cramer's rule to solve for N_α, N_β, N_γ. As α,β,γ were given in no particular order, it suffices to solve for just one :

$$N_\alpha = \frac{(s^2t-st-s+t)\beta\gamma-(t^2-1)s(\beta+\gamma)+(t^2-1)(s+t)}{(\alpha-\beta)(\alpha-\gamma)} \quad . \tag{9}$$

First, suppose that $N_\beta = N_\gamma = 0$, i.e. there is at most one index for which $N_\alpha \neq 0$. Then equations (1), (2), (3) become

$$\begin{aligned} N_\alpha &= s^2t-st-s+t, \\ \alpha N_\alpha &= t^2s-s, \\ \alpha(\alpha-1)N_\alpha &= t^3-t. \end{aligned} \tag{10}$$

Eliminating α and N_α we find that $(t-1)(s-1)(s^2-1) = 0$, and that if $s^2 = t \neq 1$, then $\alpha = s+1$ and $N_\alpha = s(s-1)(s^2+1)$. This result was also contained in 1.2.4.

Second, suppose that $N_\gamma = 0$. By the formula for N_γ we know the following :

$$(s^2t-st-s+t)\alpha\beta-(t^2-1)s(\alpha+\beta)+(t^2-1)(s+t) = 0. \tag{11}$$

Here there are two cases of special interest : $\alpha = 0$ and $\alpha = 1$. If $\alpha = 0$, then $\beta = (s+t)/s$, if $t > 1$. If $\alpha = 1$, then $\beta = (t^2-1)/(st-s^2+s-1)$, which forces $s \leqslant t$ if $t \neq 1$.

Finally, consider again the general case. If $s > t > 1$, then by (5) and (7) it follows that both $N_1 > 0$ and $N_0 > 0$. So suppose $\alpha = 0$, $\beta = 1$, with $s,t,\gamma > 1$. Then $N_\gamma = t(t^2-1)/\gamma(\gamma-1)$.

The case $\gamma = t+1$ forces $s \geqslant t$ by (7) and occurs precisely when (x,y) is regular. Here $N_0 = (s-t)t(s-1)$, $N_1 = (t^2-1)(s-1)$, and $N_{t+1} = t-1$.

The case $\gamma = 2$ occurs precisely when (x,y) is antiregular, in which case $N_0 = s^2t-t^2s-st+t(t^2+1)/2$, $N_1 = (t^2-1)(s-t)$, $N_2 = t(t^2-1)/2$. Since $N_1 \geqslant 0$, we have $s \geqslant t$ if $t > 1$.

There is one last specialization we consider : $1 = \alpha < \beta < \gamma = 1+t$. Here

$$\begin{aligned} N_1 &= (1+t)(s-1)(1-t+\beta(s-1))/(\beta-1), \\ N_\beta &= t(s-1)(t+1)(t-s)/(\beta-1)(t+1-\beta), \\ N_{t+1} &= (t^2-1-\beta(st-s^2+s-1))/(t+1-\beta). \end{aligned} \tag{12}$$

Since $N_1 \geq 0$, $\beta \geq (t-1)/(s-1)$ if $s \neq 1$. Since $N_\beta \geq 0$, $t \geq s$.

1.3.6. (i) *If* $1 < s < t$, *then* (x,y) *is neither regular nor antiregular.*

(ii) *The pair* (x,y) *is regular (with* $s = 1$ *or* $s \geq t$*) iff each triad* (x,y,z) *has exactly* 0,1 *or* $t+1$ *centers. When* $s = t$ *this is iff each triad* (x,y,z) *is centric.*

(iii) *If* $s \geq t$, *then* $N_1 = 0$ *iff either* $t = 1$ *or* $s = t$ *and* (x,y) *is antiregular. Hence for* $s = t$ *the pair* (x,y) *is antiregular iff each triad* (x,y,z) *has* 0 *or* 2 *centers.*

(iv) *If* $s = t$ *and each point in* $x^\perp - \{x\}$ *is regular, then every point is regular.*

Proof. In the preceding paragraph we proved that a GQ containing a regular or antiregular pair of points satisfies $s \geq t$ if $s > 1$, $t > 1$. We remark that for $s = 1$ any pair of points is regular, and that for $t = 1$ any pair of points is regular and any noncollinear pair of points is antiregular. By the definition of regularity, the pair (x,y) is regular iff each triad (x,y,z) has exactly 0,1, or $t+1$ centers. When $s = t$ and (x,y) is regular, then $N_0 = 0$ and so each triad is centric. Conversely, if $s = t$, $s \neq 1$, and each triad (x,y,z) is centric (recall that the pair (x,y) is fixed), then by (7) $n(A) = 0$ if $2 \leq |A| < t$, i.e. each triad has 0,1, or $t+1$ centers and so (x,y) is regular.

If $t = 1$, then it is trivial that $N_1 = 0$. If $s = t$ and (x,y) is antiregular, then the paragraph preceding the theorem informs us that $N_1 = 0$. Conversely, assume $N_1 = 0$ and $s \geq t$. Then from (5) we have $t = 1$ or $s = t$ and $n(A) = 0$ for $|A| > 2$, i.e. $t = 1$ or $s = t$ and (x,y) is antiregular.

Now let $s = t$ and assume that each point in $x^\perp - \{x\}$ is regular. Let $y \not\sim x$ and $z_1, z_2 \in \{x,y\}^\perp$, $z_1 \neq z_2$. Since (z_1,z_2) is regular, clearly (x,y) is regular. Hence x is regular. To complete the proof that each point is regular, it suffices to show that if (x,u,u') is a triad, then (u,u') is regular. But since x is regular, by (ii) there is some point $z \in \{x,u,u'\}^\perp$. By the regularity of z, for any point $z' \in \{u,u'\}^\perp - \{z\}$, the pair (z,z') is regular, forcing (u,u') to be regular. □

1.4. AN APPLICATION OF THE HIGMAN-SIMS TECHNIQUE

Let $A = (a_{ij})$ denote an $n \times n$ real symmetric matrix. Suppose that

$\Delta = \{\Delta_1,\ldots,\Delta_r\}$ and $\Gamma = \{\Gamma_1,\ldots,\Gamma_u\}$ are partitions of $\{1,\ldots,n\}$, and that Γ is a refinement of Δ. Put $\delta_i = |\Delta_i|$, $\gamma_i = |\Gamma_i|$, and let

$$\delta_{ij} = \sum_{\substack{\mu\in\Delta_i\\ \nu\in\Delta_j}} a_{\mu\nu}, \quad \gamma_{ij} = \sum_{\substack{\mu\in\Gamma_i\\ \nu\in\Gamma_j}} a_{\mu\nu} .$$

Define the following matrices :

$$A^\Delta = (\delta_{ij}/\delta_i)_{1\leq i,j\leq r} \text{ and } A^\Gamma = (\gamma_{ij}/\gamma_i)_{1\leq i,j\leq u} .$$

If $\mu_1,\ldots,\mu_r$, with $\mu_1\leq\ldots\leq\mu_r$, are the characteristic roots of A^Δ and $\lambda_1,\ldots,\lambda_u$, with $\lambda_1\leq\ldots\leq\lambda_u$, are the characteristic roots of A^Γ, then by a theorem of C.C. Sims (cf. p. 144 of [76]; the details are in S.E. Payne [134] and are considerably generalized in W. Haemers [66]) it must be that $\lambda_1 \leq \mu_1 \leq \mu_r \leq \lambda_u$. Moreover, if $\bar{y} = (y_1,\ldots,y_r)^T$ satisfies $A^\Delta\bar{y} = \lambda_1\bar{y}$ (so $\lambda_1 = \mu_1$), then $A^\Gamma\bar{x} = \lambda_1\bar{x}$, where $\bar{x} = (\ldots,x_k,\ldots)^T$ is defined by $x_k = y_i$ whenever $\Gamma_k \subset \Delta_i$.

We give the following important application.

1.4.1. (*S.E. Payne* [134]). *Let* $X = \{x_1,\ldots,x_m\}$, $m \geq 2$, *and* $Y = \{y_1,\ldots,y_n\}$, $n \geq 2$, *be disjoint sets of pairwise noncollinear points of the GQ* $S = (P,B,I)$ *of order* (s,t), $s > 1$, *and suppose that* $X \subset Y^\perp$. *Then* $(m-1)(n-1) \leq s^2$. *If equality holds, then one of the following must occur :*

(i) $m = n = 1+s$, *and each point of* $Z = P-(X \cup Y)$ *is collinear with precisely two points of* $X \cup Y$.

(ii) $m \neq n$. *If* $m < n$, *then* $s|t$, $s < t$, $n = 1+t$, $m = 1+s^2/t$, *and each point of* $P-X$ *is collinear with either* 1 *or* $1+t/s$ *points of* Y *according as it is or is not collinear with some point of* X .

Proof. Let $P = \{w_1,\ldots,w_v\}$ and let B be the $(0,1)$-matrix (b_{ij}) over $\mathbb{R}$ defined by $b_{ij} = 1$ if $w_i \not\sim w_j$ and $b_{ij} = 0$ otherwise. So $B = J-A-I$, where A,J,I are as in the proof of 1.2.2, and it readily follows from that proof that B has eigenvalues s^2t, t, $-s$. Let $\{\Delta_1,\Delta_2,\Delta_3\}$ be the partition of $\{1,\ldots,v\}$ determined by the partition $\{X,Y,Z\}$ of P. Put $\delta_{ij} = \sum_{\substack{k\in\Delta_i\\ \ell\in\Delta_j}} b_{k\ell}$, $\delta_i = |\Delta_i|$, and define the 3×3 matrix $B^\Delta = (\delta_{ij}/\delta_i)_{1\leq i,j\leq 3}$. Clearly

$\delta_1 = m$, $\delta_2 = n$, $\delta_3 = v-(m+n)$, $\delta_{11} = (m-1)m$, $\delta_{12} = 0$, $\delta_{13} = (s^2t-(m-1))m$, $\delta_{21} = 0$, $\delta_{22} = (n-1)n$, $\delta_{23} = (s^2t-(n-1))n$, $\delta_{31} = (s^2t-(m-1))m$, $\delta_{32} = (s^2t-(n-1))n$, $\delta_{33} = s^2t\delta_3-\delta_{31}-\delta_{32}$. Hence

$$B^{\Delta} = \begin{pmatrix} m-1 & 0 & s^2t-m+1 \\ 0 & n-1 & s^2t-n+1 \\ \frac{(s^2t-m+1)m}{v-m-n} & \frac{(s^2t-n+1)n}{v-m-n} & s^2t - \frac{(s^2t-m+1)m+(s^2t-n+1)n}{v-m-n} \end{pmatrix}$$

If s^2t, θ_1, θ_2 (with $\theta_1 \leqslant \theta_2$) are the eigenvalues of B^{Δ} then $\theta_1+\theta_2 = \mathrm{tr}(B^{\Delta})-s^2t = ((m+n)(st+s+2)-2v-2mn)/(v-m-n)$ and $\theta_1\theta_2 = (\det B^{\Delta})/s^2t = ((1+s+st)(2mn-m-n)+v-mnv)/(v-m-n)$. By the theorem of C.C. Sims the eigenvalues of B^{Δ} belong to the closed interval determined by the smallest and largest eigenvalue of B. Hence $-s \leqslant \theta_1 \leqslant \theta_2 \leqslant s^2t$. But θ_1 and θ_2 are the roots of the equation $f(x) = 0$ with $f(x) = x^2-(\theta_1+\theta_2)x+\theta_1\theta_2$, so that $f(-s) \geqslant 0$. Writing this out with the values of $\theta_1+\theta_2$ and $\theta_1\theta_2$ given above yields $(s-1)(st+1)(s^2-1-mn+m+n) \geqslant 0$, i.e. $s^2 \geqslant (m-1)(n-1)$. In case of equality, i.e. $-s = \theta_1$, then $\bar{y} = (y_1,y_2,y_3)^T$ satisfies $B^{\Delta}\bar{y} = (-s)\bar{y}$, if we put $y_1 = (m-1-s^2t)/(s+m-1)$, $y_2 = (n-1-s^2t)/(s+n-1)$, $y_3 = 1$. Hence it must be that $B\bar{x} = (-s)\bar{x}$, where $\bar{x} = (\ldots,x_k,\ldots)^T$ is defined by $x_k = y_i$ whenever $k \in \Delta_i$. Let us now assume, without loss of generality, that $X = \{w_1,\ldots,w_m\}$ and $Y = \{w_{m+1},\ldots,w_{m+n}\}$. Then $\bar{x} = (\underbrace{y_1,\ldots,y_1}_{m\text{ times}}, \underbrace{y_2,\ldots,y_2}_{n\text{ times}}, \underbrace{1,\ldots,1}_{v-m-n\text{ times}})^T$. For the first m+n rows of B this yields no new information. But consider the point w_i, $i > m+n$. Suppose w_i is not collinear with t_1 points of x, is not collinear with t_2 points of Y, and hence is not collinear with $s^2t-t_1-t_2$ points of Z. Then the product of the ith row of B with $\bar{x}$, which must equal $-s$, is actually $t_1y_1+t_2y_2+s^2t-t_1-t_2 = -s$. This becomes

$$t_1/(s+m-1)+t_2/(s+n-1) = 1. \tag{13}$$

If w_i lies on a line joining a point of X and a point of Y, then $t_1 = m-1$ and $t_2 = n-1$, and eq. (13) gives no information. On the other hand, if w_i is not on such a line, then either $t_1 = m$ or $t_2 = n$. Suppose $t_1 = m$, so w_i is collinear with no point of X. Using eq. (13) we find that the number of points of Y collinear with w_i is $n-t_2 = 1+(n-1)/s$. If $m = n = s+1$, this says each point not on a line joining a point of X with a point of Y must be collinear with two points of X and none of Y or with two of Y and none of X. If $1 < m < s+1$, so $1+(m-1)/s$ is not an integer, then each point of P is collinear with some point of Y. This implies that each point w_i of Z is either on a line joining points of X and Y or is collinear with $1+(n-1)/s$ points of Y. Suppose $n < 1+t$. Then there is some line L incident with some point of X but not incident with any point of Y. But then any point w_i on L, $w_i \notin X$, cannot be collinear with any point of Y, a contradiction. Hence it must be that $n = 1+t$, from which it follows that $m = 1+s^2/t$. This essentially completes the proof of the theorem. □

This result has several interesting corollaries.

1.4.2. *Let* x_1, x_2 *be noncollinear points.*

(i) *By putting* $X = \{x_1,x_2\}$ *and* $Y = \{x_1,x_2\}^{\perp}$ *we obtain the inequality of D.G. Higman. If also* $t = s^2$, *part of the corollary* 1.2.4 *of R. C. Bose and S.S. Shrikhande is obtained.*

(ii) *Put* $X = \{x_1,x_2\}^{\perp\perp}$ *and* $Y = \{x_1,x_2\}^{\perp}$. *If* $|X| = p+1$ (*and* $s > 1$) *it follows that* $pt \leq s^2$. *Moreover, if* $pt = s^2$ *and* $p < t$, *then each point* $w_i \notin cl(x_1,x_2)$ *is collinear with* $1+t/s = 1+s/p$ *points of* $\{x_1,x_2\}^{\perp}$. (*This inequality and its interpretation in the case of equality were first discovered by J.A. Thas* [194]. *Moreover, using an argument analogous to that of P.J. Cameron in the proof of* 1.2.3, *he proves that if* $p < t$ *and if each triad* (w_i,x_1,x_2), $w_i \notin cl(x_1,x_2)$, *has the same number of centers, then* $pt = s^2$).

(iii) *Let* $s^2 = t$, $s > 1$, *and suppose that the triad* (x_1,x_2,x_3) *is* 3-*regular. Put* $X = \{x_1,x_2,x_3\}^{\perp\perp}$ *and* $Y = \{x_1,x_2,x_3\}^{\perp}$. *Then* $|X| = |Y| = s+1$, *so that by* 1.4.1 *each point of* $P-(X \cup Y)$ *is collinear with precisely two points of* $X \cup Y$. (*This lemma was first discovered by J.A. Thas* [192] *using the trick of Bose-Cameron.*)

1.5. REGULARITY REVISITED

Let $S = (P,B,I)$ be a GQ of order (s,t), $s > 1$ and $t > 1$.

1.5.1. (i) *If* (x,y) *is antiregular with* $s = t$, *then* s *is odd* [213].

(ii) *If* S *has a regular point* x *and a regular pair* (L_0,L_1) *of noncon-*

concurrent lines for which x *is incident with no line of* $\{L_0,L_1\}^\perp$, *then* s = t *is even* [185] .

(iii) *If* x *is coregular, then the number of centers of any triad* (x,y,z) *has the same parity as* 1+t [144] .

(iv) *If each point is regular, then* $(t+1)|(s^2-1)s^2$.

Proof. Let (x,y) be antiregular with s = t > 1 and $\{x,y\}^\perp = \{u_0,\ldots,u_s\}$. For i = 0,1, let x I L_i I u_i I M_i I y, and let $K \in \{L_0,M_1\}^\perp$, $L_1 \neq K \neq M_0$. The points of K not collinear with x or y are denoted $v_2,\ldots,v_s$. Let $u_i \sim v_j$ for some $i \geqslant 2$. Then (x,y,v_j) is a triad with center u_i, and hence, by 1.3.6 (iii) , with exactly one other center $u_{i'}$. It follows that $u_2,\ldots,u_s$ occur in pairs of centers of triads of the form (x,y,v_j), each pair being uniquely determined by either of its members. Hence s-1 is even, and (i) is proved.

Next suppose that x and (L_0,L_1) satisfy the hypotheses of (ii), so that by 1.3.6 (i) we have s = t. If $\{L_0,L_1\}^{\perp\perp} = \{L_0,\ldots,L_s\}$, then let y_i be defined by $x \sim y_i$ I L_i, i = 0,...,s. By 1.3.1 the elements $x,y_0,y_1,\ldots,y_s$ are s+2 points of the projective plane π_x of order s defined by x. It is easy to see that each line of π_x through x contains exactly one point of the set $\{y_0,\ldots,y_s\}$. Suppose that the points y_i,y_j,y_k, with i,j,k distinct, are collinear in the plane π_x. Then the triad (y_i,y_j,y_k) has s+1 centers. Let u_j (resp., u_k) be the point incident with L_j (resp., L_k) and collinear with y_i. Then $u_k \in \{y_i,y_k\}^\perp$, hence $u_k \sim y_j$, giving a triangle with vertices y_j,u_k,u_j. Consequently $\{y_0,\ldots,y_s\}$ is an oval [49] of the plane π_x. Since the s+1 tangents of that oval concur at x, the order s of π_x is even [49].

Now assume that x is coregular. Let $u_1,\ldots,u_m$ be all the centers of a triad (x,y,z) with $\{x,y\}^\perp = \{u_1,\ldots,u_m,u_{m+1},\ldots,u_{t+1}\}$. We may suppose m < t+1. For i > m, let L_i be the line through x and u_i and M_i the line through y and u_i. Let K be the line through z meeting L_i and N the line through z meeting M_i. Met M be the line through y meeting K, and L the line through x meeting N. Since the line L_i through x is regular, the pair (L_i,N) must be regular, and it follows that M must meet L in some point $u_{i'} \in \{x,y\}^\perp$, $m+1 \leqslant i' \leqslant t+1$, $i' \neq i$. In this way with each point $u_i \in \{u_{m+1},\ldots,u_{t+1}\}$ there corresponds a point $u_{i'} \in \{u_{m+1},\ldots,u_{t+1}\}$, $i \neq i'$, and clearly this correspondence is involutory. Hence the number of points of $\{x,y\}^\perp$ that are not centers of (x,y,z) is even, proving (iii).

Finally, assume that each point is regular. The number of hyperbolic lines of S equals $(1+s)(1+st)s^2t/(t+1)t$. Hence $(t+1)|(1+s)(1+st)s^2$. Since $(1+s)(1+st)s^2 = (1+s)(1+s(t+1)-s)s^2$, this divisibility condition is equivalent to $(t+1)|(1+s)(1-s)s^2$, proving (iv). □

We collect here several useful consequences of 1.3.6 and 1.5.1 , always with $s > 1$ and $t > 1$.

1.5.2. (i) *If* S *has a regular point* x *and a regular line* L *with* $x \not\mathrm{I}\, L$, *then* $s = t$ *is even* [185].

(ii) *If* $s = t$ *is odd and if* S *contains two regular points, then* S *is not self-dual* [185].

(iii) *If* x *is coregular and* t *is odd, then* $|\{x,y\}^{\perp\perp}| = 2$ *for all* $y \notin x^{\perp}$ [144].

(iv) *If* x *is coregular and* $s = t$, *then* x *is regular iff* s *is even* [127, 144] .

(v) *If* x *is coregular and* $s = t$, *then* x *is antiregular iff* s *is odd* [144] .

Proof. If S has a regular point x and a regular line L, $x \not\mathrm{I}\, L$, then it is easy to construct a line Z, $Z \not\sim L$, such that x is incident with no line of $\{L,Z\}^{\perp}$. Then from 1.5.1 (ii) it follows that $s = t$ is even. Now suppose that $s = t$ is odd, and that S contains two regular points x and y. If S admits an anti-automorphism θ, then x^{θ} and y^{θ} are regular lines. Since at least one of x^{θ} and y^{θ} is not incident with at least one of x and y, an application of part (i) finishes the proof of (ii).

For the remainder of the proof suppose that x is coregular, and y is an arbitrary point not collinear with x. If $z \in \{x,y\}^{\perp\perp} - \{x,y\}$, and if $z' \mathrm{I}\, zu$, $z' \notin \{z,u\}$, for some $u \in \{x,y\}^{\perp}$, then u is the unique center of (x,y,z'). Hence t is even by 1.5.1 (iii), proving (iii). Now assume $s = t$. If x is regular, then any triad (x,y,z) has 1 or 1+s centers by 1.3.6, implying s is even. Conversely, if s is even, then by 1.5.1 (iii) any triad (x,y,z) is centric, hence by 1.3.6 x is regular, proving (iv). To prove (v), first note that if $s = t$ and x is antiregular then s is odd by 1.5.1 (i). Conversely, suppose that $s = t$ is odd and let (x,y,z) be any triad containing x. By 1.5.1 (iii) the number of centers of (x,y,z) must be even. Hence from 1.3.6 (iii) it follows that x must be antiregular. □

1.6. SEMIREGULARITY AND PROPERTY (H)

Throughout this section $S = (P,B,I)$ will denote a GQ of order (s,t), and the notation of Section 1.3 will be used freely.

Let x,y be fixed, noncollinear points. Each point $u \in \{x,y\}^\perp$ is the unique center of $(s-1)n(\{x,y\}^\perp)$ triads (x,y,z) with $z \in cl(x,y)$. It follows that (x,y,z) can be a unicentered triad only for $z \in cl(x,y)$ precisely when $N_1 = (t+1)(s-1)n(\{x,y\}^\perp)$, which proves the first part of the following theorem.

1.6.1. (i) *Each point of* S *is semiregular iff* $N_1 = (t+1)(s-1)N_{t+1}$ *for each pair* (x,y) *of noncollinear points.*

(ii) *If* $s = 1$ *or* $t = 1$ *then each point is semiregular and hence satisfies property* (H).

(iii) *If* $s = t$ *and* $u \in P$ *is regular or antiregular, then* u *is semiregular.*

(iv) *If* $s > t$, *then* $u \in P$ *is regular iff* u *is semiregular.*

Proof. Parts (i) and (ii) are easy. Suppose $u \in P$ is regular. If u is a center of the triad (x,y,z), then (x,y) is regular. But $|\{x,y\}^{\perp\perp}| = 1+t$ implies that $u^\perp \subset cl(x,y)$. Hence $z \in cl(x,y)$, implying that u is semiregular. Conversely, suppose that $u \in P$ is both semiregular and not regular. Then there must be a pair (x,y), $x \not\sim y$, $x, y \in u^\perp$, with $|\{x,y\}^{\perp\perp}| < 1+t$. It follows that some line L through u is incident with no point of $\{x,y\}^{\perp\perp}$. By the semiregularity of u, the s points of L different from u must each be collinear with a distinct point of $\{x,y\}^\perp$ different from u. Hence $s \leqslant t$, proving (iv). To complete the proof of (iii), let $s = t$ and suppose u is antiregular. From 1.3.6 (iii) it follows that each triad of points in $u^\perp$ has exactly two centers, implying that u is semiregular. □

Remark: *It is now easy to see that each point of* S *is semiregular if any one of the following holds :*

(i) $s = 1$ *or* $t = 1$,

(ii) *each point of* S *is regular,*

(iii) $s = t$ *and each point is antiregular,*

(iv) $t = s^2$ (*since each triad has* $1+s$ *centers*),

(v) $|\{x,y\}^{\perp\perp}| \geqslant 1+s^2/t$ *for all points* x,y *with* $x \not\sim y$ (*use* 1.4.2 (ii)).

For $x,y \in P$, $x \not\sim y$, let $u \in \{x,y\}^\perp$ and put $T = \{x,y\}^{\perp\perp}$, so $T \subset u^\perp$. If $L_1,\ldots,L_r$ are the lines projecting T from u, $r = |T|$, put T_u =

$\{x \in P \parallel x\ I\ L_i$ for some $i = 1,\ldots,r\}$.

1.6.2. (i) *Let* u *be a point of S having property* (H). *Let* T *and* T' *be two spans of noncollinear points both contained in* $u^{\perp}$. *If* $Tu \cap T'u$ *contains two noncollinear points, then* $Tu = T'u$, *so* $|T| = |T'|$.

(ii) *Let each point of S have property* (H). *Then there is a constant* p *such that* $|\{x,y\}^{\perp\perp}| = 1+p$ *for all points* x,y *with* $x \not\sim y$.

Proof. (i) If $|T \cap T'| \geq 2$, clearly $T = T'$. So first suppose $T \cap T' = \{x\}$. By hypothesis there must be points $y \in T$, $y' \in T'$, with $y \sim y'$, $y \neq y'$. If $T' = \{x,y'\}$, then clearly $T'u \subset Tu$. Now suppose there is some point $z' \in T'-\{x,y'\}$. Since $y \in cl(x,z')$ and u has property (H), it must be that $z' \in cl(x,y)$, i.e. $z'^{\perp} \cap T \neq \phi$. Since S contains no triangles we have $z' \in Tu$, implying $T' \subset Tu$. It follows that always $T'u \subset Tu$. Similarly, $Tu \subset T'u$. Finally, suppose that $T \cap T' = \phi$, but $\{z, z'\} \subset Tu \cap T'u$, $z \not\sim z'$. Let x and x' be the points of T and T', respectively, on the line uz, and let y and y' be the points of T and T', respectively, on the line uz'. So $T = \{x,y\}^{\perp\perp}$, $T' = \{x',y'\}^{\perp\perp}$. If we put $T'' = \{x,y'\}^{\perp\perp}$, then by the previous case $Tu = T''u = T'u$.

(ii) First suppose that T and T' are both hyperbolic lines with $T \cup T' \subset u^{\perp}$ for some point u. If $Tu \cap T'u$ contains two noncollinear points, then $|T| = |T'|$ by (i). Suppose there is a point $y \neq u$ with $y \in Tu \cap T'u$. Let $y_1 \in T$, $y_1 \not\sim y$, and $y_1' \in T'$, $y_1' \not\sim y$. If $T_1 = \{y,y_1\}^{\perp\perp}$ and $T_1' = \{y,y_1'\}^{\perp\perp}$, then by (i) we have $Tu = T_1u$ and $T'u = T_1'u$. We may assume that $T_1 \neq T_1'$, and hence that $T_1 \not\subset T_1'$ and $T_1' \not\subset T_1$. Let $z \in T_1-T_1'$ and $z' \in T_1'-T_1$, where we may assume $z \not\sim z'$, since otherwise $|T| = |T_1| = |T_1'| = |T_1|$. As $z' \notin T_1$, there is a point $u' \in \{y,z\}^{\perp}$ and $u' \notin \{y,z'\}^{\perp}$. Let L be the line through z' that has a point v on u'z ($u' \neq v \neq z$), and let M be the line through y having a point w in common with L ($v \neq w \neq z'$). By (i) we know $|\{v,y\}^{\perp\perp}| = |\{z,y\}^{\perp\perp}| = |T_1|$ and $|\{v,y\}^{\perp\perp}| = |\{z',y\}^{\perp\perp}| = |T_1'|$. Hence $|T| = |T'|$ in case $Tu \cap T'u \neq \phi$. So suppose that $Tu \cap T'u = \phi$, and let $z \in T$, $z' \in T'$. From the preceding case it follows that $|T| = |\{z,z'\}^{\perp\perp}| = |T'|$. This completes the proof that $|T| = |T'|$ in case $T \cup T' \subset u^{\perp}$ for some point u.

Finally, suppose $T = \{y,z\}^{\perp\perp}$, $y \not\sim z$, $T' = \{y',z'\}^{\perp\perp}$, $y' \not\sim z'$, and $\{y,z\}^{\perp} \cap \{y',z'\}^{\perp} = \phi$. If each point of $\{y,z\}^{\perp}$ is collinear with each point of $\{y',z'\}^{\perp}$, then $T = \{y',z'\}^{\perp}$ and $T' = \{y,z\}^{\perp}$, so $|T| = |T'|$. So suppose that $u \not\sim u'$ with $u \in \{y,z\}^{\perp}$, $u' \in \{y',z'\}^{\perp}$. Let $v,w \in \{u,u'\}^{\perp}$. The points of $\{v,w\}^{\perp\perp} \cup T$ (resp., $\{v,w\}^{\perp\perp} \cup T'$) are collinear with the point u (resp., u'). Hence $|T| = |\{v,w\}^{\perp\perp}| = |T'|$, by the preceding case . □

1.6.3. *Let each point of S be semiregular and suppose* $s > 1$. *Then one of the following must occur :*

(i) $s > t$ *and each point of S is regular.*

(ii) $s = t$ *and each point of S is regular or each point is antiregular.*

(iii) $s < t$ *and* $|\{x,y\}^{\perp\perp}| = 2$ *for all* $x,y \in P$ *with* $x \not\sim y$.

(iv) *There is a constant* p, $1 < p < t$, *such that* $|\{x,y\}^{\perp\perp}| = 1+p$ *for all points* x,y *with* $x \not\sim y$.

Proof. Since each point of S is semiregular, each point of S has property (H). Hence there is a constant p, $1 \leqslant p \leqslant t$, such that $|\{x,y\}^{\perp\perp}| = 1+p$ for all points x,y with $x \not\sim y$. If $p = t$, then each point is regular and consequently $s \geqslant t$. Now assume $p = 1$ and $s \geqslant t$. If $t = 1$, then $s > t$ and each point of S is regular. For $t \neq 1$ we must show that $s = t$ and each point is antiregular. So let (x,y,z) be a triad with center u. Then $|\{x,y\}^{\perp\perp}| = 2$ implies $z \notin cl(x,y)$, so by the semiregularity of u the triad must have another center. Hence (x,y) belongs to no triad with a unique center, i.e. $N_1 = 0$. By 1.3.6 (iii) $s = t$ and (x,y) is antiregular. So each point of S is antiregular. □

1.7. TRIADS WITH EXACTLY 1+t/s CENTERS

Let x be a fixed point of the GQ S = (P,B,I) of order (s,t), $s > 1$, $t > 1$.

1.7.1. (i) *The triads* (y_1,y_2,y_3) *contained in* $x^{\perp}$ *have a constant number of centers iff the triads* (x,u_1,u_2) *containing* x *have exactly* 0 *or* α (α *a constant) centers . If one of these equivalent situations occurs, then* $(s+t) \mid s(t-1)$ *and the constants both equal* 1+t/s.

(ii) *Let* $y \in P - x^{\perp}$. *Then no triad containing* (x,y) *has more than* 1+t/s *centers iff each such triad has exactly* 0 *or* 1+t/s *centers iff no such triad has* α *centers with* $0 < \alpha < 1+t/s$. *In such a case there are* $t(s-1)(s^2-t)/(s+t)$ *acentric triads containing* x *and* y, *and* $(t^2-1)s^2/(s+t)$ *triads containing* x *and* y *with exactly* 1+t/s *centers.*

(iii) *If* $s = q^n$ *and* $t = q^m$ *with* q *a prime power, and if each triad in* $x^{\perp}$ *has* 1+t/s *centers, then there is an odd integer* a *for which* $n(a+1) = ma$.

Proof. (i) Suppose there is a constant α such that each triad (x,u_1,u_2)

containing x has 0 or α centers. By the remark following eq. (11) in 1.3, we have $\alpha = 1+t/s$. There are $d = (t^2-1)s^3t/6$ triads $T_1, T_2, \ldots, T_d$ contained in $x^\perp$. Let $1+r_i$ be the number of centers of T_i, so that $\sum_{i=1}^{d} r_i = s^2t(t+1)t(t-1)/6$. Count the ordered triples (T_i, u_1, u_2), where T_i is a triad in $x^\perp$ and (x, u_1, u_2) is an ordered triad in $T_i^\perp$, to obtain $\sum_i r_i(r_i-1) = s^2 t N_\alpha (1+t/s)(t/s)(t/s-1)/6$. Here N_α is the number of triads containing $(x, u_1), x \not\sim u_1$, and having exactly $\alpha = 1+t/s$ centers. From eq. (2) of 1.3.5 it follows that $N_\alpha = (t^2-1)s^2/(s+t)$. Hence $d(\sum r_i^2)-(\sum r_i)^2 = 0$, implying that r_i is the constant $(\sum r_i)/d = t/s$.

Conversely, assume that the number r_i+1 of centers of T_i is a constant. Then $r_i = s^2t(t+1)t(t-1)/(t^2-1)s^3t = t/s$. Fix y_1 in $x^\perp - \{x\}$. The number of triads $V_1, \ldots, V_{d'}$ containing x and having y_1 as center is $d' = t(t-1)s^2/2$. If $1+t_i$ denotes the number of centers of V_i, $1 \leqslant i \leqslant d'$, it is easy to check that $\sum_i t_i = stt(t-1)/2$ and $\sum_i t_i(t_i-1) = sts(t-1)(t/s)(t/s-1)/2$. Hence $d'(\sum t_i^2) = (\sum t_i)^2$, and t_i is the constant $(\sum t_i)/d' = t/s$. It follows immediately that each centric triad (x, u_1, u_2) has exactly $1+t/s$ centers.

Suppose that these equivalent situations occur. Fix u_1, $u_1 \not\sim x$, and let L be a line which is incident with no point of $\{x, u_1\}^\perp$ (since $s \neq 1$ such a line exists). Then the number of points $u_2, u_2 \mathrm{I} L$, for which (x, u_1, u_2) is a centric triad equals $(t-1)/(1+t/s)$. Hence $(s+t) \mid s(t-1)$, and (i) is proved.

(ii) Fix $y \in P - x^\perp$, and apply the notation and results of 1.3.

Using eq. (2) and eq. (3) we have

$$s^{-1} \sum_{\alpha=0}^{1+t} \alpha N_\alpha - t^{-1} \sum_{\alpha=0}^{1+t} \alpha(\alpha-1)N_\alpha = (t^2-1)-(t^2-1) = 0,$$

which may be rewritten as

$$N_1 = \sum_{\alpha=2}^{1+t} ((s\alpha^2 - \alpha(s+t))/t)N_\alpha. \tag{14}$$

The coefficient of N_α in (14) is nonnegative iff $\alpha \geq 1+t/s$, and equals 0 iff $\alpha = 1+t/s$. Assume that $N_\alpha = 0$ for $\alpha > 1+t/s$. Since $N_1 \geq 0$, we must have $N_\alpha = 0$ for $\alpha = 1,2,\ldots,t/s$. Hence each triad containing (x,y) has exactly 0 or $1+t/s$ centers. Conversely, assume $N_\alpha = 0$ for $0 < \alpha < 1+t/s$. Since $N_1 = 0$, we necessarily have $N_\alpha = 0$ for $\alpha > 1+t/s$. Hence each triad containing (x,y) has exactly 0 or $1+t/s$ centers. Finally, if this last condition holds, it is easy to use eq. (1) and eq. (2) to solve for N_0 and $N_{1+t/s}$, completing the proof of (ii).

(iii) Given the hypotheses of (iii), from part (i) we have $t \geq s$ and $(s+t) | s(t-1)$, from which it follows that $(1+q^{m-n}) | (q^m-1)$. Since $q^m-1 = (q^{m-n}+1)q^n-q^n-1$, there results $(1+q^{m-n}) | (1+q^n)$. Consequently $n(a+1) = ma$ with a odd. □

Remark : If the conditions of 1.7.1 (i) hold with $s = t > 1$, then s is odd and x is antiregular. Moreover, putting $s = t > 1$ in 1.7.1 (ii) yields part of 1.3.6 (iii).

For the remainder of this section we suppose that each triad contained in $x^\perp$ has exactly $1+t/s$ centers, so that each triad containing x has 0 or $1+t/s$ centers. Let $T = \{x,u_1,u_2\}$ be a fixed triad with $T^\perp = \{y_0,y_1,\ldots,y_{t/s}\}$. Each triad in $T^\perp$ also has $1+t/s$ centers. For $A \subset \{0,1,\ldots,t/s\}$, let $m(A)$ be the number of points collinear with y_i for $i \in A$, but not collinear with x,u_1,u_2 or y_i for $i \notin A$.

Note first that $\sum_A m(A) = |P-(x^\perp \cup u_1^\perp \cup u_2^\perp)|$, so

$$\sum_A m(A) = s^2t-2st-2s+3t-t/s . \tag{15}$$

Now count in different ways the number of ordered pairs (w,y_i) with $w \sim y_i$ and w not collinear with x,u_1, or u_2, to obtain

$$\sum_A |A| m(A) = (s+t)(t-2). \tag{16}$$

Next count the number of ordered triples (w,y_i,y_j) with $w \sim y_i$, $w \sim y_j$, $y_i \neq y_j$, and w not collinear with x,u_1, or u_2, to obtain

$$\sum_A |A|(|A|-1)m(A) = (s+t)t(t-2)/s^2. \tag{17}$$

Finally, count the number of ordered 4-tuples (w,y_i,y_j,y_k) with w a center of the triad (y_i,y_j,y_k), and w not collinear with x,u_1, or u_2, to obtain

$$\sum_A |A|(|A|-1)(|A|-2)m(A) = (s+t)t(t-s)(t-2s)/s^4. \tag{18}$$

For $0 \leqslant \alpha \leqslant 1+t/s$, put $M_\alpha = \Sigma^\alpha m(A)$, where Σ^α denotes the sum over all A with $|A| = \alpha$. Then eqs. (15) - (18) become

$$\sum_\alpha M_\alpha = s^2t-2st-2s+3t-t/s, \tag{19}$$

$$\sum_\alpha \alpha M_\alpha = (s+t)(t-2), \tag{20}$$

$$\sum_\alpha \alpha(\alpha-1)M_\alpha = (s+t)t(t-2)/s^2, \tag{21}$$

$$\sum_\alpha \alpha(\alpha-1)(\alpha-2)M_\alpha = (s+t)t(t-s)(t-2s)/s^4. \tag{22}$$

We conclude this section with a little result on GQ of order (s,s^2), in which each triad must have exactly $1+t/s = 1+s$ centers.

1.7.2. *Let* $S = (P,B,I)$ *be a* GQ *of order* (s,s^2), $s > 2$, *with a triad* (x_0,x_1,x_2) *for which* $\{x_0,x_1,x_2\}^\perp = \{y_0,\ldots,y_s\}$, $\{x_0,\ldots,x_{s-1}\} \subset \{x_0,x_1,x_2\}^{\perp\perp}$. *Suppose there is a point* x_s *for which* $x_s \sim y_i$, $i = 0,\ldots,s-1$ *and* $x_s \neq x_j$, $j = 0,\ldots,s-1$. *Then* $x_s \sim y_s$, *i.e.* (x_0,x_1,x_2) *is* 3-*regular. It follows immediately that any triad in a* GQ *of order* (3,9) *must be* 3-*regular.*

Proof. The number of points collinear with y_s and also with at least two points of $\{y_0,\ldots,y_{s-1}\}$ is at most $s(s-1)/2+s$, and the number of points collinear with y_s and incident with some line x_sy_i, $i = 0,1,\ldots,s-1$, is at most s. Since $s > 2$, we have $s(s-1)/2+2s < s^2+1 = t+1$. Hence there is a line L incident with y_s, but not concurrent with x_sy_i, $i = 0,1,\ldots,s-1$, and not incident with an element of $\{y_i,y_j\}^\perp$, $i \neq j$, $0 \leqslant i, j \leqslant s-1$. The point incident with L and collinear with y_i is denoted by z_i, $i = 0,\ldots,s-1$. Clearly all s points z_i are distinct. Since S has no triangles, the point x_s is not collinear with any z_i, forcing $x_s \sim y_s$. □

1.8. <u>OVOIDS, SPREADS, AND POLARITIES</u>

An *ovoid* of the GQ $S = (P,B,I)$ is a set O of points of S such that each line of S is incident with a unique point of O. A *spread* of S is a set R of lines of S such that each point of S is incident with a unique line. It is trivial that a GQ with s = 1 or t = 1 has ovoids and spreads.

1.8.1. *If* O (*resp.,* R) *is an ovoid* (*resp., spread*) *of the* GQ S *of order* (s,t), *then* $|O| = 1+st$ (*resp.,* $|R| = 1+st$).

Proof. For an ovoid O, count in different ways the number of ordered pairs (x,L) with $x \in O$ and L a line of S incident with x. Use duality for a spread. □

1.8.2. (*S.E. Payne* [116]). *If the* GQ $S = (P,B,I)$ *of order* s *admits a polarity, then* 2s *is a square. Moreover, the set of all absolute points* (*resp., lines*) *of a polarity* θ *of* S *is an ovoid* (*resp., a spread*) *of* S .

Proof. Let θ be a polarity of the GQ $S = (P,B,I)$ of order s. A point x (resp., line L) of S is an absolute point (resp., line) of θ provided $x \mathrm{I} x^\theta$ (resp., $L \mathrm{I} L^\theta$). We first prove that each line L of S is incident with at most one absolute point of θ . So suppose that x,y are distinct absolute points incident with L. Then $x \mathrm{I} x^\theta$, $y \mathrm{I} y^\theta$, and $x^\theta \sim y^\theta$ since $x \sim y$. Hence $L \in \{x^\theta, y^\theta\}$, since otherwise there arises a triangle with sides L, x^θ, y^θ. So suppose $L = x^\theta$. As $y \mathrm{I} x^\theta$, we have $x \mathrm{I} y^\theta$. Since $y \mathrm{I} y^\theta$, clearly $y^\theta = xy = L = x^\theta$, implying x = y, a contradiction. So each line of S is incident with at most one absolute point of θ . A line L is absolute iff $L \mathrm{I} L^\theta$ iff L^θ is absolute. Now assume L is not absolute, i.e. $L \not\mathrm{I} L^\theta$. If $L^\theta \mathrm{I} M \mathrm{I} u \mathrm{I} L$, then $L^\theta \mathrm{I} u^\theta \mathrm{I} M^\theta \mathrm{I} L$, hence $u^\theta = M$ and $M^\theta = u$. Consequently u and M are absolute, and we have proved that each line L is incident with at least one absolute point. It follows that the set of absolute points of θ is an ovoid. Dually, the set of all absolute lines is a spread.

Denote the absolute points of θ by $x_1, \ldots, x_{s^2+1}$. It is clear that the absolute lines of θ are the images $x_i^\theta = L_i$, $1 \leq i \leq s^2+1$. Let $P = \{x_1, \ldots, x_{s^2+1}, \ldots, x_v\}$, $B = \{L_1, \ldots, L_{s^2+1}, \ldots, L_v\}$, with $x_i^\theta = L_i$, $1 \leq i \leq v$, and let $D = (d_{ij})$ be the $v \times v$ matrix over R for which $d_{ij} = 0$ if $x_i \not\mathrm{I} L_j$, and $d_{ij} = 1$ if $x_i \mathrm{I} L_j$ (i.e., D is an incidence matrix of the structure S). Then D is symmetric and $D^2 = (1+s)I+A$, where A is the adjacency matrix of the graph (P,E) (cf. the proof of 1.2.2). By 1.2.2 D^2 has eigenvalues $(s+1)^2$, 0, and 2s, with respective multiplicities 1, $s(s^2+1)/2$, and $s(s+1)^2/2$. Since D has a constant row sum (resp., column sum) equal to s+1, it clearly has s+1 as an eigenvalue . Hence D has eigenvalues s+1, 0, $\sqrt{2s}$ and $-\sqrt{2s}$ with respective multiplicities 1, $s(s^2+1)/2$, a_1 and a_2, where $a_1+a_2 = s(s+1)^2/2$. Consequently $\mathrm{tr}D = s+1+(a_1-a_2)\sqrt{2s}$. But trD is also the number of absolute points of θ, i.e. $\mathrm{tr}D = 1+s^2$. So $s^2+1 = s+1+(a_1-a_2)\sqrt{2s}$, implying that 2s is a square. □

1.8.3. *A* GQ $S = (P,B,I)$ *of order* (s,t), *with* $s > 1$ *and* $t > s^2-s$, *has no ovoid.*

Proof. We present two proofs of this theorem. The first is due to J.A. Thas [207]; the second is the original one due to E.E. Shult [165].

(a) Let O be an ovoid of a GQ S of order (s,s^2), $s \neq 1$. Let $x,y \in O$, $x \neq y$, and count the number N of ordered pairs (z,u) with $u \in O-\{x,y\}$, $z \in \{x,y\}^\perp$, and $u \sim z$. Since any two points of O are noncollinear, we have $\{x,y\}^\perp \cap O = \phi$, and hence $N = (s^2+1)(s^2-1)$. By 1.2.4 N also equals $(|O|-2)(1+s) = (s^3-1)(1+s)$, which yields a contradiction if $s \neq 1$. Hence a GQ of order (s,s^2), $s \neq 1$, has no ovoid.

Let O be an ovoid of a GQ S of order (s,t), $s \neq 1$. Since $t \neq s^2$, by 1.2.5 $t \leqslant s^2-s$.

(b) Let O be an ovoid of the GQ $S = (P,B,I)$ of order (s,t) with $s \neq 1$. Fix a point $x \notin O$ and let $V = \{y \in O \parallel y \sim x\}$. Further, let $z \notin O$, $z \not\sim x$, and let $L_z = \{u \in V \parallel u \sim z\}$. We note that $d = |\{z \in P \parallel z \notin O \text{ and } z \not\sim x\}| = t(s^2-s+1)$. If $t_z = |L_z|$, then $\sum_z t_z = (1+t)ts$ and $\sum_z t_z(t_z-1) = (1+t)t^2$. Since $d\sum_z t_z^2-(\sum_z t_z)^2 \geqslant 0$, there arises $t(s^2-s+1)(1+t)t(t+s)-(1+t)^2t^2s^2 \geqslant 0$. Hence $(s-1)(s^2-t-s) \geqslant 0$, form which $t \leqslant s^2-s$. □

1.8.4. ([207]). *Let* $S = (P,B,I)$ *be a* GQ *of order* s, *having a regular pair* (x,y) *of noncollinear points. If* O *is an ovoid of* S, *then* $|O \cap \{x,y\}^{\perp\perp}|$, $|O \cap \{x,y\}^\perp| \in \{0,2\}$, *and* $|O \cap (\{x,y\}^\perp \cup \{x,y\}^{\perp\perp})| = 2$. *If the* GQ S *of order* s, $s \neq 1$, *contains an ovoid* O *and a regular point* z *not on* O , *then* s *is even.*

Proof. Let $O \cap (\{x,y\}^\perp \cup \{x,y\}^{\perp\perp}) = \{y_1,\ldots,y_r\}$. If $u \in P-(\{x,y\}^\perp \cup \{x,y\}^{\perp\perp})$, then u is on just one line joining a point of $\{x,y\}^\perp$ to a point of $\{x,y\}^{\perp\perp}$; if $u \in \{x,y\}^\perp \cup \{x,y\}^{\perp\perp}$, then u is on s+1 lines joining a point of $\{x,y\}^\perp$ to a point of $\{x,y\}^{\perp\perp}$. We count the number of all pairs (L,u), with L a line joining a point of $\{x,y\}^\perp$ to a point of $\{x,y\}^{\perp\perp}$ and with u a point of O which is incident with L. We obtain $(s+1)^2 = s^2+1-r+r(1+s)$. Hence r = 2.

Since no two points of O are collinear, there follows $|O \cap \{x,y\}^{\perp\perp}|$, $|O \cap \{x,y\}^{\perp}| \in \{0,2\}$.

Let O be an ovoid of the GQ *S* of order s and let z be a regular point not on O. Let $y \notin O$, $z \sim y$, $z \neq y$. The points of O collinear with y are denoted by $z_0,\ldots,z_s$, with z_0 I zy. By the first part of the theorem, for each $i = 1,\ldots,s$, $\{z,z_i\}^{\perp\perp} \cap O = \{z_i,z_j\}$ for some $j \neq i$. Hence $|\{z_1,\ldots,z_s\}|$ is even, proving the second part of the theorem. □

There is an immediate corollary.

1.8.5. *Let* $S = (P,B,I)$ *be a* GQ *of order* s,s *even, having a regular pair of noncollinear points. Then the pointset* P *cannot be partitioned into ovoids.*
Proof. Let (x,y) be a regular pair of noncollinear points of the GQ $S = (P,B,I)$ of order s. If P can be partitioned into ovoids, then by 1.8.4 $|\{x,y\}^{\perp\perp}|$ is even. Hence s is odd. □

The following is a related result for the case $s \neq t$.

1.8.6.([207]). *Let* $S = (P,B,I)$ *be a* GQ *of order* (s,t), $1 \neq s \neq t$, *and suppose that there is an hyperbolic line* $\{x,y\}^{\perp\perp}$ *of cardinality* p+1 *with* $pt = s^2$. *Then any ovoid* O *of S has empty intersection with* $\{x,y\}^{\perp\perp}$.
Proof. Suppose that the ovoid O has r+1, $r \geqslant 0$, points in common with $\{x,y\}^{\perp\perp}$. Count the number N of ordered pairs (z,u), with $z \in \{x,y\}^{\perp}$, $u \in O-\{x,y\}^{\perp\perp}$, and $u \sim z$. Since any two points of O are noncollinear, we have $O \cap \{x,y\}^{\perp} = \phi$. Hence $N = (t+1)(t-r) = (s^2/p+1)(s^2/p-r)$. The number of points of $O-\{x,y\}^{\perp\perp}$ collinear with a point of $\{x,y\}^{\perp\perp}$ equals $(p-r)(t+1) = (p-r)(s^2/p+1)$. Each of these points of O is collinear with exactly one point of $\{x,y\}^{\perp}$. Further, by 1.4.2 (ii), any point of O-cl(x,y) is collinear with exactly 1+s/p points of $\{x,y\}^{\perp}$. Consequently N also equals $(p-r)(s^2/p+1)+(s^3/p-r-(p-r)(s^2/p+1))(s+p)/p$. Comparing these two values for N, we find $r = p/s$. Hence both $p|s$ and $s|p$, implying $p = s$ and $r = 1$. From $pt = s^2$ it follows that $t = s$, a contradiction. □

Remark : Putting $t = s^2$, $s \neq 1$, in the above result we find that a GQ of order (s,s^2), $s \neq 1$, has no ovoid.

1.9. AUTOMORPHISMS

Let $S = (P,B,I)$ be a GQ of order (s,t) with $P = \{x_1,\ldots,x_v\}$ and $B = \{L_1,\ldots,L_b\}$. Further, let $D = (d_{ij})$ be the $v \times b$ matrix over C for which $d_{ij} = 0$ or 1 according as $x_i \not{I}\, L_j$ or x_i I L_j (i.e. D is an incidence matrix of the structure S). Then $DD^T = A+(t+1)I$, where A is an adjacency matrix of the point graph of S (cf. 1.2.2). If $M = DD^T$, then by the proof of 1.2.2, M has eigenvalues $\tau_0 = (1+s)(1+t)$, $\tau_1 = 0$, $\tau_2 = s+t$, with respective multiplicities $m_0 = 1$, $m_1 = s^2(1+st)/(s+t)$, $m_2 = st(1+s)(1+t)/(s+t)$.

Let θ be an automorphism of S and let $Q = (q_{ij})$ (resp., $R = (r_{ij})$) be the $v \times v$ matrix (resp., $b \times b$ matrix) over C, with $q_{ij} = 1$ (resp., $r_{ij} = 1$) if $x_i^\theta = x_j$ (resp., $L_i^\theta = L_j$) and $q_{ij} = 0$ (resp., $r_{ij} = 0$) otherwise. Then Q and R are permutation matrices for which $DR = QD$. Since $Q^T = Q^{-1}$ and $R^T = R^{-1}$ for permutation matrices, there arises $QM = QDD^T = DRD^T = DRR^TD^T(Q^{-1})^T = DD^TQ = MQ$. Hence $QM = MQ$.

1.9.1. (*C.T. Benson* [10], *cf. also* [142]). *If* f *is the number of points fixed by the automorphism* θ *and if* g *is the number of points* x *for which* $x^\theta \neq x \sim x^\theta$, *then*

$$\mathrm{tr}(QM) = (1+t)f+g \ \textit{and}\ (1+t)f+g \equiv 1+st \pmod{s+t}.$$

Proof. Suppose that θ has order n, so that $(QM)^n = Q^nM^n = M^n$. It follows that the eigenvalues of QM are the eigenvalues of M multiplied by the appropriate roots of unity. Since $MJ = (1+s)(1+t)J$ (J is the $v \times v$ matrix with all entries equal to 1), we have $(QM)J = (1+s)(1+t)J$, so $(1+s)(1+t)$ is an eigenvalue of QM. From $m_0 = 1$ it follows that this eigenvalue of QM has multiplicity 1. Further, it is clear that 0 is an eigenvalue of QM with multiplicity $m_1 = s^2(1+st)/(s+t)$. For each divisor d of n, let U_d denote the sum of all primitive d th roots of unity. Then U_d is an integer [87]. For each divisor d of n, the primitive d th roots of unity all contribute the same number of times to eigenvalues θ of QM with $|\theta| = s+t$. Let a_d denote the multiplicity of $\xi_d(s+t)$ as an eigenvalue of QM, with $d|n$ and ξ_d a primitive d th root of unity. Then we have $\mathrm{tr}(QM) = \sum_{d|n} a_d(s+t)U_d + (1+s)(1+t)$. Hence $\mathrm{tr}(QM) \equiv 1+st \pmod{s+t}$. Let f and g be as given in the theorem. Since the entry on the i th row and i th column of QM is the number of lines incident with x_i and x_i^θ, we have $tr(QM) = (1+t)f+g$, completing the proof. □

1.9.2. *If* f_P *(resp.,* f_B*) is the number of points (resp., lines) fixed by the automorphism* θ*, and if* g_P *(resp.,* g_B*) is the number of points* x *(resp., lines* L*) for which* $x^\theta \neq x \sim x^\theta$ *(resp.,* $L^\theta \neq L \sim L^\theta$*), then*

$$tr(QDD^T) = (1+t)f_P+g_P = (1+s)f_B+g_B = tr(RD^TD).$$

Proof. The last equality is just the dual of the first, which was established in 1.9.1. To obtain the middle equality, count the pairs (x,L) for which $x \in P$, $L \in B$, x I L, $x^\theta \sim x$, $L^\theta \sim L$. This number is given by $(1+t)f_P+g_P+N/2 = (1+s)f_B+g_B+N/2$, where N is the number of pairs (x,L) for which x I L, $x^\theta \sim x$, $x \neq x^\theta$, $L^\theta \sim L$, $L^\theta \neq L$. The desired equality follows. □

1.10. A SECOND APPLICATION OF HIGMAN - SIMS

Let $S = (P,B,I)$ be a GQ of order (s,t), $P = \{w_1,\ldots,w_v\}$. Let $\Delta = \{\Delta_1,\Delta_2\}$ be any partition of $\{1,\ldots,v\}$. Put $\delta_1 = |\Delta_1|$, $\delta_2 = |\Delta_2| = v-\delta_1$.
Let δ_{ij} be the number of ordered pairs (k,ℓ) for which $k \in \Delta_i, \ell \in \Delta_j$ and $w_\ell \not\sim w_k$. Here we recall the notation of 1.4. So for the matrix B of 1.4, the resulting B^Δ is

$$B^\Delta = \begin{pmatrix} e & s^2t-e \\ \delta_1(s^2t-e)/\delta_2 & s^2t-\delta_1(s^2t-e)/\delta_2 \end{pmatrix}, \text{ with } e = \delta_{11}/\delta_1.$$

One eigenvalue of B^Δ is clearly s^2t, so the other is $\bar{t} = tr(B^\Delta)-s^2t = e-\delta_1(s^2t-e)/\delta_2$. By the result of C.C. Sims as applied in 1.4, we have $-s \leqslant e-\delta_1(s^2t-e)/\delta_2$, with equality holding iff $\delta_1-e = s+\delta_1/(1+s)$. If equality holds $(\delta_1-v, \delta_1)^T$ is an eigenvector of B^Δ associated with the eigenvalue -s, hence it must be that $\bar{x} = (\underbrace{\delta_1-v,\ldots,\delta_1-v}_{\delta_1 \text{ times}}, \underbrace{\delta_1,\ldots,\delta_1}_{\delta_2 \text{ times}})$ is an eigenvector of B associated with -s. It is straightforward to check that this holds iff each point of Δ_1 is collinear with exactly $\delta_1-e = s+\delta_1/(1+s)$ points of Δ_1, and each point of Δ_2 is collinear with exactly $\delta_2+e-s^2t+s =$

$s+\delta_2/(1+s)$ points of Δ_2. The following theorem is obtained.

1.10.1. (*S.E. Payne* [125]). *Let* X_1 *be any nonempty, proper subset of points of the* GQ S *of order* (s,t), $|X_1| = \delta_1$. *Then the average number* $\bar{e}$ *of points of* X_1 *collinear in* S *with a fixed point of* X_1 *satisfies* $\bar{e} \leqslant s+\delta_1/(1+s)$, *with equality holding iff each point of* X_1 *is collinear with exactly* $s+\delta_1/(1+s)$ *points of* X_1, *iff each point of* $X_2 = P-X_1$ *is collinear with exactly* $\delta_1/(1+s)$ *points of* Δ_1.

2 Subquadrangles

2.1. DEFINITIONS

The GQ $S' = (P',B',I')$ of order (s',t') is called a *subquadrangle* of the GQ $S = (P,B,I)$ of order (s,t) if $P' \subset P$, $B' \subset B$, and if I' is the restriction of I to $(P' \times B') \cup (B' \times P')$. If $S' \neq S$, then we say that S' is a *proper* subquadrangle of S .

From $|P| = |P'|$ it follows easily that $s = s'$ and $t = t'$, hence if S' is a proper subquadrangle then $P \neq P'$, and dually $B \neq B'$. Let $L \in B$. Then precisely one of the following occurs : (i) $L \in B'$, i.e. L belongs to S'; (ii) $L \notin B'$ and L is incident with a unique point x of P', i.e. L is *tangent to* S' *at* x; (iii) $L \notin B'$ and L is incident with no point of P', i.e. L is *external* to S'. Dually, one may define *external* points and *tangent* points of S'. From the definition of a GQ it easily follows that *no tangent point may be incident with a tangent line.*

2.2. THE BASIC INEQUALITIES

2.2.1. *Let S' be a proper subquadrangle of S, with notation as above. Then either* $s = s'$ *or* $s \geq s't'$. *If* $s = s'$, *then each external point is collinear with the* $1+st'$ *points of an ovoid of* S'; *if* $s = s't'$, *then each external point is collinear with exactly* $1+s'$ *points of* S'. *The dual holds, similarly.*

Proofs. (a) ([190]). Let V be the set of the points external to S'. Then $|V| = d = (1+s)(1+st)-(1+s')(1+s't')-(1+t')(1+s't')(s-s')$.

If $t = t'$, then from $d \geq 0$ there arises $(s-s')t(s-s't) \geq 0$, implying $s = s'$ or $s \geq s't$.

We now assume $t > t'$. Let $V = \{x_1,\ldots,x_d\}$ and let t_i be the number of points of P' which are collinear with x_i. We count in two different ways the ordered pairs (x_i,z), $x_i \in V$, $z \in P'$, $x_i \sim z$, and we obtain $\sum_i t_i = (1+s')(1+s't')(t-t')s$. Next we count in different ways the ordered triples (x_i,z,z'), $x_i \in V$, $z \in P'$, $z' \in P'$, $z \neq z'$, $x_i \sim z$, $x_i \sim z'$, and we obtain

$\sum_i t_i(t_i-1) = (1+s')(1+s't')s'^2t'(t-t')$. Hence $\sum_i t_i^2 = (1+s')(1+s't')\times$ $(t-t')(s+s'^2t')$. As $d\sum_i t_i^2-(\sum_i t_i)^2 \geqslant 0$, we obtain $(1+s')(1+s't')(t-t')\times$ $(s-s')(s-s't')(st+s'^2t'^2) \geqslant 0$. Since $t > t'$, it must be that $s = s'$ or $s \geqslant s't'$. Further, we note that $t_i = (\sum_i t_i)/d$ for all $i \in \{1,\ldots,d\}$ iff $d\sum_i t_i^2 -(\sum_i t_i)^2 = 0$, i.e. iff $s = s'$ or $s = s't'$. If $s = s'$, then $t_i = 1+st'$ for all i. Hence in such a case each external point is collinear with the $1+st'$ points of an ovoid of S'. If $s = s't'$, then $t_i = 1+s'$ for all i. □

(b) ([126]). Refer to the *proof* of 1.10.1, and let Δ_1 be the indices of the points of S', Δ_2 the set of remaining indices. Then $\delta_1 = (1+s')(1+s't')$, and each point indexed by an element of Δ_1 is collinear with exactly $1+s'+s't'$ such points. Hence $1+s'+s't' \leqslant s+(1+s')(1+s't')/(1+s)$. This is equivalent to $0 \leqslant (s-s't')(s-s')$, and when equality holds each external point is collinear with exactly $(1+s')(1+s't')/(1+s)$ points of S'. When $s = s'$, this number is $1+s't'$. When $s = s't'$, this number is $1+s'$. □

The next results are easy consequences of 2.2.1, although they first appeared in J.A. Thas [180].

2.2.2. *Let* $S' = (P',B',I')$ *be a proper subquadrangle of* $S = (P,B,I)$, *with* S *having order* (s,t) *and* S' *having order* (s,t'), *i.e.* $s = s'$ *and* $t > t'$. *Then we have :*

(i) $t \geqslant s$; *if* $t = s$, *then* $t' = 1$.

(ii) *If* $s > 1$, *then* $t' \leqslant s$; *if* $t' = s \geqslant 2$, *then* $t = s^2$.

(iii) *If* $s = 1$, *then* $1 \leqslant t' < t$ *is the only restriction on* t'.

(iv) *If* $s > 1$ *and* $t' > 1$, *then* $\sqrt{s} \leqslant t' \leqslant s$, *and* $s^{3/2} \leqslant t \leqslant s^2$.

(v) *If* $t = s^{3/2} > 1$ *and* $t' > 1$, *then* $t' = \sqrt{s}$.

(vi) *Let* S' *have a proper subquadrangle* S'' *of order* (s,t''), $s > 1$. *Then* $t'' = 1$, $t' = s$, *and* $t = s^2$.

Proofs. These are all easy consequences of 2.2.1, along with the inequality of D.G. Higman. We give two examples. (ii) Suppose that $s > 1$. By Higman's inequality we have $t \leqslant s^2$. Hence using the dual of 2.2.1 also, we have $t' \leqslant t/s \leqslant s$, implying $t' \leqslant s$. If $t' = s$, then $s = t' = t/s$, implying $t = s^2$. (vi) Let S' have a proper subquadrangle S'' of order (s,t''), $s > 1$. Then $t' \leqslant s$ and $t'' \leqslant t'/s$. Hence $t'' \leqslant t'/s \leqslant s/s$, implying $t'' = 1$ and $t' = s$. By

(ii) we have $t = s^2$. □

2.3. RECOGNIZING SUBQUADRANGLES

A theorem which will appear very useful for several characterization theorems is the following [180].

2.3.1. *Let* $S' = (P',B',I')$ *be a substructure of the* GQ $S = (P,B,I)$ *of order* (s,t) *for which the following conditions are satisfied :*

(i) *If* $x,y \in P'$ $(x \neq y)$ *and* x I L I y, *then* $L \in B'$.

(ii) *Each element of* B' *is incident with* 1+s *elements of* P'.

Then there are four possibilities :

(a) *S' is a dual grid* (*and then* s = 1).

(b) *The elements of* B' *are lines which are incident with a distinguished point of* P, *and* P' *consists of those points of* P *which are incident with these lines.*

(c) $B' = \phi$ *and* P' *is a set of pairwise noncollinear points of* P.

(d) *S' is a subquadrangle of order* (s,t').

Proof. Suppose that $S' = (P',B',I')$ satisfies (i) and (ii) and is not of type (a), (b) or (c). Then $B' \neq \phi$, $P' \neq \phi$ and $s > 1$. If $L' \in B'$, then there exists a point $x' \in P'$ such that x' $\not{I}$ L'. Let x and L be defined by x' I L I x I L'. By (i) and (ii) we have $x \in P'$ and $L \in B'$. Hence S' satisfies (iii) in the definition of GQ. Clearly S' satisfies (ii) and we now show that S' satisfies (i) of that definition. Consider a point $x' \in P'$ and suppose that x' is incident with t'+1 lines of B'. Since $B' \neq \phi$, $t' \geqslant 0$. Let $y' \in P'$ be a point which is not collinear with x' and suppose it is incident with t"+1 lines of B'. By (iii) in the definition of GQ it is clear that t' = t". Hence t'+1 is the number of lines of B' which are incident with any point not collinear with at least one of the points x' or y'. So we consider a point $z' \in P'$ which is in $\{x',y'\}^{\perp}$.

First suppose that t' = 0. Let x'z' = L', y'z' = L", and $L \in B'-\{L',L"\}$. Then x' $\not{I}$ L and y' $\not{I}$ L. Since there exists a line of B' which is incident with x' (resp., y') and concurrent with L, it follows that L and L' (resp., L and L") are concurrent. Hence z' I L, and S' is of type (b), a contradiction.

Now assume $t' > 0$. Consider a line $L \in B'$ for which x' I L and z' $\not{I}$ L. On L there is a point u', with $u' \not\sim y'$ and $u' \not\sim z'$. Then the number of lines of B' which are incident with z' equals the number of lines of B' which are incident with u', which equals t"+1 since $y' \not\sim u'$, and hence equals t'+1.

We conclude that each point of P' is incident with t'+1 (≥ 2) lines of B', which proves the theorem. □

2.4. AUTOMORPHISMS AND SUBQUADRANGLES

Let θ be an automorphism of the GQ $S = (P,B,I)$ of order (s,t).

2.4.1. *The substructure* $S_\theta = (P_\theta, B_\theta, I_\theta)$ *of the fixed elements of* θ *must be given by at least one of the following :*

(i) $B_\theta = \phi$ *and* P_θ *is a set of pairwise noncollinear points.*

(i)' $P_\theta = \phi$ *and* B_θ *is a set of pairwise nonconcurrent lines.*

(ii) P_θ *contains a point* x *such that* $x \sim y$ *for every point* $y \in P_\theta$ *and each line of* B_θ *is incident with* x .

(ii)' B_θ *contains a line* L *such that* $L \sim M$ *for every line* $M \in B_\theta$, *and each point of* P_θ *is incident with* L .

(iii) S_θ *is a grid.*

(iii)' S_θ *is a dual grid.*

(iv) S_θ *is a subquadrangle of order* (s',t'), $s' \geq 2$ *and* $t' \geq 2$.

Proof. Suppose S_θ is not of type (i), (i)', (ii), (ii)', (iii), (iii)'. Then $P_\theta \neq \phi \neq B_\theta$. If $x,y \in P_\theta$, $x \neq y$, $x \sim y$, then $(xy)^\theta = x^\theta y^\theta = xy$, so the line xy belongs to B_θ. Dually, if $L, M \in B_\theta$, $L \neq M$, $L \sim M$, then the point common to L and M belongs to P_θ. Next, let $L \in B_\theta$ and consider a point $x \in P_\theta$, with $x \not I L$. Further, let y and M be defined by x I M I y I L. Then $x^\theta \, I \, M^\theta \, I \, y^\theta \, I \, L^\theta$, i.e. $x \, I \, M^\theta \, I \, y^\theta \, I \, L$. Hence $M = M^\theta$ and $y = y^\theta$, i.e. $M \in B_\theta$, $y \in P_\theta$. It follows that S_θ satisfies (iii) in the definition of GQ. Now parts (i) and (ii) in the definition of GQ are easily obtained by making a variation on the proof of 2.3.1. □

2.5. SEMIREGULARITY, PROPERTY (H) AND SUBQUADRANGLES

The following result will be recognized later as a major step in the proofs of certain characterizations of the classical GQ.

2.5.1. *Let each point of the* GQ $S = (P,B,I)$ *of order* (s,t) *have property* (H). *Then one of the following must occur :*

(i) *Each point is regular.*

(ii) $|\{x,y\}^{\perp\perp}| = 2$ *for all* $x,y \in P$, $x \not\sim y$.

(iii) *There is a constant* p, $1 < p < t$, *such that* $|\{x,y\}^{\perp\perp}| = 1+p$ *for all* $x,y \in P$, $x \not\sim y$, *and* $s = p^2$, $t = p^3$. *Moreover, if* L *and* M *are nonconcurrent lines of S, then* $\{x \in P \parallel x \mathrm{I} L\} \cup \{y \in P \parallel y \mathrm{I} M\}$ *is contained in the pointset of a subquadrangle of order* $(s,p) = (p^2,p)$.

Proof. Let each point of *S* have property (H). By 1.6.2 there is a constant p such that $|\{x,y\}^{\perp\perp}| = 1+p$ for all points x,y with $x \not\sim y$. If $p = t$, then all points of *S* are regular, and we have case (i). If $p = 1$, we have case (ii). So assume $1 < p < t$, so that necessarily $s \neq 1$. For $L \in B$, put $L^* = \{x \in P \parallel x \mathrm{I} L\}$. Now consider two nonconcurrent lines L and M, and denote by P' the union of the sets $\{x,y\}^{\perp\perp}$, with $x \in L^*$ and $y \in M^*$. First we shall prove that each common point of the distinct sets $\{x,y\}^{\perp\perp}$ and $\{x',y'\}^{\perp\perp}$, with $x, x' \in L^*$ and $y,y' \in M^*$, belongs to $L^* \cup M^*$. If $\{x,y\}^{\perp\perp}$ and $\{x',y'\}^{\perp\perp}$ are the pointsets of distinct lines of *S*, then evidently $\{x,y\}^{\perp\perp} \cap \{x',y'\}^{\perp\perp} = \phi$. Now let $x \sim y$ and $x' \not\sim y'$. Suppose that $z \in \{x,y\}^{\perp\perp} \cap \{x',y'\}^{\perp\perp}$, with $z \notin \{x,y\}$. Since $x \in \{z,x'\}^{\perp} = \{y',x'\}^{\perp}$, we have $x \sim y'$, a contradiction as there arises a triangle xyy'. Finally, let $x \not\sim y$, $x' \not\sim y'$, and $z \in \{x,y\}^{\perp\perp} \cap \{x',y'\}^{\perp\perp}$ with $z \notin \{x,y\}$. The point which is incident with L and collinear with z is denoted by u. Since $u \in \{z,x\}^{\perp} = \{y,x\}^{\perp}$ and $u \in \{z,x'\}^{\perp} = \{y',x'\}^{\perp}$, we have $y \sim u \sim y'$, which is clearly impossible.

Now consider a point $x \in L^*$ and define V and y by $x \mathrm{I} V \mathrm{I} y \mathrm{I} M$. If $z_1 \mathrm{I} V$, $z_1 \neq y$, and $z_2 \mathrm{I} M$, $z_2 \neq y$, then by 1.6.2 the set $\mathrm{cl}(z_1,z_2) \cap y^{\perp}$, i.e. Ty if $T = \{z_1,z_2\}^{\perp\perp}$, is independent of the choice of the points z_1,z_2. That set will be denoted by xM^*. By 1.6.2, any span having at least two points in common with xM^* must be contained in xM^*. If $u \in xM^*$, $u \not\sim x$, then $\{u,x\}^{\perp\perp} \cap M^* \neq \phi$. Hence $u \in P'$, and it follows that $xM^* \subset P'$.

Next let N be a line whose points belong to the set xM^*, where $N \neq M$ and $N \neq V$. We shall prove that the union P" of the spans $\{z,u\}^{\perp\perp}$, $z \in N^*$ and $u \in L^*$, coincides with P'. First we note that the spans $\{z,u\}^{\perp\perp}$ with $z \in M^* \cap N^*$ are contained in P'. Now consider an hyperbolic line $\{z,x\}^{\perp\perp}$ with $z \in N^* - M^*$. Evidently $\{z,x\}^{\perp\perp}$ has a point in common with M^*, and $\{z,x\}^{\perp\perp} \subset xM^* \subset P'$. Finally, consider a span $\{z,u\}^{\perp\perp}$, with $z \in N^*-M^*$ and $u \in L^*-\{x\}$. The hyperbolic line $\{x,z\}^{\perp\perp}$ has a point v in common with M^*. From the preceding paragraph we have $vL^* \subset P'$. But $\{x,z\}^{\perp\perp} = \{x,v\}^{\perp\perp} \subset vL^*$, so $\{z,u\}^{\perp\perp}$ has two points in common with vL^* and hence must be contained in vL^*. This shows $P" \subset P'$. Interchanging the roles of P' and P" shows $P' = P"$.

(Or $|P''| = |P'| = (s+1)(sp+1)$ and $P'' \subset P'$ imply $P'' = P'$.)

The next step is to show that for any two distinct collinear points $x,y \in P'$, the span $\{x,y\}^{\perp\perp}$ $(= (xy)^*)$ is contained in P'. The case $\{x,y\} \subset L^* \cup M^*$ is trivial. So suppose that $\{x,y\} \not\subset L^* \cup M^*$ and that $x \in L^* \cup M^*$. Assume that $x \in L^*$ and $y \in \{y_1,y_2\}^{\perp\perp}$, with $y_1 \in L^*$ and $y_2 \in M^*$. In such a case $\{x,y\}^{\perp\perp} \subset y_2L^* \subset P'$. So suppose that $\{x,y\} \cap (L^* \cup M^*) = \phi$. Evidently we may also assume that $\{x,y\}^{\perp\perp} \cap (L^* \cup M^*) = \phi$. Let $x \in \{x_1,x_2\}^{\perp\perp}$ (resp., $y \in \{y_1,y_2\}^{\perp\perp}$), with $x_1 \in L^*$ and $x_2 \in M^*$ (resp., $y_1 \in L^*$ and $y_2 \in M^*$). First, we suppose that $\{x_1,x_2\}^{\perp\perp}$ is an hyperbolic line (i.e. $x_1 \not\sim x_2$). Then let z_1 and N be defined by x I N I z_1 I L. Clearly $N^* \subset x_2L^*$. Since we have proved that the union P'' of the sets $\{z,u\}^{\perp\perp}$, $z \in N^*$ and $u \in M^*$, coincides with P', the point y belongs to P''. By a preceding case, $\{x,y\}^{\perp\perp} \subset P''$, so $\{x,y\}^{\perp\perp} \subset P'$. Second, we suppose $x_1 \sim x_2$, and without loss of generality that $y_1 \sim y_2$. Since $x \sim y$, clearly $x_1 \neq y_1$ and $x_2 \neq y_2$. Let u be a point of the hyperbolic line $\{x_2,y_1\}^{\perp\perp}$, $x_2 \neq u \neq y_1$, and let P'' be the union of the sets $\{v,w\}^{\perp\perp}$, $w \in M^*$ and $v \in N^* = \{x_1,u\}^{\perp\perp}$. As $P' = P''$, y is contained in a set $\{v,w\}^{\perp\perp}$, $w \in M^*$ and $v \in N^*$. Evidently $v \not\sim w$, so that by a previous case $\{x,y\}^{\perp\perp} \subset P'' = P'$. We conclude that for any distinct collinear points $x,y \in P'$, the span $\{x,y\}^{\perp\perp}$ is contained in P'.

Now let B' be the set of all lines of S which are incident with at least two points of P', and let $I' = I \cap ((P' \times B') \cup (B' \times P'))$. Then by 2.3.1 the structure $S' = (P',B',I')$ is a subquadrangle of order (s,t') of S. Since $|P'| = (s+1)(sp+1) = (s+1)(st'+1)$, we have $t' = p$. By the inequality of D.G. Higman we have $s \leq p^2$, and by 2.2.1 we have $t \geq sp$. Moreover, by 1.4.2 (ii) we have $pt \leq s^2$. There results $sp^2 \leq pt \leq s^2$, implying $s \geq p^2$. Hence $s = p^2$. It now also follows easily that $t = p^3$, which completes the proof of the theorem. □

The preceding theorem is essentially contained in J.A. Thas [194]. We now have the following easy corollary.

2.5.2. *Let each point of* S *be semiregular and suppose* $s > 1$. *Then one of the following must occur :*

(i) $s > t$ *and each point of* S *is regular.*

(ii) $s = t$ *and each point of* S *is regular or each point is antiregular.*

(iii) $s < t$ *and* $|\{x,y\}^{\perp\perp}| = 2$ *for all* $x,y \in P$ *with* $x \not\sim y$.

(iv) *The conclusion of* 2.5.1 (iii) *holds.*

Proof. Immediate from 1.6.3 and 2.5.1. (Recall that any semiregular point has property (H).) □

2.6. 3-REGULARITY AND SUBQUADRANGLES

2.6.1. ([209]). *Let* (x,y,z) *be a* 3*-regular triad of the* GQ $S = (P,B,I)$ *of order* (s,s^2), $s > 1$, *and let* P' *be the set of all points incident with lines of the form* uv, $u \in \{x,y,z\}^{\perp} = X$ *and* $v \in \{x,y,z\}^{\perp\perp} = Y$. *If* L *is a line which is incident with no point of* $X \cup Y$ *and if* k *is the number of points in* P' *which are incident with* L, *then* $k \in \{0,2\}$ *if* s *is odd and* $k \in \{1,s+1\}$ *if* s *is even.*

Proof. Let L be a line which is incident with no point of $X \cup Y$. If $w \in X = \{x,y,z\}^{\perp}$, if $w \mathrm{I} M \mathrm{I} m \mathrm{I} L$, and if M is not a line of the form uv, $u \in X$ and $v \in Y = \{x,y,z\}^{\perp\perp}$, then there is just one point $w' \in X-\{w\}$ which is collinear with m. Hence the number r of lines uv, $u \in X$ and $v \in Y$, which are concurrent with L, has the parity of $|X| = s+1$. Clearly r is also the number of points in P' (P' is the set of all points incident with lines of the form uv) which are incident with L.

Let $\{L_1,L_2,\ldots\} = \mathcal{L}$ be the set of all lines which are incident with no point of $X \cup Y$, and let r_i be the number of points in P' which are incident with L_i. We have $|\mathcal{L}| = s^3(s^2-1)$ and $|P'-(X \cup Y)| = (s+1)^2(s-1)$. Clearly $\sum_i r_i = (s+1)^2(s-1)s^2$, and $\sum_i r_i(r_i-1)$ is the number of ordered triples $(uv,u'v',L_i)$, with u,u' distinct points of X, with v,v' distinct points of Y, and with $uv \sim L_i \sim u'v'$ where u,v,u',v' are not incident with L_i. Hence $\sum_i r_i(r_i-1) = (s+1)^2s^2(s-1)$.

Let s be odd. Then r_i is even, and so $\sum_i r_i(r_i-2) \geqslant 0$ with equality iff $r_i \in \{0,2\}$ for all i. Since $\sum_i r_i(r_i-2) = (s+1)^2s^2(s-1)-(s+1)^2(s-1)s^2 = 0$, we have indeed $r_i \in \{0,2\}$ for all i.

Let s be even. Then r_i is odd, and so $\sum_i (r_i-1)(r_i-(s+1)) \leqslant 0$ with equality iff $r_i \in \{1,s+1\}$ for all i. Since $\sum_i (r_i-1)(r_i-(s+1)) = (s+1)^2s^2(s-1)-(s+1)(s+1)^2(s-1)s^2+(s+1)s^3(s^2-1) = 0$, we have indeed $r_i \in \{1,s+1\}$ for all i. □

2.6.2. ([209]). *Let* (x,y,z) *be a 3-regular triad of the* GQ $S = (P,B,I)$ *of order* (s,s^2), s *even. If* P' *is the set of all points incident with lines of the form* uv, $u \in X = \{x,y,z\}^\perp$ *and* $v \in Y = \{x,y,z\}^{\perp\perp}$, *if* B' *is the set of all lines in* B *which are incident with at least two points of* P', *and if* I' *is the restriction of* I *to* $(P' \times B') \cup (B' \times P')$, *then* $S' = (P',B',I')$ *is a subquadrangle of order* s . *Moreover* (x,y) *is a regular pair of* S' , *with* $\{x,y\}^{\perp'} = \{x,y,z\}^\perp$ *and* $\{x,y\}^{\perp'\perp'} = \{x,y,z\}^{\perp\perp}$.

Proof. We have $|P'| = (s+1)^2(s-1)+2(s+1) = (s+1)(s^2+1)$. Let L be a line of B'. If L is incident with some point of $X \cup Y$, then clearly L is of type uv, with $u \in X$ and $v \in Y$. Then all points incident with L are in P'. If L is incident with no point of $X \cup Y$, then by 2.6.1 L is again incident with s+1 points of P'. Now by 2.3.1 $S' = (P',B',I')$ is a subquadrangle of order (s,t'). Since $|P'| = (s+1)(st'+1)$ we have $t' = s$, and so S' is a subquadrangle of order s. Since $X \cup Y \subset P'$, $|X| = |Y| = s+1$, and each point of X is collinear with each point of Y, we have $\{x,y\}^{\perp'} = \{x,y,z\}^\perp$ and $\{x,y\}^{\perp'\perp'} = \{x,y,z\}^{\perp\perp}$. □

2.7. k-ARCS AND SUBQUADRANGLES

Let $S = (P,B,I)$ be a GQ of order (s,t), $s > 1$, $t > 1$. A k-*arc* of S is a set of k pairwise noncollinear points. A k-arc $\mathcal{O}$ is *complete* provided it is not contained in a (k+1)-arc.

Let $\mathcal{O}$ be an $(st-\rho)$-arc of S, for some integer ρ. Let B' be the set of lines of S incident with no point of $\mathcal{O}$. An easy calculation shows that $|B'| = (1+t)(1+\rho)$, implying that $\rho \geq -1$. Evidently $\mathcal{O}$ is an ovoid precisely when $\rho = -1$. For the remainder of this section we assume that $\rho \geq 0$. Let L be a fixed line of B' incident with points $y_0,\ldots,y_s$, and let t_i be the number of lines ($\neq$ L) of B' incident with y_i, $i = 0,\ldots,s$.

$$\sum_{i=0}^{s} t_i = (1+s)t-(st-\rho) = t+\rho. \tag{1}$$

Eq. (1) says that each line of B' is concurrent with $t+\rho$ other lines of B'. Put $t_i = \theta_i+\rho$. It follows that each line M of B' incident with y_i is concurrent with exactly $t-\theta_i$ lines ($\neq$ M) of B' at points different from y_i. Count the lines of B' concurrent with lines of B' through y_i (including the latter) to obtain $(\theta_i+\rho+1)(1+t-\theta_i)$. Clearly this number is bounded above by $|B'| = (1+t)(1+\rho)$, from which we obtain the following :

$$\theta_i((t-\rho)-\theta_i) \leqslant 0, \tag{2}$$

with equality holding iff each line of B' is concurrent with some line of B' through y_i.

Clearly $t_i \leqslant t$, so $\theta_i \leqslant t-\rho$. And $\theta_i = t-\rho$ iff $\mathcal{O} \cup \{y_i\}$ is also an arc. From now on suppose that $\mathcal{O}$ is a complete $(st-\rho)$-arc, $\rho \geqslant 0$. Then $\theta_i < t-\rho$, so that by eq. (2) we have

$$\theta_i \leqslant 0. \tag{3}$$

The average number of lines of B' through a point of L is

$1+(t+\rho)/(1+s) = \sum_{i=0}^{s} (\theta_i+\rho+1)/(1+s) \leqslant \rho+1$. Hence $\rho \geqslant t/s$, with equality holding iff $\theta_i = 0$ for $i = 0,\ldots,s$, iff each point of L is on exactly $\rho+1$ lines of B'. (4)

2.7.1. *Any* $(st-\rho)$*-arc of* S *with* $0 \leqslant \rho < t/s$ *is contained in a uniquely defined ovoid of* S. *Hence if* S *has no ovoid, then any* k*-arc of* S *necessarily has* $k \leqslant st-t/s$.

Proof. By the preceding paragraph it is clear that any $(st-\rho)$-arc $\mathcal{O}$ of S with $0 \leqslant \rho < t/s$ is contained in an $(st-\rho+1)$-arc $\mathcal{O}'$. If $\rho = 0$, then $\mathcal{O}'$ is an ovoid. If $\rho > 0$, then $0 \leqslant \rho-1 < t/s$, and $\mathcal{O}'$ is contained in an $(st-\rho+2)$-arc $\mathcal{O}''$, etc. Finally, $\mathcal{O}$ can be extended to an ovoid. Now assume that $\mathcal{O}$ is contained in distinct ovoids $\mathcal{O}_1$ and $\mathcal{O}_2$. Let $x \in \mathcal{O}_1-\mathcal{O}_2$. Then each of the t+1 lines incident with x is incident with a unique point of $\mathcal{O}_2-\mathcal{O}_1$. Hence $|\mathcal{O}_2-\mathcal{O}_1| \geqslant t+1$, implying that $|\mathcal{O}_2-\mathcal{O}| \geqslant t+1$, i.e. $\rho+1 \geqslant t+1$ which is an impossibility. □

2.7.2. *Let* $\mathcal{O}$ *be a complete* $(st-t/s)$*-arc of* S. *Let* B' *be the set of lines incident with no point of* $\mathcal{O}$; *let* P' *be the set of points on (at least) one line of* B'; *and let* I' *be the restriction of* I *to points of* P' *and lines of* B'. *Then* $S' = (P',B',I')$ *is a subquadrangle of order* $(s,\rho) = (s,t/s)$.

Proof. Use (2) and (4). □

Putting s = t in 2.7.2 yields the following corollary.

2.7.3. *Any* GQ *of order* s *having a complete* (s^2-1)*-arc must have a regular pair of lines.*

2.7.4. *Let S be a* GQ *of order* s, $s > 1$, *with a regular point* x. *Then S has a complete* $(2s+1)$*-arc.*

Proof. Let T_1 and T_2 be two distinct hyperbolic lines containing x. By 1.3.1 there is a point y for which $T_1^\perp \cap T_2^\perp = \{y\}$. Let z I xy , $z \neq x$, $z \neq y$. Then $(T_1-\{x\}) \cup (T_2^\perp-\{y\}) \cup \{z\}$ is a complete $(2s+1)$-arc of *S*. □

2.7.5. *Let S be a* GQ *of order* s, $s > 1$, *having an ovoid* O *and a regular point* $x, x \notin O$ (*so* s *is even by* 1.8.4). *Then* $(O-x^\perp) \cup \{x\}$ *is a complete* (s^2-s+1)*-arc.*

Proof. Clearly $O' = (O-x^\perp) \cup \{x\}$ is an (s^2-s+1)-arc. If $O' \cup \{y\}$ is an arc, then $\{x,y\}^\perp = O \cap y^\perp = O \cap x^\perp$, contradicting 1.8.4. □

3 The known generalized quadrangles and their properties

3.1. DESCRIPTION OF THE KNOWN GQ

We start by giving a brief description of three families of examples known as the classical GQ, all of which are associated with classical groups and were first recognized as GQ by J. Tits [49].

3.1.1. *The classical* GQ, *embedded in* PG(d,q), $3 \leqslant d \leqslant 5$, *may be described as follows* :

(i) Consider a nonsingular quadric Q of projective index 1 [80] of the projective space PG(d,q), with d = 3, 4 or 5. Then the points of Q together with the lines of Q (which are the subspaces of maximal dimension on Q) form a GQ Q(d,q) with parameters

$$s = q,\ t = 1,\ v = (q+1)^2,\ b = 2(q+1),\ \text{when } d = 3,$$
$$s = t = q,\ v = b = (q+1)(q^2+1),\ \text{when } d = 4,$$
$$s = q,\ t = q^2,\ v = (q+1)(q^3+1),\ b = (q^2+1)(q^3+1),\ \text{when } d = 5.$$

Since Q(3,q) is a grid, its structure is trivial. Further, recall that the quadric Q has the following canonical equation :

$$x_0x_1+x_2x_3 = 0,\ \text{when } d = 3,$$
$$x_0^2+x_1x_2+x_3x_4 = 0,\ \text{when } d = 4,$$
$$f(x_0,x_1)+x_2x_3+x_4x_5 = 0,$$

where f is an irreducible binary quadratic form when d = 5.

(ii) Let H be a nonsingular hermitian variety of the projective space $PG(d,q^2)$, d = 3 or 4. Then the points of H together with the lines on H form a GQ $H(d,q^2)$ with parameters

$$s = q^2,\ t = q,\ v = (q^2+1)(q^3+1),\ b = (q+1)(q^3+1),\ \text{when } d = 3,$$
$$s = q^2,\ t = q^3,\ v = (q^2+1)(q^5+1),\ b = (q^3+1)(q^5+1),\ \text{when } d = 4.$$

Recall that H has the canonical equation

$$x_0^{q+1}+x_1^{q+1}+\dots+x_d^{q+1} = 0 .$$

(iii) The points of PG(3,q), together with the totally isotropic lines with respect to a symplectic polarity, form a GQ W(q) with parameters

$$s = t = q,\ v = b = (q+1)(q^2+1).$$

Recall that the lines of W(q) are the elements of a linear complex of lines of PG(3,q), and that a symplectic polarity of PG(3,q) has the following canonical bilinear form :

$$x_0y_1-x_1y_0+x_2y_3-x_3y_2 = 0.$$

The earliest known non-classical examples of GQ were discovered by J. Tits and first appeared in P. Dembowski [49].

3.1.2. *For each oval or ovoid* O *in* PG(d,q), d = 2 *or* 3, *there is a* GQ T(O) *constructed as follows :*
Let d = 2 (resp., d = 3) and let O be an oval [49] (resp., an ovoid [49]) of PG(d,q). Further, let PG(d,q) = H be embedded as an hyperplane in PG(d+1,q) = P. Define points as (i) the points of P-H, (ii) the hyperplanes X of P for which $|X \cap O| = 1$, and (iii) one new symbol (∞). Lines are defined as (a) the lines of P which are not contained in H and meet O (necessarily in a unique point), and (b) the points of O. Incidence is defined as follows : A point of type (i) is incident only with lines of type (a); here the incidence is that of P. A point of type (ii) is incident with all lines of type (a) contained in it and with the unique element of O in it. The point (∞) is incident with no line of type (a) and all lines of type (b). It is an easy exercise to show that the incidence structure so defined is a GQ with parameters

$$s = t = q,\ v = b = (q+1)(q^2+1), \text{ when } d = 2,$$
$$s = q,\ t = q^2,\ v = (q+1)(q^3+1),\ b = (q^2+1)(q^3+1), \text{ when } d = 3.$$

If $d = 2$, the GQ is denoted by $T_2(O)$; if $d = 3$, the GQ is denoted by $T_3(O)$. If no confusion is possible, these quadrangles are also denoted by $T(O)$.

For each prime power q, R.W. Ahrens and G. Szekeres [1] constructed GQ with order $(q-1, q+1)$. For q even, these examples were found independently by M. Hall, Jr. [70] . Then a construction was found by S.E. Payne [119] which included all these examples and for q even produced some additional ones (cf. [120, 123]). These examples yield the only known cases (with $s \neq 1$ and $t \neq 1$) in which s and t are not powers of the same prime.

3.1.3. ([1, 70]). *Associated with any complete oval* O *in* $PG(2,2^h)$ *there is a* GQ $T_2^*(O)$ *of order* $(q-1,q+1)$, $q = 2^h$.

Proof. Let O be a complete oval, i.e. a $(q+2)$-arc [80], of the projective plane $PG(2,q)$, $q = 2^h$, and let $PG(2,q) = H$ be embedded as a plane in $PG(3,q) = P$. Define an incidence structure $T_2^*(O)$ by taking for points just those points of $P-H$ and for lines just those lines of P which are not contained in H and meet O (necessarily in a unique point). The incidence is that inherited from P. It is evident that the incidence structure so defined is a GQ with parameters $s = q-1$, $t = q+1$, $v = q^3$, $b = q^2(q+2)$. □

3.1.4. ([119]). *To each regular point* x *of the* GQ $S = (P,B,I)$ *of order* s, $s > 1$, *there is associated a* GQ $P(S,x)$ *of order* $(s-1,s+1)$.

Proof. Let x be a regular point of the GQ $S = (P,B,I)$ of order s, $s > 1$. Then P' is defined to be the set $P-x^{\perp}$. The elements of B' are of two types : the elements of type (a) are the lines of B which are not incident with x; the elements of type (b) are the hyperbolic lines $\{x,y\}^{\perp\perp}$, $y \not\sim x$. Now we define the incidence I'. If $y \in P'$ and $L \in B'$ is of type (a), then $y\ I'\ L$ iff $y\ I\ L$; if $y \in P'$ and $L \in B'$ is of type (b), then $y\ I'\ L$ iff $y \in L'$. We now show that the incidence structure $S' = (P',B',I')$ is a GQ of order $(s-1,s+1)$.

It is clear that any two points of S' are incident (with respect to I') with at most one line of S'. Moreover, any point of P' is incident with $s+2$ lines of B', and any line of B' is incident with s points of P'. Consider a point $y \in P'$ and a line L of type (a), with $y \not{I}'\ L$. Let z be defined by $x \sim z$ and $z\ I\ L$. If $y \sim z$, then no line of type (a) is incident with y and concurrent with L. But then, by the regularity of x, there is a point of P' which is incident with the line $\{x,y\}^{\perp\perp}$ of type (b) and the line L. If $y \not\sim z$, then there is just one line of type (a) which is incident with y

and concurrent with L. By the regularity of x, the line of type (b) containing y is not concurrent with L. Finally, consider a point $y \in P'$ and a line $L = \{x,u\}^{\perp\perp}$, $x \not\sim u$, of type (b) with $y \notin L$. It is clear that no line of type (b) is incident with y and concurrent with L. If y is collinear with at least two points of L, then by the regularity of x we have $y \sim x$, i.e. $y \notin P'$, a contradiction. Hence y is collinear with at most one point of L. If $u \not\sim y$, then by 1.3.6 the triad (x,y,u) has a center v, and consequently the line of type (a) defined by v I M I y is incident with y and concurrent with L. □

The GQ $S' = (P',B',I')$ of order $(s-1,s+1)$ will be denoted by $P(S,x)$.

A quick look at the examples of order s in 3.1.1 and 3.1.2 reveals that regular points and regular lines arise in the following cases (for more details and proofs see 3.3) : all lines of Q(4,q) are regular; the points of Q(4,q) are regular iff q is even; all points of W(q) are regular; the lines of W(q) are regular iff q is even; the unique point (∞) of type (iii) of $T_2(O)$ is regular iff q is even; all lines of type (b) of $T_2(O)$ are regular . The corresponding GQ of S.E. Payne will be considered in detail in 3.2.

3.1.5. ([1]). *For each odd prime power* q *there is a* GQ AS(q) *of order* (q-1,q+1).

Proof. An incidence structure AS(q) = (P,B,I), q an odd prime power, is to be constructed as follows. Let the elements of P be the points of the affine 3-space AG(3,q) over GF(q). Elements of B are the following curves of AG(3,q) :

(i) $x = \sigma$, $y = a$, $z = b$,

(ii) $x = a$, $y = \sigma$, $z = b$,

(iii) $x = c\sigma^2 - b\sigma + a$, $y = -2c\sigma + b$, $z = \sigma$.

Here the parameter σ ranges over GF(q) and a,b,c are arbitrary elements of GF(q). The incidence I is the natural one. It remains to show that AS(q) is indeed a GQ of order (q-1,q+1).

It is clear that $|P| = q^3$, that $|B| = q^2(q+2)$, and that each element of B is incident with q elements of P. For each value of c there are q^2 curves of type (iii), and these curves have no point in common. For suppose the curves corresponding to (a,b,c) and (a',b',c) intersect. Then for some σ we have $c\sigma^2 - b\sigma + a = c\sigma^2 - b'\sigma + a'$ and $-2c\sigma + b' = -2c\sigma + b$, which clearly implies

$b = b'$ and $a = a'$. Similarly, no two curves of the form (i) (or of the form (ii)) intersect. Thus we have $q+2$ families of nonintersecting curves, q^2 curves in each family and q points on each curve. Hence each point of P is incident with exactly $q+2$ elements of B, one from each family.

Now we shall show that two curves in different families meet in at most one point. This is clear if one of the curves is of type (i) or (ii), and we only need to consider two curves of type (iii). Suppose the curve corresponding to (a,b,c) meets the curve corresponding to (a',b',c') at two different parameter values, say σ and τ. Then we have $-2c\sigma+b = -2c'\sigma+b'$ and $-2c\tau+b = -2c'\tau+b'$. Hence $c(\tau-\sigma) = c'(\tau-\sigma)$, with $\tau \neq \sigma$. Consequently $c = c'$ and the two curves coincide.

Finally, we shall show that axiom (iii) in the definition of GQ is satisfied. It is sufficient to prove that AS(q) does not contain triangles. For indeed, if AS(q) has no triangles, then the number of points collinear with at least one point of a line L equals $q+q(q+1)(q-1) = q^3 = |P|$, which proves (iii) in the definition of GQ. We must consider the following possibilities for L_1,L_2,L_3 to form a triangle.

(a) L_1 of type (i), L_2 of type (ii), L_3 of type (iii). Let L_1 be $x = \sigma$, $y = a_1$, $z = b_1$; let L_2 be $x = a_2$, $y = \sigma$, $z = b_2$; let L_3 be $x = c_3\sigma^2-b_3\sigma+a_3$, $y = -2c_3\sigma+b_3$, $z = \sigma$. Since L_1 and L_2 meet, we must have $b_1 = b_2$. But then both L_1 and L_2 meet L_3 at the same point, with parameter value $\sigma = b_1 = b_2$, and there is no triangle.

(b) L_1 of type (i) and L_2,L_3 of type (iii). Let L_1 be $x = \sigma$, $y = a_1$, $z = b_1$; let L_2 and L_3 be, respectively, $x = c_2\sigma^2-b_2\sigma+a_2$, $y = -2c_2\sigma+b_2$, $z = \sigma$, and $x = c_3\sigma^2-b_3\sigma+a_3$, $y = -2c_3\sigma+b_3$, $z = \sigma$. The line L_1 meets both L_2 and L_3 at points with parameter value b_1. We have $a_1 = -2c_2b_1+b_2 = -2c_3b_1+b_3$. If L_2,L_3 meet at the point with parameter value $\sigma \neq b_1$, then $-2c_2\sigma+b_2 = -2c_3\sigma+b_3$, which with the previous equation gives $2c_2(\sigma-b_1) = 2c_3(\sigma-b_1)$, implying $c_2 = c_3$. Hence L_2 and L_3 do not meet, a contradiction.

(c) L_1 of type (ii) and L_2,L_3 of type (iii). Let L_1 be $x = a_1$, $y = \sigma$, $z = b_1$; and let L_2 and L_3 be as given in (b). The line L_1 meets both L_2 and L_3 at points with parameter value b_1. We now have $a_1 = c_2b_1^2-b_2b_1+a_2$

and $a_1 = c_3b_1^2-b_3b_1+a_3$. If L_2,L_3 meet at the point with parameter value $\sigma \neq b_1$, then $c_2\sigma^2-b_2\sigma+a_2 = c_3\sigma^2-b_3\sigma+a_3$ and $-2c_2\sigma+b_2 = -2c_3\sigma+b_3$, which with the previous equations give $c_2(\sigma+b_1)-b_2 = c_3(\sigma+b_1)-b_3$ and $(b_1-\sigma)(c_2-c_3) = 0$, from which $c_2 = c_3$. Hence L_2 and L_3 do not meet, a contradiction.

(d) L_1,L_2,L_3 are of type (iii). Let L_i be $x = c_i\sigma^2-b_i\sigma+a_i$, $y = -2c_i\sigma+b_i$, $z = \sigma$, $i = 1,2,3$. Suppose that L_i,L_j, $i \neq j$, meet each other at the point with parameter value $\sigma_{ij} = \sigma_{ji}$, where σ_{12}, σ_{23}, σ_{31} are distinct. Then

$$c_i\sigma_{ij}^2-b_i\sigma_{ij}+a_i = c_j\sigma_{ij}^2-b_j\sigma_{ij}+a_j \tag{1}$$

and

$$-2c_i\sigma_{ij}+b_i = -2c_j\sigma_{ij}+b_j, \tag{2}$$

giving

$$-b_i\sigma_{ij}+2a_i = -b_j\sigma_{ij}+2a_j . \tag{3}$$

By (2) we have

$$\begin{aligned}&\sigma_{23}\sigma_{31}(b_1-b_2)+\sigma_{31}\sigma_{12}(b_2-b_3)+\sigma_{12}\sigma_{23}(b_3-b_1) = \\ &2\sigma_{23}\sigma_{31}\sigma_{12}(c_1-c_2)+2\sigma_{31}\sigma_{12}\sigma_{23}(c_2-c_3)+2\sigma_{12}\sigma_{23}\sigma_{31}(c_3-c_1) = 0 .\end{aligned} \tag{4}$$

By (3) we have

$$\sigma_{12}(b_1-b_2)+\sigma_{23}(b_2-b_3)+\sigma_{31}(b_3-b_1) = 0 . \tag{5}$$

Eliminating b_1 from (4) and (5), we obtain $(\sigma_{12}-\sigma_{23})(\sigma_{12}-\sigma_{31})(\sigma_{31}-\sigma_{23})(b_2-b_3) = 0$. Hence $b_2 = b_3$, and by (2) $\sigma_{23}(c_2-c_3) = 0$. Since $c_2 \neq c_3$, we have $\sigma_{23} = 0$. Analogously $\sigma_{31} = \sigma_{12} = 0$. So $\sigma_{12} = \sigma_{23} = \sigma_{31}$, a contradiction.

It follows that AS(q) has no triangles and consequently is a GQ. □

In their paper [1], R.W. Ahrens and G. Szekeres also note that the incidence structure (P^*, B^*, I^*), with $P^* = B^* = B$ and $L \, I^* \, M$, $L \in P^*$, $M \in B^*$, iff $L \sim M$ and $L \neq M$ in (P,B,I), is a symmetric 2-$(q^2(q+2), q(q+1), q)$ design. These designs were new. (See Section 3.6 for a further study of symmetric designs arising from GQ.) They also remark that for q = 3 there arises a GQ with 27 points and 45 lines, whose dual can also be obtained as follows : lines of the GQ are the 27 lines on a general cubic surface V in PG(3,C) [4], points of the GQ are the 45 tritangent planes [4] of V, and incidence is inclusion.

The only known family of GQ remaining to be discussed was discovered by W.M. Kantor [88] while studying generalized hexagons and the family $G_2(q)$ of simple groups. We now give the method by which Kantor used the hexagons to construct the GQ. In 10.6 , using the theory of GQ as group coset geometries, we shall give a self-contained algebraic presentation that was directly inspired by W.M. Kantor's original paper.

3.1.6. ([88]). *For each prime power* q, $q \equiv 2 \pmod 3$, *there is a* GQ K(q) *of order* (q, q^2) *which arises from the generalized hexagon* H(q) *of order* q *associated with the group* $G_2(q)$.

Construction : *A generalized hexagon* [122] of order q (≥ 1) is an incidence structure $S = (P,B,I)$, with a symmetric incidence relation satisfying the following axioms :

(i) each point (resp., line) is incident with q+1 lines (resp., points);

(ii) $|P| = |B| = 1+q+q^2+q^3+q^4+q^5$;

(iii) 6 is the smallest positive integer k such that S has a circuit consisting of k points and k lines.

There is a natural metric defined on $P \cup B$: an object is at distance 0 from itself, an incident point and line are at distance 1, etc. Clearly the maximum distance between any two objects in $P \cup B$ is 6. The generalized hexagon of order 1 is the ordinary hexagon. Up to duality only one generalized hexagon of order q, q a prime power , is known.

This generalized hexagon arises from the group $G_2(q)$ and was introduced by J. Tits in his celebrated paper on triality [216]. One of the two dual choices of this generalized hexagon has a nice representation in PG(6,q) [216] : its points are the points of a nonsingular quadric Q; its lines are (some, but not all of the) lines of Q; incidence is that of PG(6,q). The generalized hexagon with that representation will be denoted by H(q).

Let $S = (P,B,I)$ be a generalized hexagon of order q. Define an incidence structure $S^* = (P^*,B^*,I^*)$ as follows. Let L be a fixed line of S. The points of S^* will be the points of L and the lines of S at distance 4 from L. Lines of S^* are L, the points of S at distance 3 from L and the lines of S at distance 6 from L. We now define the incidence I^*: a point of L (in S) is defined to be incident (in S^*) with L and with the lines of S^* which are at distance 2 from it in S ; a line at distance 4 from L (considered as a point of S^*) is defined to be incident (in S^*) with the lines of S^* which are at distance 1 or 2 (in S) from it. For the incidence structure S^* so defined, the following properties are easy to check : each point is incident with $1+q^2$ lines, each line is incident with 1+q points, any two points are incident with at most one line, and both $|P^*|$ and $|B^*|$ have the correct values for S^* to be a GQ of order (q,q^2). It is easy to discover a simple geometric configuration whose absence from S is necessary and sufficient for S^* to be a GQ. Recently, using mainly projective geometry techniques in PG(6,q), J.A. Thas [84, 207] proved that this configuration is absent from H(q) when $q \equiv 2 \pmod 3$. In this work we shall give a group theoretical proof directly inspired by W.M. Kantor's paper, and hence we defer it until 10.6.2 , where GQ as group coset geometries are introduced.

3.2. ISOMORPHISMS BETWEEN KNOWN GQ

We start off by considering GQ of order q, q > 1, for which the known examples are Q(4,q), W(q), $T_2(O)$, and their duals.

3.2.1. Q(4,q) *is isomorphic to the dual of* W(q). *Moreover,* Q(4,q) (*or* W(q)) *is self-dual iff* q *is even.*

Proof. Let Q^+ be the Klein quadric of the lines of PG(3,q) [4] . Then Q^+ is an hyperbolic quadric of PG(5,q). The image of W(q) on Q^+ is the intersection of Q^+ with a nontangent hyperplane PG(4,q) of PG(5,q). The nonsingular quadric $Q^+ \cap PG(4,q)$ of PG(4,q) is denoted by Q. The lines of W(q) which are incident with a given point form a flat pencil of lines, hence their images on Q^+ form a line of Q. Now it easily follows that W(q) is isomorphic to the dual of Q(4,q).

Now consider the nonsingular quadric Q of PG(4,q). Let L_0 and L_1 be nonconcurrent lines of Q(4,q). Then the 3-space L_0L_1 intersects Q in an hyperbolic quadric having reguli $\{L_0,L_1,\dots,L_q\}$ and $\{M_0,M_1,\dots,M_q\}$. In

$Q(4,q)$ we have $L_i \sim M_j$, $i, j = 0,\dots,q$, so (L_0,L_1) is a regular pair of lines of $Q(4,q)$. Hence each point of $Q(4,q)$ is coregular. From 1.5.2 it follows that each point of $Q(4,q)$ is regular or antiregular according as q is even or odd. Thus for q odd $Q(4,q)$ (and also $W(q)$) is not self-dual. So let q be even. The tangent 3-spaces of Q all meet in one point n, the nucleus of Q [80] . From n we project Q onto a $PG(3,q)$ not containing n. This yields a bijection of the pointset of $Q(4,q)$ onto $PG(3,q)$, mapping the $(q+1)(q^2+1)$ lines of $Q(4,q)$ onto $(q+1)(q^2+1)$ lines of $PG(3,q)$. Since the $q+1$ lines of $Q(4,q)$ which are incident with a given point are contained in a tangent 3-space of Q, they are mapped onto the elements of a flat pencil of lines of $PG(3,q)$. Hence the images of the lines of $Q(4,q)$ constitute a linear complex of lines [159] of $PG(3,q)$, i.e. they are the totally isotropic lines with respect to a symplectic polarity of $PG(3,q)$. It follows that $Q(4,q) \cong W(q)$, and consequently $Q(4,q)$ (and $W(q)$) are self-dual. □

Remark : In [217] J. Tits proves that $W(q)$ is self-polar iff $q = 2^{2h+1}$, $h \geq 0$. Let θ be a polarity of $W(q)$, $q = 2^{2h+1}$, $h \geq 1$. By 1.8.2 the set of all absolute points (resp. , lines) of θ is an ovoid O (resp., a spread V) of $W(q)$. It is easily seen that O (resp., V) is an ovoid [49] (resp., spread [49]) of $PG(3,q)$. J. Tits proves that O is not a quadric and that the associated inversive plane admits the Suzuki group $Sz(q)$ as automorphism group. Finally, the spread V is the Lüneburg-spread giving rise to the non-desarguesian Lüneburg-plane [100, 181].

3.2.2. *The GQ* $T_2(O)$ *is isomorphic to* $Q(4,q)$ *iff* O *is an irreducible conic; it is isomorphic to* $W(q)$ *iff* q *is even and* O *is a conic.*

Proof. Let Q be a nonsingular quadric of $PG(4,q)$ and let $x \in Q$. Project Q from x onto a $PG(3,q)$ contained in $PG(4,q)$ but not containing x. Then there arises a bijection θ from the set of all points of $Q(4,q)$ not collinear with x, onto the pointset $PG(3,q)-PG(2,q)$, where $PG(2,q)$ is the intersection of $PG(3,q)$ and the tangent 3-space of Q at x . In other words, if O is the conic $Q \cap PG(2,q)$, then we have a bijection θ from the set of all points of $Q(4,q)$ not collinear with x, onto the set of all points of $T_2(O)$ not collinear with (∞). Now we extend θ in the following way : if y is a point of $Q(4,q)$ with $y \sim x$ and $y \neq x$, then define y^θ to be the intersection of $PG(3,q)$ and the tangent 3-space of Q at y, i.e. y^θ is the projection of that tangent 3-space from x onto PG(3,q); define x^θ to be (∞); if L is a line of Q(4,q), define L^θ to be the projection of L onto PG(3,q) (from x). If L does not contain x, then L^θ is a line of PG(3,q) containing a point of O; if L contains x, then L^θ is a point of O. Now it is clear that θ is an isomorphism of Q(4,q) onto $T_2(O)$. Hence, if O is an irreducible conic, then $T_2(O) \cong Q(4,q)$.

Conversely, suppose that $T_2(O) \cong Q(4,q)$. Then by an argument in the proof of the previous theorem, all pairs of lines of $T_2(O)$ are regular. Let L_0 and L_1 be nonconcurrent lines of type (a) of $T_2(O)$, and suppose they define distinct points x_0 and x_1 of O. If $\{L_0,L_1\}^\perp = \{M_0,M_1,\ldots,M_q\}$ and $\{L_0,L_1\}^{\perp\perp} = \{L_0,L_1,\ldots,L_q\}$, then $L_0,\ldots,L_q$, $M_0,\ldots,M_q$ are lines of type (a) of $T_2(O)$. Moreover, in $T_2(O)$ and also in PG(3,q) L_i is concurrent with M_j, i, j = 0,...,q. Hence $\{L_0,L_1\}^\perp$ and $\{L_0,L_1\}^{\perp\perp}$ are the reguli of an hyperbolic quadric Q^+ of PG(3,q) [80] . Clearly O is the intersection of Q^+ with a nontangent plane, so O is an irreducible conic.

If q is even and O is a conic, then $T_2(O) \cong Q(4,q) \cong W(q)$. Conversely, suppose that $T_2(O) \cong W(q)$. As all points of W(q) are regular and the lines of $T_2(O)$ through (∞) are regular, by 1.5.2 q must be even. In this case $T_2(O) \cong W(q) \cong Q(4,q)$, implying that O is a conic. (In the case q is odd another pleasant argument is as follows : by B. Segre's theorem [158] the oval O is a conic. Hence $T_2(O) \cong Q(4,q)$, implying $Q(4,q) \cong W(q)$, a contradiction since q is odd.) □

<u>Remark</u> : If q is odd, then the oval O is a conic, implying $T_2(O) \cong Q(4,q)$. In such a case $T_2(O)$ is not self-dual. If q is even and O is a conic, then $T_2(O)$, which is isomorphic to Q(4,q), is self-dual. The problem of determining all ovals for which $T_2(O)$ is self-dual has been solved (cf. M. Eich and S.E. Payne [56], S.E. Payne and J.A. Thas [143], and also Chapter 12). A complete classification of $T_2(O)$ would also entail a complete classification of the ovals, a problem which at present seems hopeless.

We now consider the known GQ of order (q,q^2). For q = 2, the GQ of order (q,q^2) are also the GQ of order (q,q+2). But in 5.3.2 we shall prove that up to isomorphism there is only one GQ of order (2,4). For q > 2, the known examples are Q(5,q), the dual of $H(3,q^2)$, $T_3(O)$, and K(q).

3.2.3. $Q(5,q)$ *is isomorphic to the dual of* $H(3,q^2)$.

Proof. Let Q be an elliptic quadric, i.e. a nonsingular quadric of projective index 1, in $PG(5,q)$. Extend $PG(5,q)$ to $PG(5,q^2)$. Then the extension of Q is an hyperbolic quadric Q^+, i.e. a nonsingular quadric of projective index 2, in $PG(5,q^2)$. Hence Q^+ is the Klein quadric of the lines of $PG(3,q^2)$. So to Q in Q^+ there corresponds a set V of lines in $PG(3,q^2)$. To a given line L of the GQ $Q(5,q)$ there correspond $q+1$ lines of $PG(3,q^2)$ that all lie in a plane and pass through a point x. Let H be the set of points on the lines of V. Then with each point of $Q(5,q)$ there corresponds a line of V, and with each line L of $Q(5,q)$ there corresponds a point x of H. With distinct lines L, L' of $Q(5,q)$ correspond distinct points x,x' of H (a plane of Q^+ contains at most one line of Q). Since a point y of $Q(5,q)$ is on q^2+1 lines of $Q(5,q)$, these q^2+1 lines are mapped onto the q^2+1 points of the image of y. Hence we obtain an anti-isomorphism from $Q(5,q)$ onto the structure (H,V,I), where I is the natural incidence relation. So (H,V,I) is a GQ of order (q^2,q) embedded in $PG(3,q^2)$. But now by a celebrated result of F. Buekenhout and C. Lefèvre [29], which will be proved in the next chapter, the GQ (H,V,I) must be $H(3,q^2)$. □

The proof just given is in J.A. Thas and S.E. Payne [213]. An algebraic proof of the same theorem was given by A.A. Bruen and J.W.P. Hirschfeld [24]

3.2.4. $T_3(O)$ *is isomorphic to* $Q(5,q)$ *iff* O *is an elliptic quadric of* $PG(3,q)$

Proof. Let Q be a nonsingular quadric of projective index 1 of $PG(5,q)$, and let $x \in Q$. Project Q from x onto a $PG(4,q) \subset PG(5,q)$ not containing x. Then there arises a bijection θ from the set of all points of $Q(5,q)$ not collinear with x, onto the pointset $PG(4,q)-PG(3,q)$, where $PG(3,q)$ is the intersection of $PG(4,q)$ and the tangent 4-space of Q at x. In other words, if O is the elliptic quadric $PG(3,q) \cap Q$, then we have a bijection θ from the set of all points of $Q(5,q)$ not collinear with x, onto the set of all points of $T_3(O)$ not collinear with (∞). We extend θ in the following way : if y is a point of $Q(5,q)$ with $x \neq y \sim x$, then define y^θ to be the intersection of $PG(4,q)$ and the tangent 4-space of Q at y, i.e. y^θ is the projection of that tangent 4-space from x onto $PG(4,q)$ (note that $y^\theta \cap PG(3,q)$ is a tangent plane of O); define x^θ to be (∞); if L is a line of $Q(5,q)$, define L^θ to be the projection of L onto $PG(4,q)$ (from x). If L does not contain x, then L^θ is a line of PG(3,q) which contains a point of O; if L contains x, then L^θ is a point of O. Now it is clear that θ is an isomorphism of Q(5,q) onto $T_3(O)$. Hence, if O is an elliptic quadric of PG(3,q), then $T_3(O) \cong Q(5,q)$.

Conversely, suppose that $T_3(O) \cong Q(5,q)$. Since the 3-space defined by any pair of nonconcurrent lines of Q(5,q) intersects Q in an hyperbolic quadric, it is clear that any pair of lines of Q(5,q) is regular. Hence any pair of lines of $T_3(O)$ is regular.

Let L_0 and L_1 be nonconcurrent lines of type (a) of $T_3(O)$, and suppose they define distinct points x_0 and x_1 of O. If $\{L_0,L_1\}^\perp = \{M_0,M_1,\ldots,M_q\}$ and $\{L_0,L_1\}^{\perp\perp} = \{L_0,L_1,\ldots,L_q\}$ then $L_0,\ldots,L_q,M_0,\ldots,M_q$ are lines of type (a) of $T_3(O)$. Moreover, in $T_3(O)$ and also in PG(4,q) L_i is concurrent with M_j, i, j = 0,...,q. Hence $L_0,\ldots,L_q,M_0,\ldots,M_q$ are contained in a three dimensional space P, and moreover $\{L_0,L_1\}^\perp$ and $\{L_0,L_1\}^{\perp\perp}$ are the reguli of an hyperbolic quadric Q^+ of P [80]. If PG(3,q) is the three dimensional space containing O, then clearly $Q^+ \cap O = Q^+ \cap PG(3,q) = P \cap O$. Hence $P \cap O$ is an irreducible conic. It follows that for any 3-space P of PG(5,q) with $P \not\subset PG(4,q)$ and $|P \cap O| > 1$, the oval $P \cap O$ is an irreducible conic. Since all ovals on O are conics, the ovoid O is an elliptic quadric by a result of A. Barlotti [5]. □

3.2.5. *For* $q \equiv 2 \pmod 3$ *and* $q > 2$ *the* GQ K(q) *is never isomorphic to a* $T_3(O)$.

Proof. For a complete proof of this theorem we refer to 10.6.2, where it is shown that K(q) has a unique regular line if q > 2, whereas the point (∞) of $T_3(O)$ is always coregular. □

We now turn to isomorphisms between the known GQ of order (q-1,q+1). For the case q = 3 see the remarks preceding 3.2.3. For q > 3, the known examples are $T_2^*(O)$, P(*S*,x) (resp., P(*S*,L)) with x (resp., L) a regular point (resp., line) of the GQ *S* of order q, and AS(q). In choosing the regular point x or regular line L in some GQ *S* of order q, by 3.2.1 and 3.2.2 we may restrict ourselves to the GQ $T_2(O)$.

For q odd, every oval O is an irreducible conic by B. Segre's theorem [158] and hence by 3.2.2 $T_2(O) \cong Q(4,q)$. So in that case all lines of $T_2(O)$ are regular and all points are antiregular, and moreover $T_2(O)$ is homogeneous in its points (resp., lines). Consequently for q odd there arises only one GQ of S.E. Payne. Perhaps the nicest model of that GQ is obtained by considering $P(W(q),x)$: points of the GQ are the points of $PG(3,q) - PG(2,q)$, with $PG(2,q)$ the polar plane of x with respect to the symplectic polarity θ defining $W(q)$; lines of the GQ are the totally isotropic lines of θ which do not contain x, and also all lines of $PG(3,q)$ which contain x and are not contained in $PG(2,q)$.

Now assume that q is even. Here the structure $T_2(O)$ depends, naturally, on the nature of the oval O. In general the point (∞) and all lines incident with it are regular. If some additional point or line is regular then $T_2(O)$ must belong to a completely determined list of examples (cf. 3.3 and Chapter 12 for the details). And for $q = 2^h \geq 8$, there are examples of O for which there is a unique line L_∞ of regular points. For any one of these regular points x different from (∞), the GQ $P(T_2(O),x)$ was shown by S.E. Payne [123] not to be isomorphic to any $T_2^*(O)$. However, as we show below, both $T_2^*(O)$ and AS(q) do arise as special cases of the general construction $P(\mathcal{S},x)$. This underscores the importance of this general method of construction, and it strongly suggests that a complete classification of the GQ $P(\mathcal{S},x)$ and $P(\mathcal{S},L)$, for q even, is hopeless.

3.2.6. ([120]). *The* GQ $T_2(O)$ *and* AS(q) *are isomorphic to the respective* GQ $P(T_2(O'),(\infty))$, *with* $O' = O - \{x\}$ *and* $x \in O$, *and* $P(W(q),y)$.
Proof. Consider $T_2^*(O)$, with O a complete oval of $PG(2,q)$, $q = 2^h$. Let $O' = O-\{x\}$, with x some point of O. Then O' is an oval with nucleus x [80]. Now consider the GQ $T_2(O')$. The point (∞) is a regular point of $T_2(O')$ (which may be considered to follow from the fact that all tangent lines of O' meet at x, or from the fact that (∞) is coregular and q is even). It is easy to see that the GQ $P(T_2(O'),(\infty))$ coincides with the GQ $T_2^*(O)$. Hence $T_2^*(O)$ is a GQ of S.E. Payne.

Now consider the GQ AS(q), q odd, of R.W. Ahrens and G. Szekeres. Recall that the elements of P are the points of AG(3,q) and that the elements of B are the following curves of AG(3,q) :

(i) $x = \sigma$, $y = a$, $z = b$; denoted $[-,a,b]$.

(ii) $x = a$, $y = \sigma$, $z = b$; denoted $[a,-,b]$.

(iii) $x = c\sigma^2-b\sigma+a$, $y = -2c\sigma+b$, $z = \sigma$; denoted $[c,b,a]$.

Here the parameter σ ranges over the elements of GF(q), and a,b,c are fixed but arbitrary elements of GF(q). The set of q lines of type (ii) with fixed b will be denoted by (b); the set of q lines of type (iii) of AS(q) with fixed c and b will be denoted by (c,b). Further, we introduce the notation $[c] = \{(c,b) \parallel b \in GF(q)\}$ and $[\infty] = \{(b) \parallel b \in GF(q)\}$. Then we define a new incidence structure $S' = (P',B',I')$ in the following way. The elements of P' are of four types : a symbol (∞), the elements (b) and (c,b), and the points of P. The elements of B' are the lines of type (ii) and (iii) of B, the elements [c], and $[\infty]$. Further, we define I' by (∞) I' $[\infty]$, (∞) I [c] for all $c \in GF(q)$, (b) I' $[\infty]$ for all $b \in GF(q)$, (b) I' [a,-,b] for all $a,b \in GF(q)$, (c,b) I' [c] for all $b,c \in GF(q)$, (c,b) I' [c,b,a] for all $a,b,c \in GF(q)$, u I' L iff u I L for all $u \in P$ and all lines L of type (ii) or (iii) of B. It is easily checked that each point of P' is incident with q+1 lines of B', and each line of B' is incident with q+1 points of P'. Now using the fact that for each value of c there are q^2 mutually disjoint lines of type (iii) in AS(q), and after checking that in S' two lines L,M of type (ii) or (iii) concur at a point (b) or (c,b) iff in AS(q) the q lines of $\{L,M\}^{\perp}$ are of type (i), it is not difficult to show that S' is a GQ of order q.

Next we show that all points of S' are regular. By 1.3.6 it is sufficient to prove that any triad of points of S' is centric. There are several cases according to the types of the points in the triad. In the following a point $(x,y,z) \in P$ will be called a type I point, a point (c,b) a type II point, a point (b) a type III point, and the point (∞) the type IV point. There are many cases to consider, but several of them are easy. We present the details only for the least trivial of the cases.

First of all we consider the case (IV,I,I). Let u and v be the points of type I. Further, assume that L is the line of type (i) of AS(q) incident with u and that M is the line of AS(q) which contains v and is concurrent with L. If w is defined by u I' N I' w I' M, then in S' the point w is collinear with (∞). Hence in S' the triad $((\infty),u,v)$ is centric, so that the point (∞) is regular.

Before starting with the other cases we remark that in S' the points (x_0,y_0,z_0) and (x_1,y_1,z_1) (resp., (c,b) and (x,y,z)) are collinear iff $(y_0+y_1)(z_1-z_0) = 2(x_0-x_1)$ (resp., $y = -2cz+b$).

Consider now the case (I,I,I), and suppose (u_0,u_1,u_2), $u_i = (x_i,y_i,z_i)$, is a triad of points. For subscripts reduced mod 3 to one of 0,1,2, this means that $(y_i+y_{i+1})(z_{i+1}-z_i) \neq 2(x_i-x_{i+1})$, i = 0,1,2. We then wish to find a point $u_3 = (x_3,y_3,z_3)$ of type I such that $(y_i+y_3)(z_3-z_i) = 2(x_i-x_3)$, $i = 0,1,2$, or a point (c,b) of type II such that $b = y_i+2cz_i$, $i = 0,1,2$, or a point (b) of type III such that $b = z_i$, $i = 0,1,2$. A point $u_3 = (x_3,y_3,z_3)$ satisfying the above conditions can be found iff the following system of linear equations in y_3,z_3 has a solution :

$$(z_1-z_0)y_3+(y_0-y_1)z_3 = y_0z_0-y_1z_1+2(x_0-x_1),$$
$$(z_2-z_0)y_3+(y_0-y_2)z_3 = y_0z_0-y_2z_2+2(x_0-x_2).$$

The determinant of this system is $\Delta = z_0(y_2-y_1)+z_1(y_0-y_2)+z_2(y_1-y_0)$. Hence if $\Delta \neq 0$ we can solve for a u_3 of type I. On the other hand, if $\Delta = 0$ and $z_i \neq z_j$ for some $i \neq j$, then it is easily verified that the system $b = y_i+2cz_i$, $i = 0,1,2$, has a solution in b,c. Finally, if $\Delta = 0$ and $z_0 = z_1 = z_2$, then (z_0) is collinear with u_0,u_1,u_2. This completes case (I,I,I). The other cases that are not trivial are for triads (III,II,I), (III,I,I), (II,II,I), and (II,I,I). But even there the computations are somewhat simpler than, and in the same spirit as the ones just presented.

So we have proved that all points of S' are regular. Clearly we have $AS(q) = P(S',(\infty))$. We finally prove that $S' \cong W(q)$. For that purpose we introduce the incidence structure $S'' = (P'',B'',I'')$, with $P'' = P'$, B'' the set of spans (in S') of all points-pairs of P', and I'' the natural incidence. By 1.3.1 and using the fact that any triad of points of S' is centric, it follows that any three noncollinear points of S'' generate a projective plane. Since $|P''| = q^3+q^2+q+1$, S'' is the design of points and lines of the projective 3-space $PG(3,q)$ over $GF(q)$. Clearly all spans (in S') of collinear point-pairs containing a given point x, form a flat pencil of lines in $PG(3,q)$. It follows immediately that the set of all spans of collinear point-pairs is a linear complex of lines of $PG(3,q)$ [159], i.e. is the set of all totally isotropic lines for some symplectic polarity. Hence $S' \cong W(q)$ and the theorem is proved. □

Remark : In [15] it is proved that $T_2^*(O_1) \cong T_2^*(O_2)$ iff there is an isomorphism θ of the plane π_1 of O_1 onto the plane π_2 of O_2 for which $O_1^\theta = O_2$.

3.3. COMBINATORIAL PROPERTIES : REGULARITY, ANTIREGULARITY, SEMIREGULARITY AND PROPERTY (H)

In this section we consider the pure combinatorics of the known GQ. Many of the properties in the following theorems will be seen to be of fundamental importance for a large variety of characterizations of the known GQ. We start by considering the classical GQ, and by 3.2.1 it is sufficient to consider $Q(3,q)$, $Q(4,q)$, $Q(5,q)$, and $H(4,q^2)$. Of course, the structure of $Q(3,q)$ is trivial.

3.3.1. (i) *Properties of* $Q(4,q)$ *: all lines are regular; all points are regular iff* q *is even; all points are antiregular iff* q *is odd; all points and lines are semiregular and have property* (H).

(ii) *Properties of* $Q(5,q)$ *: all lines are regular; all points are 3-regular; all points and lines are semiregular and have property* (H).

(iii) *Properties of* $H(4,q^2)$ *: for any two noncollinear points* x,y *we have* $|\{x,y\}^{\perp\perp}| = q+1$*; for any two nonconcurrent lines* L,M *we have* $|\{L,M\}^{\perp\perp}| = 2$*, but* (L,M) *is not antiregular; all points are semiregular and have property* (H)*; all lines have property* (H) *but no line is semiregular.*

Proof. (i) This is an immediate corollary of 1.6.1 and the proof of 3.2.1.

(ii) It was observed in the proof of 3.2.4 that all lines of $Q(5,q)$ are regular. So consider a triad (x_0,x_1,x_2). Since $t = s^2$ we have $|\{x_0,x_1,x_2\}^{\perp}| = q+1$. It is clear that $\{x_0,x_1,x_2\}^{\perp} = Q \cap \pi^{\perp}$, where $\pi^{\perp}$ is the polar plane of the plane $\pi = x_0x_1x_2$ with respect to the quadric Q. Since π and $\pi^{\perp}$ are mutually polar, each point of $Q \cap \pi$ is collinear in $Q(5,q)$ with each point of $Q \cap \pi^{\perp}$. Hence $|\{x_0,x_1,x_2\}^{\perp\perp}| = |Q \cap \pi| = q+1$, and (x_0,x_1,x_2) is 3-regular. It follows that all points are 3-regular. Since all lines are regular, by 1.6.1 they are semiregular and hence satisfy property (H). Since no triad (of points) has a unique center, also all points are semiregular and satisfy property (H).

(iii) Consider two noncollinear points x,y of $H(4,q^2)$. Then $\{x,y\}^{\perp} = H \cap \pi$, where π is the polar plane of the line $L = xy$ (of $PG(4,q^2)$) with respect to the hermitian variety H. The set of all points of H that are collinear with all points of $H \cap \pi$ is clearly $L \cap H$. Hence $|\{x,y\}^{\perp\perp}| = |H \cap L| = q+1$. Further, consider two nonconcurrent lines L,M of $H(4,q^2)$. If $PG(3,q^2)$ is the 3-space containing L and M, then

$PG(3,q^2) \cap H = H'$ is a nonsingular hermitian variety of $PG(3,q^2)$. Moreover, the trace (resp., span) of (L,M) in $H(4,q^2)$ coincides with the trace (resp., span) of (L,M) in $H'(3,q^2)$. Hence $|\{L,M\}^{\perp\perp}| = 2$. Since $t > s$ in $H(4,q^2)$, the pair (L,M) is not antiregular.

Now we shall show that all points are semiregular. Suppose that u is the unique center of the triad (x,y,z). Since $(|\{x,y\}^{\perp\perp}|-1)t = s^2$, part (ii) of 1.4.2 tells us that $z \in cl(x,y)$. Hence u is semiregular. It follows also that all points satisfy property (H).

From $|\{L,M\}^{\perp\perp}| = 2$ for each pair (L,M) of nonconcurrent lines, it follows that all lines have property (H). Finally, we show that no line is semiregular. Consider three lines L,M,V of $H(4,q^2)$ with $L \sim V \sim M \not\sim L$. Further, let N be a line of $H(4,q^2)$ for which $N \sim V$, $L \not\sim N \not\sim M$, and which is not contained in the 3-space $PG(3,q^2)$ defined by L and M. Then V is the unique center of the triad (L,M,N), but $N \notin cl(L,M)$. Hence V is not semiregular, and the proof of (iii) is complete. □

We now turn our attention to the GQ $T(O)$.

3.3.2. (i) *All lines of type* (b) *of* $T_2(O)$ *are regular. The point* (∞) *is regular or antiregular according as* q *is even or odd.*

(ii) *All lines of type* (b) *of* $T_3(O)$ *are regular , and the point* (∞) *is 3-regular.*

Proof. (i) Let $x \in O$ be a line of type (b). We shall prove that x is regular. Consider a line L of type (a) which is not concurrent with x. The intersection of O and L is denoted by y, $x \neq y$. It is clear that $\{x,L\}^{\perp}$ contains y and the q lines of the plane xL which contain x but not y. And $\{x,L\}^{\perp\perp}$ contains x and the q lines of the plane xL which contain y but not x. Hence $|\{x,L\}^{\perp\perp}| = q+1$ and (x,L) is regular. It follows that (∞) is coregular and the proof of (i) is complete by 1.5.2.

(ii) An argument analogous to that in (i) shows that all lines of type (b) of $T_3(O)$ are regular. It remains only to prove that (∞) is 3-regular. Let $((\infty),x,y)$ be a triad, so that x and y are points of type (i). The 3-space $PG(3,q)$ which contains O and the line xy of $PG(4,q)$ have a point $z \notin O$ in common. Exactly q+1 tangent planes $\pi_1,\dots,\pi_{q+1}$ of O contain z. It is clear that $\{(\infty),x,y\}^{\perp}$ consists of the q+1 3-spaces $x\pi_1,\dots,x\pi_{q+1}$. And $\{(\infty),x,y\}^{\perp\perp}$ contains (∞) and the q points of $xy-\{z\}$. Hence $|\{(\infty),x,y\}^{\perp\perp}| = q+1$, and consequently (∞) is 3-regular. □

There is a kind of converse.

3.3.3. (i) *If* $T_2(O)$ *has even one regular pair of nonconcurrent lines of type* (a) *defining distinct points of* O, *then* O *is a conic and* $T_2(O)$ *is isomorphic to* $Q(4,q)$.

(ii) *If* $T_2(O)$ *has a regular point of type* (i), *then* q *is even,* O *is a conic and* $T_2(O)$ *is isomorphic to* $Q(4,q)$.

(iii) *If* $T_3(O)$ *has a regular line of type* (a), *then* O *is an elliptic quadric and* $T_3(O)$ *is isomorphic to* $Q(5,q)$.

(iv) *If* $T_3(O)$ *has a 3-regular point other than* (∞), *then* O *is an elliptic quadric and* $T_3(O)$ *is isomorphic to* $Q(5,q)$.

Proof. (i) Suppose that (L,M) is a regular pair of nonconcurrent lines of $T_2(O)$ of type (a) defining distinct points of O. Then all elements of $\{L,M\}^{\perp}$ and $\{L,M\}^{\perp\perp}$ are of type (a), and in PG(3,q) each line of $\{L,M\}^{\perp}$ has a point in common with each line of $\{L,M\}^{\perp\perp}$. Hence $\{L,M\}^{\perp}$ and $\{L,M\}^{\perp\perp}$ are the reguli of some hyperbolic quadric Q^+ of PG(3,q). Evidently O is a plane intersection of Q^+, and thus O is a conic and by 3.2.2 the proof of (i) is complete.

(ii) Suppose that some type (i) point of $T_2(O)$ is regular. Since the translations of PG(3,q)-PG(2,q), $O \subset PG(2,q)$, induce a group of automorphisms of $T_2(O)$ which is transitive on points of type (i), clearly all points of type (i) are regular. It follows easily that all points are regular, and then by 1.5.2 that q is even since lines of type (b) are regular. Then by 1.5.2 again all lines are regular, so an appeal to part (i) completes the proof of (ii).

(iii) Suppose that $T_3(O)$ has a regular line L of type (a). The point of O defined by L is denoted by x. By an argument analogous to that used in the proof of (i) it follows that all ovals on O containing x are conics. So if $x' \in O-\{x\}$, the ovals on O containing x and x' are conics. If q is even, by a theorem of O. Prohaska and M. Walker [147] the ovoid O is an elliptic quadric, i.e. $T_3(O) \cong Q(5,q)$. If q is odd, by a result of A. Barlotti [5] O must be an elliptic quadric, so that by 3.2.4 $T_3(O) \cong Q(5,q)$.

(iv) Finally, suppose that $T_3(O)$ has a 3-regular point x of type (i) or (ii). In Step 3 of the proof of 5.3.1 we shall show that x is coregular. Now by part (iii) the proof is complete. □

Note : If q is odd, any oval (resp., ovoid) is necessarily a conic (resp., an elliptic quadric), so that $T_2(O) \cong Q(4,q)$ (resp., $T_3(O) \cong Q(5,q)$). For q even, $q > 4$, there are always ovals O for which $T_2(O)$ has a unique line of regular points (cf. Chapter 12 for more details). And for $q = 2^h$, h odd, $h > 2$, the Tits ovoids provide examples of $T_3(O)$ not isomorphic to $Q(5,q)$.

In 10.6.2 we shall prove that for $q \equiv 2 \pmod 3$, $q > 2$, L (as used in the construction given in 3.1.6) is the unique regular line of K(q). Hence by Step 3 of the proof of 5.3.1 K(q), $q \neq 2$, has no 3-regular point. This has the following interesting corollary : *if* $q = 2^{2h+1}$, $h \geq 1$, *there are at least three pairwise nonisomorphic* GQ *of order* (q,q^2).

We now turn to the known GQ of order (q-1,q+1). In 5.3.2 we shall prove that every GQ of order (2,4) is isomorphic to Q(4,2). Hence all lines of the GQ of order (2,4) are regular, and all its points are 3-regular. Note that a GQ of order (q-1,q+1), $q > 2$, has no regular pair of noncollinear points since $s < t$.

3.3.4. *The pair* (L,M) *of nonconcurrent lines of* $T_2^*(O)$ *is regular iff* L *and* M *define the same point of* O .

Proof. Let L and M be distinct lines of $T_2^*(O)$ which define the same point y of the complete oval O. The plane LM of PG(3,q) intersects O in two points y and z. It is clear that $\{L,M\}^\perp$ consists of the q lines distinct from yz which are contained in the plane LM and pass through the point z. The span $\{L,M\}^{\perp\perp}$ consists of the q lines distinct from yz which are contained in the plane LM and pass through the point y. Hence the pair (L,M) is regular.

Further, let L and M be nonconcurrent lines of $T_2^*(O)$ which define different points y and z of O. If (L,M) is regular, then $\{L,M\}^\perp$ (resp., $\{L,M\}^{\perp\perp}$) defines q different points $u_1,\ldots,u_q$ (resp., $y = y_1$, $z = y_2,\ldots,y_q$) of O. Moreover, $u_i \neq y_j$ for all $i,j = 1,\ldots,q$. Hence $|O| \geq 2q$, which is impossible since $q > 2$. □

In determining all regular elements of P(W(q),x) we may restrict ourselves to the case q odd, since otherwise $P(W(q),x) \cong T_2^*(O)$ where O is a conic. Note that in the case q odd $P(W(q),x) \cong AS(q)$.

3.3.5. *The pair* (L,M), $L \not\sim M$, *of* P(W(q),x),$q > 3$, *is regular iff one of the following holds :* (i) L *and* M *are hyperbolic lines of* W(q) *which contain* x, *or* (ii) *in* W(q) L *and* M *are concurrent lines.*

Proof. Let L and M be distinct lines of P(W(q),x) which are both of type (b), i.e. which are hyperbolic lines of W(q) through x. Let π be the polar plane of x with respect to the symplectic polarity θ of PG(3,q) defining W(q). If y is the pole of the plane LM with respect to θ, then $\{L,M\}^{\perp}$ consists of the q lines of PG(3,q) distinct from xy which are contained in the plane LM and pass through the point y. All lines of $\{L,M\}^{\perp}$ are of type (a). The span $\{L,M\}^{\perp\perp}$ consists of the q lines of PG(3,q) distinct from xy, which are contained in the plane LM and pass through the point x. All lines of $\{L,M\}^{\perp\perp}$ are of type (b). Since $|\{L,M\}^{\perp\perp}| = q$, the pair (L,M) is regular.

Now let L and M be nonconcurrent lines of P(W(q),x) which are both of type (a) and are concurrent in W(q). Then $\{L,M\}^{\perp}$ consists of q lines of type (b), and hence by the preceding paragraph (L,M) is regular.

Finally, suppose that (L,M) is a regular pair of nonconcurrent lines with L of type (b) and M of type (a) or with both L and M of type (a) but not concurrent in W(q). Then $\{L,M\}^{\perp}$ and $\{L,M\}^{\perp\perp}$ each contain at least q-1 lines of type (a). Let $Z_1,\ldots,Z_{q-1}$ (resp., $V_1,\ldots,V_{q-1}$) be lines of type (a) contained in $\{L,M\}^{\perp}$ (resp., $\{L,M\}^{\perp\perp}$). Then in W(q) the line Z_i is concurrent with the line V_j, for all i, j = 1,...,q-1. Since q > 3 and by 3.3.1 all lines of W(q), q odd, are antiregular, and we have a contradiction. □

3.4. OVOIDS AND SPREADS OF THE KNOWN GQ

As usual we consider the classical case first.

3.4.1. (i) *The* GQ Q(4,q) *always has ovoids. It has spreads iff* q *is even, but in that case has no partition into ovoids or spreads by* 1.8.5.

(ii) *The* GQ Q(5,q) *has spreads but no ovoids.*

(iii) *The* GQ $H(4,q^2)$ *has no ovoid. For* q = 2 *it has no spread (A.E. Brouwer* [21]) . *For* q > 2, *whether or not it has a spread seems to be an open problem.*

Proof. (i) Let us consider the GQ Q(4,q). In PG(4,q) consider a hyperplane PG(3,q) for which PG(3,q) ∩ Q is an elliptic quadric Q^-. Then Q^- is an ovoid of Q(4,q). If q is even, then Q(4,q) is self-dual, and hence Q(4,q) has spreads. If q is odd, since all lines of Q(4,q) are regular, the dual of 1.8.4 guarantees that Q(4,q) has no spread.

(ii) Let H be a nonsingular hermitian variety in $PG(3,q^2)$. Then any hermitian curve on H, i.e. any nonsingular plane intersection of H , is an ovoid of the GQ $H(3,q^2)$. Hence $H(3,q^2)$ has ovoids, implying that $Q(5,q)$ has spreads. By 1.8.3 $Q(5,q)$ has no ovoid.

(iii) Here we propose two proofs.

(a) Suppose $H(4,q^2)$ did have an ovoid O, and let $\{x,y\} \subset O$. Then $\{x,y\}^{\perp\perp}$ has cardinality q+1. Since $qt = s^2$, by 1.8.6 O has an empty intersection with $\{x,y\}^{\perp\perp}$, a contradiction.

(b) Again suppose that O is an ovoid of $H(4,q^2)$, and consider the intersection of O with a hyperplane $PG(3,q^2)$ of $PG(4,q^2)$. If $H \cap PG(3,q^2)$ is a nonsingular hermitian variety H' of $PG(3,q^2)$, then $O \cap PG(3,q^2) = O'$ is an ovoid of the GQ $H'(3,q^2)$. Hence $|O'| = q^3+1$. If $H \cap PG(3,q^2)$ has a singular point p, then $|O \cap PG(3,q^2)| = 1$ if $p \in O$ and $|O \cap PG(3,q^2)| = q^3+1$ if $p \notin O$. So for any $PG(3,q^2)$ we have $|O \cap PG(3,q^2)| \in \{1,q^3+1\}$. From J.A. Thas [183] it follows that O is a line of $PG(4,q^2)$, a contradiction.

By an exhaustive search A.E. Brouwer [21] showed that $H(4,4)$ has no spread. We do not know whether or not $H(4,q^2)$ has a spread when $q > 2$. □

For q an even power of 2, only one type of ovoid of $Q(4,q)$ is known. But for $q = 2^{2h+1}$, $h \geqslant 1$, two types of ovoids of $Q(4,q)$ are known. Their projections from the nucleus of Q onto a $PG(3,q)$ are the elliptic quadric and the Tits ovoid [217] in $PG(3,q)$. On the other hand, the corresponding spreads of $W(q)$ are the regular spread [49] and the Lüneburg-spread [100] of $PG(3,q)$. Details and proofs are in J.A. Thas [181].

Recently W.M. Kantor [90] proved that for odd values of q there exist ovoids of $Q(4,q)$ which are not contained in some $PG(3,q)$, i.e. which are not obtained in the way described in the proof of the first part of 3.4.1. One of the classes constructed by W.M. Kantor is the following. Consider in $PG(4,q)$, q odd, the nonsingular quadric Q with equation $x_2^2+x_0x_4+x_3x_1 = 0$. Let $\sigma \in \mathrm{Aut}\, GF(q)$ and let $-k$ be a nonsquare in $GF(q)$. Then $\{(0,0,0,0,1)\} \cup \{(1,x_1,x_2,kx_1^\sigma,-x_2^2-kx_1^{\sigma+1}) \| x_1,x_2 \in GF(q)\}$ is an ovoid O of $Q(4,q)$. (It is easy to check that O is contained in some $PG(3,q)$ iff σ is the identity permutation). Moreover, he showed that the corresponding spread of $W(q)$ gives rise to a Knuth semifield plane [49].

Without using the duality between $H(3,q^2)$ and $Q(5,q)$ it is possible to give a short proof that $Q(5,q)$ has a spread. Indeed, over a quadratic extension of $GF(q)$ we consider two mutually skew and conjugated planes π and

π' on the extension Q^* of Q. For each point $p \in Q$, let L be the line containing p and intersecting π and π'. Since L contains at least three points of Q^*, L is a line of Q^*. As L is a line of PG(5,q), L is a line of Q. The set of all such lines L evidently is a spread of the GQ Q(5,q).

Further, we show that $H(3,q^2)$ has different types of ovoids. Let H' be an hermitian curve on H. If $x,y \in H'$, $x \neq y$, then $(H'-\{x,y\}^{\perp\perp}) \cup \{x,y\}^{\perp}$ is also an ovoid. For more information about the spreads of Q(5,q) we refer to J.A. Thas [206, 207].

Finally, we remark that the second part of (ii) was first proved by A.A. Bruen and J.A. Thas [25] using a method analogous to that used in proof (b) of (iii).

3.4.2. (i) *The* GQ $T_2(O)$ *always has an ovoid, but for q even it has no partition into ovoids or spreads by* 1.8.5.

(ii) *The* GQ $T_3(O)$ *has no ovoid but always has spreads.*

Proof. (i) Let π be a plane which has no point in common with O. The q^2 points of type (i) in π together with the point (∞) clearly constitute an ovoid of $T_2(O)$. If the oval O of PG(2,q) is contained in some ovoid O' of PG(3,q), $PG(2,q) \subset PG(3,q)$, then an ovoid of $T_2(O)$ may also be obtained as follows. Let $O = \{x_0,\ldots,x_q\}$ and let π_i be the tangent plane of O' at x_i, $i = 0,\ldots,q$. Then the set $(O'-O) \cup \{\pi_0,\pi_1,\ldots,\pi_q\}$ is an ovoid of $T_2(O)$.

(ii) By 1.8.3 the GQ $T_3(O)$ has no ovoid. Finally, we show that $T_3(O)$ always has spreads. Let $x \in O$, let π be a plane of $PG(3,q) \supset O$ for which $x \notin \pi$, and let L be the intersection of π and the tangent plane of O at x. Further, let V be a threespace through π which is distinct from PG(3,q), and let W be a line spread of V containing L as an element. The elements of W are denoted by $L, L_1,\ldots,L_{q^2}$. Since $L \cap L_i = \phi$, the plane L_ix has exactly two points in common with O, say x and x_i. Notice that $\{x,x_i\} = O \cap xy_i$, with $\{y_i\} = \pi \cap L_i$. Clearly $O = \{x,x_1,\ldots,x_{q^2}\}$. The lines distinct from xx_i which join x_i to the points of L_i are denoted by $M_{i1},M_{i2},\ldots,M_{iq}$. Now we show that $\{M_{11},M_{12},\ldots,M_{q^2q},x\}$ is a spread of $T_3(O)$.

Clearly the lines M_{ij} and $M_{ij'}$ of $T_3(O)$, $j \neq j'$, are not incident with a common point of $T_3(O)$ of type (i), and since $M_{ij}M_{ij'} \cap PG(3,q) = xx_i$ is not a tangent line of O, they are not incident with a common point of $T_3(O)$

of type (ii). It is also clear that M_{ij} and $M_{i'j'}$, $i \neq i'$, are not incident with a common point of type (ii), and since M_{ij}, $M_{i'j'}$ and x generate a four dimensional space, the lines M_{ij} and $M_{i'j'}$ of $T_3(O)$ cannot be incident with a common point of type (i). Finally, the lines x and M_{ij} of $T_3(O)$ are not incident with a common point of $T_3(O)$. Since $|\{M_{11},\ldots,M_{q^2q},x\}| = q^3+1$, we conclude that $\{M_{11},\ldots,M_{q^2q},x\}$ is a spread of $T_3(O)$. □

3.4.3. *The* GQ P(*S*,x) *always has spreads. It has an ovoid iff S has an ovoid containing* x.

Proof. Consider a GQ *S* of order s, s > 1, with a regular point x. If x I L, then the s^2 lines of *S* which are concurrent with L but not incident with x constitute a spread of P(*S*,x). In addition, the set of all lines of type (b) is a spread of P(*S*,x). Further, we note that for each spread V of *S*, the set V-{L}, where L is the line of V which is incident with x, is a spread of P(*S*,x).

Let O be an ovoid of the GQ *S* with x ∈ O. It is clear that O-{x} is an ovoid of P(*S*, x) if every line of type (b) contains exactly one point of O-{x}. But this follows immediately from 1.8.4.

Conversely, suppose that O' is an ovoid of P(*S*,x). It is immediate from the construction of P(*S*, x) that O' ∪ {x} is an ovoid of *S*. □

3.4.4. *The* GQ K(q) *has spreads but no ovoid.*

Proof. By 1.8.3 K(q) has no ovoid. We sketch a proof that K(q) always has spreads. Let V be a spread of the generalized hexagon H(q), i.e. let V be a set of q^3+1 lines of H(q) every two of which are at distance 6 [36]. If the regular line L of K(q) belongs to V, then it is easy to show that V is a spread of K(q). Since H(q) always has spreads containing L [200], the GQ K(q) always has a spread. □

Note : Suppose that H(q) is constructed on the quadric Q of PG(6,q). Let PG(5,q) be a hyperplane of PG(6,q) which contains L and for which Q ∩ PG(5,q) is elliptic [80]. Then by J.A. Thas [200] the lines of H(q) which are contained in PG(5,q) constitute a spread of K(q).

3.5. SUBQUADRANGLES

Here we shall describe some of the known subquadrangles of both the classical and of the other known GQ, with the main emphasis being on large subquadrangles.

(a) Consider $Q(5,q)$, with Q a nonsingular quadric of projective index 1 in $PG(5,q)$. Intersect Q with a nontangent hyperplane $PG(4,q)$. Then the points and lines of $Q' = Q \cap PG(4,q)$ form the GQ $Q'(4,q)$. Here $s^2 = t = q^2$, $s = s' = t'$, so that $t = s't'$. Since all lines of $Q(5,q)$ (resp., $Q'(4,q)$) are regular, $Q(5,q)$ (resp., $Q'(4,q)$) has subquadrangles with $t'' = 1$ and $s'' = s' = s$.

Similarly, consider $H(4,q^2)$, with H a nonsingular hermitian variety of $PG(4,q^2)$. Intersect H with a nontangent hyperplane $PG(3,q^2)$. Then the points and lines of $H' = H \cap PG(3,q^2)$ form the GQ $H'(3,q^2)$. Here $t = s^{3/2} = q^3$, $s = s'$, $t' = \sqrt{s}$, and again $t = s't'$. Since all points of $H'(3,q^2)$ are regular, $H'(3,q^2)$ has subquadrangles with $t'' = t' = \sqrt{s}$ and $s'' = 1$.

Now consider $Q(4,q)$ and extend $GF(q)$ to $GF(q^2)$. Then Q extends to $\bar{Q}$ and $Q(4,q)$ to $\bar{Q}(4,q^2)$. Here $Q(4,q)$ is a subquadrangle of $\bar{Q}(4,q^2)$, and we have $t = s = q^2$ and $t' = s' = q$. Hence $t = s = s't'$.

(b) Consider $T_3(O)$ and let π be a plane of $PG(3,q) \supset O$ for which $O \cap \pi = O'$ is an oval. Then by considering a hyperplane $PG'(3,q)$ of $PG(4,q)$, for which $PG(3,q) \cap PG'(3,q) = \pi$, we obtain $T_2(O')$ as a subquadrangle of $T_3(O)$. Here $s^2 = t = q^2$ and $s = s' = t'$, so again $t = s't'$.

(c) Consider an irreducible conic C' of the plane $PG(2,q) \subset PG(3,q)$, where $q = 2^h$. Let $GF(q^n)$, $n > 1$, be an extension of the field $GF(q)$ and let $PG(3,q^n)$ (resp., $PG(2,q^n)$ and C) be the corresponding extension of $PG(3,q)$ (resp., $PG(2,q)$ and C'). If x is the nucleus of C', then x is also the nucleus of C, and $C' \cup \{x\} = O'$ (resp., $C \cup \{x\} = O$) is a complete oval of the plane $PG(2,q)$ (resp., $PG(2,q^n)$). Evidently $T_2^*(O')$ is a subquadrangle of $T_2^*(O)$. In this case we have $s = q^n-1$, $t = q^n+1$, $s' = q-1$, and $t' = q+1$. For $n = 2$ we have $s = s't'$.

3.6. SYMMETRIC DESIGNS DERIVED FROM GQ

3.6.1. (i) *A GQ of order* q *gives rise to a symmetric* 2-$(q^3+q^2+q+1, q^2+q+1, q+1)$ *design.*

(ii) *A GQ of order* $(q+1,q-1)$ *gives rise to a symmetric* 2-$(q^2(q+2), q(q+1), q)$ *design.*

Proof. (i) Let $S = (P,B,I)$ be a GQ of order q. Define as follows a new incidence structure $S' = (P',B',I')$: $P' = B' = P$, and $x\ I'\ y$ for $x \in P'$, $y \in B'$, iff $x \sim y$ in S . Clearly S' is a symmetric 2-$(q^3+q^2+q+1,q^2+q+1,q+1)$ design. The identity mapping of P is a bijection of P' onto B' which defines a polarity θ of S'. Moreover, all points and lines of S' are absolute for θ . We also remark that an incidence matrix of S' is given by A + I, where A is an adjacency matrix of the point graph of S .

(ii) Let $S = (P,B,I)$ be a GQ of order $(q+1,q-1)$, and let $S' = (P',B',I')$ be defined by : $P' = B' = P$ and $x\ I'\ y$, with $x \in P'$ and $y \in B'$, iff $x \neq y$ and $x \sim y$ in S . Clearly S' is a symmetric 2-$(q^2(q+2),q(q+1),q)$ design (cf. also our comments following the proof of 3.1.5). The identity mapping of P is a bijection of P' onto B', which defines a polarity θ of S'. Moreover, θ has no absolute point. Further, we note that any adjacency matrix of the point graph of S is an incidence matrix of the design S'. □

Let $S_1 = (P_1,B_1,I_1)$ and $S_2 = (P_2,B_2,I_2)$ be two GQ of order q (resp., $(q+1,q-1)$), and let S_1' and S_2' be the corresponding designs. It is straightforward to check that any isomorphism of S_1 onto S_2 induces an isomorphism of S_1' onto S_2' . In [56] M.M. Eich and S.E. Payne consider the following converse : In which cases is an isomorphism between S_1' and S_2' necessarily induced by an isomorphism of the underlying GQ? We now survey their main results.

3.6.2. *If* S_1 *and* S_2 *have order* $(q+1,q-1)$, $q \geqslant 3$, *then any isomorphism from* S_1' *onto* S_2' *is induced by a unique isomorphism from* S_1 *onto* S_2 . *For* $q = 2$ *this result does not hold.*

Proof. First suppose $q \geqslant 3$ and let τ be an isomorphism from $S_1' = (P_1',B_1',I_1')$ onto $S_2' = (P_2',B_2',I_2')$. Then τ is a pair (α,β), where α is a bijection from P_1' onto P_2' and β is a bijection of B_1' onto B_2' satisfying $x\ I_1'\ y$ iff $x^\alpha\ I_2'\ y^\beta$. Hence α and β are really bijections from P_1 onto P_2 satisfying $x \sim y$ iff $x^\alpha \sim y^\beta$ and $x^\alpha \neq y^\beta$ for distinct elements x and y. Assume that α is not an isomorphism of the point graph of S_1 onto the point graph of S_2. Then there must be distinct collinear points x and y in S_1 such that x^α and y^α are not collinear in S_2. Let $z_1,\dots,z_q$ be the remaining points incident with the line xy of S_1. Then $z_1^\beta, z_2^\beta,\dots,z_q^\beta$ must be precisely the elements of $\{x^\alpha,y^\alpha\}^\perp$. Since $x \sim y$, clearly $x^\beta \sim y^\alpha$ $(x^\beta \neq y^\alpha)$. So we may assume that y^α, x^β, and say z_1^β are collinear in S_2. But $z_i^\alpha \sim x^\beta$ $(z_i^\alpha \neq x^\beta)$ and $z_i^\alpha \sim z_1^\beta$ $(z_i^\alpha \neq z_1^\beta)$, for $i = 2,\ldots,q$. Hence $y^\alpha, x^\beta, z_1^\beta, z_2^\alpha, \ldots, z_q^\alpha$ must be the q+2 points incident with some line L of S_2. For $2 \leqslant i, j \leqslant q$, $i \neq j$, it must be that $z_i^\beta \sim z_j^\alpha$ $(z_i^\beta \neq z_j^\alpha)$, $z_i^\beta \sim y^\alpha$ $(z_i^\beta \neq y^\alpha)$, so that z_i^β is incident with L. But then $x^\alpha \sim z_i^\beta$ $(x^\alpha \neq z_i^\beta)$ for $1 \leqslant i \leqslant q$ implies x^α is incident with L, so $x^\alpha \sim y^\alpha$, a contradiction. Hence α must be an isomorphism of the point graph of S_1 onto the point graph of S_2. Again let $x \sim y$ in S_1 with $x \neq y$, and let $z_1,\ldots,z_q$ be the remaining points incident with the line xy of S_1. Then $z_1^\alpha, z_2^\alpha, \ldots, z_q^\alpha$ are the remaining points incident with the line $x^\alpha y^\alpha$ of S_2. We have $y^\beta \sim z_i^\alpha$ $(y^\beta \neq z_i^\alpha)$ and $y^\beta \sim x^\alpha$ $(y^\beta \neq x^\alpha)$. Hence $y^\beta = y^\alpha$, implying $\alpha = \beta$. It is now clear that τ is induced by a unique isomorphism from S_1 onto S_2.

Now suppose that q = 2, so $S = (P,B,I)$ is a grid. Let $P = \{x_{ij} \parallel i,j = 1,\ldots,4\}$, $B = \{L_1,\ldots,L_4, M_1,\ldots,M_4\}$, x_{ij} I L_k iff $i = k$ and x_{ij} I M_k iff $j = k$. Let α be the permutation of P defined by $x_{12}^\alpha = x_{21}$, $x_{21}^\alpha = x_{12}$, $x_{11}^\alpha = x_{22}$, $x_{22}^\alpha = x_{11}$, $x_{34}^\alpha = x_{43}$, $x_{43}^\alpha = x_{34}$, $x_{33}^\alpha = x_{44}$, $x_{44}^\alpha = x_{33}$ and $x_{ij}^\alpha = x_{ij}$ in all other cases. Then the permutation α of the pointset of the corresponding 2-(16,6,2) design S' clearly defines an automorphism of S', but α is not an automorphism of the point graph of S . □

The situation for GQ of order q requires somewhat more effort. Let $S = (P,B,I)$ be a GQ of order q, $q \neq 1$. A point x_∞ of S is called a *center of irregularity* provided the following is true : if y and z are distinct collinear points in $P - x_\infty^\perp$, then there is some point w such that $w \sim z$ and (y,w) is an irregular (i.e. not regular) pair. The following result is a key lemma in the treatment of the order q case.

3.6.3. *Suppose S has a center of irregularity. Let α be a permutation of* P *satisfying the following :*

(i) $y \sim y^\alpha$ *for all* $y \in P$,

(ii) $y \sim w$ *iff* $y^\alpha \sim w^{\alpha^{-1}}$, *for all* $y,w \in P$,

(iii) *If* (y,w) *is an irregular pair of points, then* $w \not\sim y^{\alpha^{-1}}$.

Then α is the identity.

Proof. Suppose x_∞ is a center of irregularity, and let y be a point such that $y \not\sim x_\infty$ and $y \neq y^{\alpha^{-1}}$, so $y \neq y^\alpha$. By (i) $y \sim y^{\alpha^{-1}}$. If $y^{\alpha^{-1}} \not\sim x_\infty$, there must be some point w such that $w \sim y^{\alpha^{-1}}$ and (y,w) is irregular. But this is impossible by (iii). Hence $y^{\alpha^{-1}} \sim x_\infty$. Now if $y \neq y^{\alpha^{-1}}$, then by (i) and (ii) y^α must be incident with the line $yy^{\alpha^{-1}}$. Hence $y^\alpha \not\sim x_\infty$, and the argument just applied to show $y^{\alpha^{-1}} \sim x_\infty$ now shows that $(y^\alpha)^{\alpha^{-1}} = y$ must be collinear with x_∞, a contradiction. So $y^\alpha = y^{\alpha^{-1}}$. We have proved that α^2 fixes each point in $P-x_\infty^\perp$. If $z \in P-x_\infty^\perp$, then by (ii) $x_\infty^\alpha \not\sim z^{\alpha^{-1}}$, i.e. $x_\infty^\alpha \not\sim z^\alpha$. Again by (ii) $x_\infty^{\alpha^2} \not\sim z$. Since $x_\infty^{\alpha^2} \not\sim z$ for all $z \in P-x_\infty^\perp$, we have $x_\infty^{\alpha^2} = x_\infty$. If $u \in x_\infty^\perp-\{x_\infty\}$, then for $u' \in \{x_\infty\} \cup (P-x_\infty^\perp)$ and $u' \sim u$, we have $u^\alpha \sim (u')^{\alpha^{-1}}$, i.e. $u^\alpha \sim u'^\alpha$. Again by (ii) $u^{\alpha^2} \sim u'$. It easily follows that $u^{\alpha^2} = u$. Hence α^2 is the identity permutation of P, and by (ii) α defines an automorphism π of S.

We now claim α fixes x_∞. For suppose $x_\infty^\alpha = z \neq x_\infty$. Then $z \sim x_\infty$. Let L be the line zx_∞. Since α^2 is the identity, $z^\alpha = x_\infty$, which implies that α must fix the set of all points incident with L. Also z must be a center of irregularity. It now follows for z just as it did for x_∞ that if y is a point such that $y \neq y^\alpha$ and $y \not\sim z$, then $y^\alpha \in \{y,z\}^\perp$. If $y \not\sim z$, $y \not\sim x_\infty$, $y \neq y^\alpha$, then $y^\alpha \in \{y,z\}^\perp \cap \{y,x_\infty\}^\perp$, implying y^α I L. This is impossible since y $\not\!\mathrm{I}$ L. Hence any point y with $y \not\sim z$ and $y \not\sim x_\infty$ must be fixed by α. Since each line M, $M \not\sim L$, is incident with at least two points not collinear with x_∞ or z (by 1.3.4 (iv) all points of a GQ of order 2 are regular), it is clear that $M^\pi = M$. It follows readily that π is the identity automorphism of S, which contradicts the assumption that $x_\infty^\alpha \neq x_\infty$.

Finally, since α fixes x_∞, it must leave $P-x_\infty^\perp$ invariant. Then by the first part of the proof α must fix each point of $P-x_\infty^\perp$. It follows readily that α is the identity. □

If S is a GQ of order q, $q \neq 1$, in which each pair of noncollinear points is irregular, then clearly each point is a center of irregularity and 3.6.3 applies.

3.6.4. *Let* $S_1 = (P_1,B_1,I_1)$ *and* $S_2 = (P_2,B_2,I_2)$ *be* GQ *of order* q, q > 1. *If* S_2 *has a center of irregularity, then any isomorphism from* S_1' *onto* S_2' *is induced by an isomorphism from* S_1 *onto* S_2 .

Proof. Suppose that S_1' and S_2' are isomorphic , and that S_2 has a center of irregularity. Further, assume that Q_i is an incidence matrix of S_i, i = 1,2, with points labeling columns and lines labeling rows. Then $Q_i^TQ_i = (s+1)I+A_i$, with A_i an adjacency matrix of the point graph of S_i. Hence $Q_i^TQ_i-sI = N_i$ is an incidence matrix of the design S_i'. Since $S_1' \cong S_2'$, there are permutation matrices M_1 and M_2 such that $M_1N_1M_2 = N_2$. By reordering the points of S_1 so that its new incidence matrix is $Q_1M_1^{-1}$, we may suppose $N_1 = N_2M$ for some permutation matrix M. If M = I, we are done.

So suppose M ≠ I. Since $N_1 = N_2M$ and N_2 are symmetric, we have

$$M^TN_2 = N_2M \tag{1}$$

If $P_2 = \{x_1,\ldots,x_v\}$, then let the permutation α be defined by $x_j = x_i^\alpha$ iff $(M)_{ij} \neq 0$. By (1) $x_i \sim x_j$ iff $x_i^\alpha \sim x_j^{\alpha^{-1}}$, for all points x_i, x_j of P_2. Since N_2M has only 1's on its main diagonal, we have $x_i \sim x_i^\alpha$ for all points x_i of P_2. We now prove that for any irregular pair (x_i,x_j) of points of S_2, we have $x_i \not\sim x_j^{\alpha^{-1}}$. Suppose the contrary for a particular i and j. If $P_1 = \{y_1,\ldots,y_v\}$, then from $N_1 = N_2M$ it follows that $y_n \sim y_m$ iff $x_n \sim x_m^{\alpha^{-1}}$. So in particular $y_i \sim y_j$ in S_1. Since $x_j \sim x_j^{\alpha^{-1}}$, $x_i \sim x_i^{\alpha^{-1}}$, $x_i \sim x_j^{\alpha^{-1}}$, $x_i^{\alpha^{-1}} \sim x_j$, we have $x_j^{\alpha^{-1}}, x_i^{\alpha^{-1}} \in \{x_i,x_j\}^\perp$ (notice that $x_i \not\sim x_j$ since (x_i,x_j) is irregular). Now consider a point y_k incident with the line y_iy_j. Then $x_i \sim x_k^{\alpha^{-1}}$ and $x_j \sim x_k^{\alpha^{-1}}$, implying $\{x_i,x_j\}^\perp = \{x_k^{\alpha^{-1}} \parallel y_k \text{ I } y_iy_j\}$. If $y_r \text{ I } y_iy_j$, then $x_r \sim x_k^{\alpha^{-1}}$ for all $x_k^{\alpha^{-1}} \in \{x_i,x_j\}^\perp$. Hence (x_i,x_j) is regular, a contradiction. So for any irregular pair (x_i,x_j) of points of S_2 we have $x_i \not\sim x_j^{\alpha^{-1}}$. Now by 3.6.3 α is the identity permutation of P_2. So M = I and the proof is complete. □

We are now in a position to resolve the problem of this section for at least the known GQ.

3.6.5. *Suppose* S_1 *and* S_2 *are* GQ *of order* q, q > 1, *and that* S_2 *is isomorphic to one of the known* GQ . *Then one of the following two situations must arise :*

(i) $S_2 \cong W(q)$. *If* $S_1' \cong S_2'$, *then also* $S_1 \cong W(q)$. *However, not every isomorphism from* S_1' *to* S_2' *is induced by one from* S_1 *to* S_2 .

(ii) S_2 *has a center of irregularity, so that each isomorphism from* S_1' *to* S_2' *is induced by one from* S_1 *to* S_2 .

Proof. Since all points of W(q) are regular, it has no center of irregularity The symmetric design arising from W(q) clearly is isomorphic to the well known design of points and planes of PG(3,q). Here the polarity θ of the design is essentially the symplectic polarity of PG(3,q) defining W(q). Moreover, it is an easy geometrical exercise to prove that W(q) is the only GQ of order q that gives rise to the symmetric design S' formed by the points and planes of PG(3,q). Since any element of PGL(4,q) defines an automorphism of S' and since there are always elements in PGL(4,q) that are not automorphisms of the point graph of W(q), there are automorphisms of S' not induced by automorphisms of W(q).

If $S_2 \cong Q(4,q)$, there are two cases. If q is even, then $Q(4,q) \cong W(q)$, so it is already handled . If q is odd, then each point is antiregular, and in particular is a center of irregularity.

The only remaining known GQ of order q is $T_2(O)$ and its dual, where q is even and O is a nonconical oval. In this case we now show that (∞) is a center of irregularity for $T_2(O)$ and each line of $T_2(O)$ of type (b) is a center of irregularity for the dual of $T_2(O)$.

First we prove that any line x of type (b) of $T_2(O)$ $(x \in O)$ is a center of irregularity for the dual of $T_2(O)$. Let L and M be two concurrent lines each of which is not concurrent with x (then L and M are of type (a)). Let $L \sim y$ and $M \sim z$, with y and z of type (b) (possibly $y = z$). In PG(3,q), let $u \in M - \{z\}$ and $u \notin L$. The line xu of PG(3,q) is a line N of type (a) of $T_2(O)$ for which $N \sim M$. By 3.3.3 the pair (L,N) is irregular, so we have proved that x is a center of irregularity for the dual of $T_2(O)$.

Finally, we prove that the point (∞) is a center of irregularity for $T_2(O)$. By 3.3.3 no point of type (i) is regular. Let x and y be two collinear points of type (i). Since x is not regular, there is some point z for which (x,z) is irregular. The point which is collinear with z and incident with xy is denoted by u. First let u be of type (i). The perspectivity of $PG(3,q)$ with center x and axis $PG(2,q) \supset O$ which maps u onto y is denoted by σ. Since σ induces an automorphism of $T_2(O)$ and since (x,z^σ) is irregular, where $z^\sigma \sim y$, we are done. So suppose u is of type (ii). Let $\{x,z\}^\perp = \{u,u_1,\ldots,u_q\}$ and let $\{u_1,u_i\}^\perp = \{x,z,u_i^1,\ldots,u_i^{q-1}\}$, $i = 2,\ldots,q$. Clearly (x,u_i^j) is irregular. Now suppose $u_i^j \sim u$ for all $i = 2,\ldots,q$ and $j = 1,\ldots,q-1$. Then $u \in \{u_1,u_i\}^{\perp\perp}$ and equivalently $u_i \in \{u,u_1\}^{\perp\perp}$, $i = 2,\ldots,q$. Hence (u,u_1) is regular, a contradiction. It follows that there is some u_i^j for which $u_i^j \not\sim u$. Now by a preceding argument there is a point u' for which $u' \sim y$ and (x,u') is irregular. □

4 Generalized quadrangles in finite projective spaces

4.1. PROJECTIVE GENERALIZED QUADRANGLES

A *projective* GQ S = (P,B,I) is a GQ for which P is a subset of the pointset of some projective space PG(d,K) (of dimension d over a field K), B is a set of lines of PG(d,K), P is the union of all members of B considered as sets of points, and the incidence relation I is the one induced by that of PG(d,K). If PG(d',K) is the subspace of PG(d,K) generated by all points of P, then we say PG(d',K) is the *ambient space* of S.

All finite projective GQ were first determined by F. Buekenhout and C. Lefèvre in [29] with a proof most of which is valid in the infinite case. Independently, D. Olanda [110, 111] has given a typically finite proof and J.A. Thas and P. De Winne [212] have given a different combinatorial proof under the assumption that the case d = 3 is already settled. More recently, K.J. Dienst [51, 52] has settled the infinite case. The main goal of this chapter is to give the proof of F. Buekenhout and C. Lefèvre. However, because the GQ in this book are finite, we have modified their presentation somewhat.

The definition of a GQ used by F. Buekenhout and C. Lefèvre was a little more general and included grids. However, a routine exercise shows that a projective grid consists of a pair of opposite reguli in some PG(3,K) (and hence is a GQ). Until further notice we shall suppose S = (P,B,I) to be a finite projective GQ of order (s,t), $s \geqslant 2$, $t \geqslant 2$, with ambient space PG(d,s), $d \geqslant 3$.

For the subspace of PG(d,s) generated by the pointsets or points $P_1,\ldots,P_k$, we shall frequently use the notation $\langle P_1,\ldots,P_k\rangle$.

4.2. THE TANGENT HYPERPLANE

4.2.1. *If* W *is a subspace of* PG(d,s) *and if* W ∩ B *denotes the set of all lines of* S *in* W , *then for the substructure* W ∩ S = (W ∩ P, W ∩ B,∈) *we have one of the following :* (a) *The elements of* W ∩ B *are lines which are incident with a distinguished point of* P , *and* W ∩ P *consists of the points of* P *that are incident with these lines;* (b) $W \cap B = \phi$ *and* $W \cap P$ *is a set of pairwise noncollinear points of S;* (c) $W \cap S$ *is a projective subquadrangle of S . If* W *is a hyperplane of* PG(d,s) , *then* $W \cap P$ *generates* W.

Proof. By 2.3.1 and since $s \neq 1$, it is immediate that we have one of (a), (b), (c). So suppose W is a hyperplane of PG(d,s). By definition there is a point $p \in P-(W \cap P)$. It suffices to show that an arbitrary line L of B is in $\langle W \cap P,p \rangle$. We may suppose that L meets W in some point q. If $p \in L$, the required conclusion is obvious. So suppose $p \notin L$ and let L' be a line of B through p ($L' \neq L$) meeting W in a point q', with $q' \neq q$. Clearly L' is in $\langle W \cap P,p \rangle$. There must be a point r' of L', $r' \neq q'$, such that the line M of B through r' intersecting L meets L in a point r different from q. Then M has two distinct points in $\langle W \cap P,p \rangle$: the point r' of L' and the point $M \cap W$. Hence r is in $\langle W \cap P,p \rangle$, so that L has two points of $\langle W \cap P,p \rangle$. □

If $p \in P$, a *tangent to S at* p is any line L through p such that either $L \in B$ or $L \cap P = \{p\}$. The union of all tangents to S at p will be called the *tangent set of S at* p, and we denote it by S(p). The relation between S(p) and $p^{\perp}$ is : $p^{\perp} = P \cap S(p)$. A line L of PG(d,s) is a *secant* to S if L intersects P in at least two points but is not a member of B.

4.2.2. *If* p *and* q *are collinear points of S, then* $p^{\perp} \cap q^{\perp}$ *is the line* $\langle p,q \rangle$.

Proof. Clear. □

4.2.3. *For each* $p \in P$, $\langle p^{\perp} \rangle \subset S(p)$.

Proof. We must show that for each line L through p in $\langle p^{\perp} \rangle$ either $L \in B$ or L intersects P exactly in p. So suppose that $p \in L \notin B$, $L \subset \langle p^{\perp} \rangle$. First, suppose that there is some line L_1 of B through p and a second tangent L_2 to S at p for which the plane $\alpha = \langle L_1,L_2 \rangle$ contains L . If L were not a tangent at p it would contain some point q, $p \neq q \in P$. There would be a unique line $M \in B$ through q and intersecting L_1 in p_1, $p_1 \neq p$. As M is contained in α , M meets L_2 in a point p_2, $p_2 \neq p$. Then p, $p_2 \in L_2$ implies $L_2 \in B$, since L_2 is a tangent to S containing two points of S . But then L_1 and L_2 are two lines of S through p intersecting M, contradicting the assumption that S is a GQ. Hence L must be a tangent.

Second, as PG(d,s) is finite dimensional there is an integer k such that $\langle p^{\perp} \rangle$ is generated by k lines $L_1,\ldots,L_k$ of S through p. Let $X_i = \langle L_1 \cup \ldots \cup L_i \rangle$, $i = 2,\ldots,k$. By the first case we know $X_2 \subset S(p)$. Now we use induction on i.

Assume $X_i \subset S(p)$, and let L be some line of X_{i+1} through p. We may suppose $L \neq L_{i+1}$ and $L \not\subset X_i$. Then the plane $\alpha = \langle L, L_{i+1} \rangle$ intersects X_i along a line L'. By the induction hypothesis L' is a tangent to S at p, so that $\alpha = \langle L_{i+1}, L' \rangle$ satisfies the hypothesis of the first case. Hence L is a tangent to S at p, and it follows that $X_{i+1} \subset S(p)$. □

4.2.4. $\langle p^\perp \rangle$ *is a hyperplane of* PG(d,s).

Proof. Consider a point $q \in P - \langle p^\perp \rangle$. By 4.2.1 $\langle p^\perp, q \rangle \cap S$ is a subquadrangle of S. Clearly this subquadrangle has order (s,t), so it must coincide with S. Hence $\langle p^\perp, q \rangle = PG(d,s)$, i.e. $\dim \langle p^\perp \rangle = d-1$. □

4.2.5. *The hyperplane* $\langle p^\perp \rangle$ *is the tangent set* $S(p)$ *to* S *at* p, *and is called the tangent hyperplane to* S *at* p.

Proof. By the preceding results we know that $\langle p^\perp \rangle$ is a hyperplane contained in $S(p)$. If equality did not hold, there would be some tangent line L at p not in $\langle p^\perp \rangle$. We use induction on the dimension of PG(d,s) to obtain the desired contradiction. First suppose d = 3. Let L_1 be a line of S through p and let α be the plane $\langle L, L_1 \rangle$. If there were a point $q \in \alpha \cap P$ with $q \notin L_1$, there would be a line M of B through q meeting L_1 in a point not p. But M would be in α and hence meet L in a point ($\neq$ p) of P, an impossibility. Hence each point of $\alpha \cap P$ is on L_1. But every line of S intersects α, implying every line of S meets L_1, an impossibility. So the result holds for d = 3. Suppose d > 3 and consider two lines L_1 and L_2 of S through p. Let H be a hyperplane containing $\langle L, L_1, L_2 \rangle$. As L is not in $\langle p^\perp \rangle$, H is not $\langle p^\perp \rangle$. By 4.2.1 $\langle H \cap P \rangle = H$, and either $H \cap P$ is the pointset of a GQ in H or $H \cap P \subset p^\perp$. If $H \cap P \subset p^\perp$, then $H = \langle H \cap P \rangle \subset \langle p^\perp \rangle$, or $H = \langle p^\perp \rangle$, a contradiction. So $H \cap S$ is a subquadrangle of S. But then using the induction hypothesis in the ambient space of $H \cap S$ we reach a contradiction. □

4.2.6. *Let* p,q,r *be three distinct points of* S *on a line of* PG(d,s). *Then the intersections* $S(p) \cap S(q)$, $S(q) \cap S(r)$, *and* $S(r) \cap S(p)$ *coincide.*

Proof. First suppose that $\langle p,q,r \rangle$ is not a line of S, and let w be any point of $p^\perp \cap q^\perp$. Then $p,q \in w^\perp$, and $r \in \langle p,q \rangle \subset \langle w^\perp \rangle = S(w)$. Since $r \in P$ and $r \in S(w)$, clearly $r \in w^\perp$. Hence any point of $p^\perp \cap q^\perp$ also belongs to $r^\perp$. We claim $\langle p^\perp \rangle \cap \langle q^\perp \rangle = \langle p^\perp \cap q^\perp \rangle$. Indeed $\langle p^\perp \rangle \cap \langle q^\perp \rangle$ must be a (d-2)-dimensional subspace containing $\langle p^\perp \cap q^\perp \rangle$. But it easily follows that $\langle p^\perp \cap q^\perp, p \rangle = \langle p^\perp \rangle$, so that $\langle p^\perp \cap q^\perp \rangle$ is at least (d-2)-dimensional. Hence

$\langle p^{\perp}\rangle \cap \langle q^{\perp}\rangle = \langle p^{\perp} \cap q^{\perp}\rangle$. Then $S(p) \cap S(q) = \langle p^{\perp}\rangle \cap \langle q^{\perp}\rangle = \langle p^{\perp} \cap q^{\perp}\rangle \subset \langle r^{\perp}\rangle = S(r)$, completing the proof. Now suppose $\langle p,q,r\rangle$ is a line of S, so by 4.2.2 $p^{\perp} \cap q^{\perp} = \langle p,q\rangle$ and $S(p) \cap S(q) \cap P = \langle p,q\rangle$. Let w be any point of $S(p) \cap S(q)$ not on $\langle p,q\rangle$. If r' is on the line $\langle r,w\rangle$, $r' \neq r$ and $r' \neq w$, then $\langle p,r'\rangle$ is in $S(p)$. Since $\langle p,r'\rangle \cap \langle q,w\rangle$ is not a point of S, $\langle p,r'\rangle$ is not a line of S, so $r' \notin P$. Hence the line $\langle r,w\rangle$ intersects P at the unique point r, implying that each point w of $S(p) \cap S(q)$ not on $\langle p,q\rangle$ belongs to $S(r)$. This completes the proof. □

4.2.7. *Let* L *be a secant containing three distinct points* p,a,a' *of* P. *Then the perspectivity* σ *of* PG(d,s) *with center* p *and axis* $S(p)$ *mapping* a *onto* a' *leaves* P *invariant.*

Proof. Clearly σ fixes all points of $S(p)$ and thus fixes $p^{\perp}$. Let $b \in P-p^{\perp}$. First suppose b is not on L and let α be the plane $\langle p,a,b\rangle$. Consider the line $M = \langle a,b\rangle$. Then M intersects $S(p)$ at a point c, fixed by σ . Hence $M^{\sigma} = \langle a',c\rangle$.

If M is a line of S, then $c \in P$ so the tangent line $\langle p,c\rangle$ is a line of S. Thus the plane $\langle p,a,c\rangle = \alpha$ is in the tangent hyperplane $S(c)$. Hence, since $a' \in \alpha$, it follows that $a' \sim c$ and M^{σ} is a line of S and b^{σ} is a point of S.

If M is not a line of S, suppose there is a point $u \in P-S(p)$ with $u \in a^{\perp} \cap b^{\perp}$. The argument of the previous paragraph, with u in the role of b, shows that $u^{\sigma} \in P$. Then with u and u^{σ} playing the roles of a and a', respectively, it follows that $b^{\sigma} \in P$. On the other hand, suppose $a^{\perp} \cap b^{\perp} \subset S(p)$. Consider points $u, u' \in P-S(p)$ with $a \sim u \sim u' \sim b$. Then consecutive applications of the previous paragraph show that u^{σ}, u'^{σ}, and finally b^{σ} are all in P.

Second, suppose b is on L, and use the fact that if u is any point of P not on L then $u^{\sigma} \in P$. It follows readily that $b^{\sigma} \in P$. □

4.2.8. *All secant lines contain the same number of points of* S .

Proof. Let L and L' be secant lines. First suppose L and L' have a point p of P in common, and let M be any secant line through p. If some M is incident with more than two points of P, by 4.2.7 we may consider the nontrivial group G of all perspectivities with center p and axis $S(p)$, leaving P invariant. The group G is regular on the set of points of M in P but different from p, for each M. Hence each secant through p has $1+|G|$ points of P, so that L and L' have the same number of points of S . If no M is incident with more than two points of P, then clearly L and L' contain two points of S .

Secondly, suppose L and L' do not have any point of P in common, and choose points p,p' of P on L,L', respectively. If $p \not\sim p'$, then $\langle p,p' \rangle$ is a secant, so meets P in the same number of points as do L and L', by the previous paragraph. If $p \sim p'$, choose a point $q \in P$ with $p \not\sim q \not\sim p'$, and apply the previous paragraph to the secant lines L, $\langle p,q \rangle$, $\langle p',q \rangle$, L'. □

4.3. EMBEDDING S IN A POLARITY : PRELIMINARY RESULTS

The goal of this section and the next is to extend the mapping $p \mapsto S(p)$ to a polarity of PG(d,s), i.e. to construct a mapping π such that

(a) for each point x of PG(d,s), $\pi(x)$ is a hyperplane of PG(d,s),

(b) for each $p \in P$, $\pi(p) = S(p)$,

(c) $x \in \pi(y)$ implies $y \in \pi(x)$.

For a point x of PG(d,s), the *collar* S_x of S for x is the set of all points p of S such that p = x or the line $\langle p,x \rangle$ is a tangent to S at p. For example, if $x \in P$, S_x is just $x^\perp$. If $x \notin P$, the collar S_x is the set of points p of P such that $\langle p,x \rangle \cap P = \{p\}$.

For all $x \in PG(d,s)$ the *polar* $\pi(x)$ of x with respect to S is the subspace of PG(d,s) generated by the collar S_x, i.e. $\pi(x) = \langle S_x \rangle$. In particular, if $x \in P$, then $\pi(x) = S(x)$ (cf. 4.2.5).

4.3.1. *For any point* x, *let* p_1 *and* p_2 *be distinct points of* S_x. *Then* $P \cap \langle p_1,p_2 \rangle \subset S_x$.

Proof. Suppose $p \in P \cap \langle p_1,p_2 \rangle$, $p_1 \neq p \neq p_2$. Since $x \in S(p_1) \cap S(p_2)$, by 4.2.6 also $x \in S(p)$, hence $p \in S_x$. □

4.3.2. *Each line* L *of* S *intersects the collar* S_x *for each point* x *of* PG(d,s) , *in exactly one point, unless each point of* L *is in* S_x .

Proof. The result is clearly true if $x \in P$, so suppose $x \notin P$. Put $\alpha = \langle L,x \rangle$. If $\alpha \cap P$ is the set of points on L, then each point of L is in S_x. So suppose $y \in \alpha \cap P$, $y \notin L$. Then $y \sim p$ for a unique point p of L. By 4.2.5 each line of $\alpha = \langle L,y \rangle$ through p is a tangent at p, and hence $p \in S_x$. Moreover by 4.3.1 p is the unique point of L in S_x unless each point of L is in S_x . □

4.3.3. *Either* $\pi(x) = \langle S_x \rangle$ *is a hyperplane or* $\pi(x) = PG(d,s)$.

Proof. Again we may assume that $x \notin P$. If the assertion is false for some point x, then $\pi(x)$ is contained in some subspace X of codimension 2 in PG(d,s). Each line $L \in B$ intersects X by 4.3.2. Therefore if p is a point of S not on X, S_p is contained in $\langle X, p \rangle$ and as $\langle S_p \rangle$ is a hyperplane, $\langle X,p \rangle = \langle S_p \rangle = S(p)$. Any line L' of S through p must contain a second point q of P not in X. Then $S(p) = \langle X,p \rangle = \langle X,q \rangle = S(q)$, an obvious impossibility. □

4.3.4. *If* $\pi(x)$ *is a hyperplane, then* $S_x = P \cap \pi(x)$.

Proof. Clearly $S_x \subset P \cap \pi(x)$. Suppose there were a point p of $P \cap \pi(x)$ not in S_x. Then either some line L of S through p does not lie in $\pi(x)$, or $\pi(x) = S(p)$. In the first case L intersects $\pi(x)$ exactly in p. Then as $p \notin S_x$, L is on no point of S_x, contradicting 4.3.2. In the second case, as $p \notin S_x$, each line of B through p has exactly one point in S_x. So on any line of B through p there is a point p', $p' \neq p$, of $S(p)-S_x$, and there is a line L of B through p' but not in $\pi(x) = S(p)$, leading back to the first case. □

4.3.5. *Let* x *be a point of* PG(d,s) *and* a , a' *distinct points of* P *different from* x *and not in* $\pi(x)$, *which are collinear with* x . *Then the perspectivity* σ *of* PG(d,s) *with center* x *and axis* $\pi(x)$ *mapping* a *onto* a' *leaves* P *invariant.*

Proof. If $x \in P$, the result is known by 4.2.7, since $\langle x,a,a' \rangle$ is a secant line. So suppose $x \notin P$, and note that σ fixes all points of $P \cap \pi(x)$. Let b be a point of $P-\pi(x)$ not on $\langle a,a' \rangle$. Let α be the plane $\langle x,a,b \rangle$ and M the line $\langle a,b \rangle$. If $M \cap \pi(x) = \{c\}$, then $M^\sigma = \langle a',c \rangle$. By an argument similar to that used in the proof of 4.2.7 we may assume $M \in B$.
Then $c \in P \cap \pi(x) = S_x$, by 4.3.4, so $\langle c,x \rangle$ and M, and hence $\alpha = \langle x,a,c \rangle$ are in the tangent hyperplane $S(c)$. Then $a' \in \alpha \subset S(c)$, forcing $M^\sigma = \langle a',c \rangle \in B$, i.e. $b^\sigma \in P$. Finally, if b is a point of $P-\pi(x)$ on the line $\langle a,a' \rangle$, it follows readily that $b^\sigma \in P$. □

4.3.6. *Suppose that secant lines to* S *have at least three points of* P. *If* $\pi(x)$ *is a hyperplane, then either* $y \in \pi(x)$ *implies* $x \in \pi(y)$, *or there is a point* z *with* $\pi(z) = PG(d,s)$ *and* $S_z \neq P$.

Proof. Clearly we may suppose $\pi(y)$ to be a hyperplane. Consider a nontrivial perspectivity σ with center x and axis $\pi(x)$ and leaving P invariant

(σ exists by 4.3.5). Since $y \in \pi(x)$, σ fixes y and by definition of $\pi(y)$ must leave $\pi(y)$ invariant. But the invariant hyperplanes of a nontrivial perspectivity are its axis and all hyperplanes through its center. First suppose $\pi(y)$ is the axis of σ, i.e. $\pi(x) = \pi(y)$. If $x \in \pi(x)$, there is nothing to show. So suppose $x \notin \pi(x)$. Let p_1 and p_2 be two points of P-{y} on a secant L through y, and let $q \in P$, $p_1 \sim q \sim p_2$. Then $S(p_1)$ and $S(p_2)$ do not contain y, but $S(q)$ does contain y because it contains p_1 and p_2 on L, i.e. $q \in \pi(y) = \pi(x)$. Since $S(p_1) \cap S(p_2) = \langle p_1^\perp \cap p_2^\perp \rangle$, $S(p_1) \cap S(p_2)$ is in $\pi(x) = \pi(y)$ and does not contain y. Furthermore, as p_1 and p_2 are not in $\pi(x)$, $S(p_1)$ and $S(p_2)$ do not contain x. Consequently, as $x \notin \pi(x)$, $S(p_1)$ and $S(p_2)$ intersect the line $\langle x,y \rangle$ in two distinct points z_1 and z_2, different from x and y. But the line $\langle x,y \rangle$ is in the tangent hyperplane of each point of $S_x = S_y = P \cap \pi(x)$. Hence S_{z_1} contains $S_x = S_y$ and the point p_1, but not the point p_2. So $\pi(z_1) = PG(d,s)$ and $S_{z_1} \neq P$. Consequently, if there is no z such that $\pi(z) = PG(d,s)$ and $S_z \neq P$, then $\pi(y)$ is not the axis of σ and $\pi(y)$ must contain x, completing the proof. □

4.4. THE FINITE CASE

Throughout this book attention is concentrated on finite GQ. The arguments given in the first three sections of this chapter hold also in the case of a projective space of finite dimension $d \geq 3$ over an infinite field. For the remainder of this chapter, however, finiteness is essential. Recall that S has order (s,t), $s \geq 2$, $t \geq 2$, and denote by $\ell+1$ the constant number (cf. 4.2.8) of points of S on a secant line. If $\ell = 1$, P is a quadratic set in the sense of F. Buekenhout [27] and by his results S is formed by the points and lines on a nonsingular quadric of projective index 1 in PG(d,s), d = 4 or 5. Hence we assume that $\ell > 1$ and proceed to establish (a), (b), (c) of 4.3.

4.4.1. $\ell = t/s^{d-3}$, *and* d = 3 *or* 4.

Proof. The secant lines through a point $p \in P$ are the s^{d-1} lines of PG(d,s) through p which do not lie in the tangent hyperplane $S(p)$. Hence the total number of points of P is $\ell s^{d-1} + |p^\perp| = (1+s)(1+st)$, implying $\ell = t/s^{d-3}$. By Higman's inequality we know that $t \leq s^2$, so that $2 \leq \ell \leq s^2/s^{d-3}$, implying d = 3 or 4. □

A subset E of P is called *linearly closed* in P if for all $x,y \in E$, $x \neq y$, the intersection $\langle x,y\rangle \cap P$ is contained in E. Thus any subset X of P generates a *linear closure* $\bar{X}$ in P.

4.4.2. *Let* $d = 3$, *and suppose* a_0,a_1,a_2 *are three points of* P *noncollinear in* PG(3,s). *Then* $\overline{\{a_0,a_1,a_2\}} = P \cap \langle a_0,a_1,a_2\rangle$.

Proof. If the plane $\alpha = \langle a_0,a_1,a_2\rangle$ contains a line of S, the lemma is trivial. Hence suppose α contains no line of S. As $d = 3$, any secant line intersects P in exactly t+1 points. Take a point p ($\neq a_0$) of P on the secant line $\langle a_0,a_1\rangle$. The t+1 secant lines $\langle p,q\rangle$, where q is a point of $P \cap \langle a_0,a_2\rangle$, intersect P in points which are in the linear closure $\overline{\{a_0,a_1,a_2\}}$. As each of these lines $\langle p,q\rangle$ intersects P in t+1 points, there are t(t+1)+1 points of P on these lines. Hence $|\overline{\{a_0,a_1,a_2\}}| \geq t^2+t+1$. If the claim of 4.4.2 were false, there would be a point $r \in (P \cap \alpha)-\overline{\{a_0,a_1,a_2\}}$. Then every line of α through r contains at most one point of $\overline{\{a_0,a_1,a_2\}}$, so there are at least t^2+t+1 lines of α through r which are secant to P. Therefore, we obtain $(t^2+t+1)(t-1)+1 = t^3$ points of P in α not belonging to $\overline{\{a_0,a_1,a_2\}}$. Hence $|\alpha \cap P| \geq t^3+t^2+t+1$. Since no two points of $\alpha \cap P$ are collinear in S, and $s \leq t^2$, we have $t^3+t^2+t+1 \leq 1+st \leq 1+t^3$, an impossibility that completes the proof. □

4.4.3. *Let* $d = 4$, *and suppose* a_0,a_1,a_2 *are three points of* P *noncollinear in* PG(4,s). *Then* $\overline{\{a_0,a_1,a_2\}} = P \cap \langle a_0,a_1,a_2\rangle$.

Proof. As before we may suppose that $\alpha = \langle a_0,a_1,a_2\rangle$ contains no line of S. Fix a point $p \in P \cap \alpha$ and a line $L \in B$ incident with p. Put $Q = \langle a_0,a_1,a_2,L\rangle$. Then for $Q \cap S$ there are the following two possibilities : (a) The elements of $Q \cap B$ are lines which are incident with a distinguished point of P, and $Q \cap P$ consists of the points of P which are incident with these lines, and (b) $Q \cap S$ is a projective subquadrangle of S . If we have (b), then by the preceding result $\alpha \cap P$ is the linear closure of $\{a_0,a_1,a_2\}$ in P, as desired. If we have (a) , two cases are possible. (i) There exists a line $L' \in B$ through a point of α such that $\langle \alpha,L'\rangle$ intersects S in a subquadrangle. Then 4.4.2 still applies. (ii) For each line $L \in B$ intersecting α, $B' = \langle \alpha,L\rangle \cap B$ is a set of lines through a point b_i of L not on α , and $P' = \langle \alpha,L\rangle \cap P$ is the set of all points on the lines of B'. Here $\langle \alpha,L\rangle$ is the tangent hyperplane at b_i. Hence α contains 1+t points of P : $a_0,\ldots,a_t$. Clearly $a_j \sim b_i$

for all i and j. Furthermore, by the definition of the points b_i, the lines of B through a given point a_j are the lines $\langle a_j, b_i \rangle$. Hence there are exactly 1+t points b_i. This means S has two (disjoint) sets $\{a_j\}$, $\{b_i\}$ of 1+t (pairwise noncollinear) points with $a_j \sim b_i$, $0 \leqslant i, j \leqslant t$. By Payne's inequality 1.4.1, $t^2 \leqslant s^2$, i.e. $t \leqslant s$. But $\ell = t/s > 1$ makes this impossible, completing the proof. □

4.4.4. *Let* $\{a_i\}$ *be a family of points of* P *. Then the linear closure of* $\{a_i\}$ *in* P *is* $P \cap \langle a_i \rangle$.

Proof. First note that if s = 2 (and $\ell > 1$, $\langle P \rangle = PG(d,s)$ by assumption), then any line containing at least two points of P is entirely contained in P Hence all points of PG(d,s) are points of P, so the lemma is trivial. Hence we assume $s > 2$. Also it is clear that the result holds if $\langle a_i \rangle$ is a point, line, or plane. Further, we may assume that the points a_i are linearly independent in PG(d,s). As PG(d,s) is finite, we may apply induction as follows : suppose the result is true for k points $a_0, \ldots, a_{k-1}$, $3 \leqslant k$, indexed so that $\langle a_0, \ldots, a_{k-1} \rangle$ is not contained in $S(a_0)$, and let $a_k \in P - \langle a_0, \ldots, a_{k-1} \rangle$. We show that the result holds for $\{a_0, \ldots, a_k\}$. Put $L_i = \langle a_0, a_i \rangle$, i = 1,...,k, and let β be any plane through L_k contained in $\langle L_1, \ldots, L_k \rangle$. Clearly β intersects $\langle L_1, \ldots, L_{k-1} \rangle$ in a line L. We show that $P \cap \langle L_k, L \rangle \subset \overline{\{a_0, \ldots, a_k\}}$, from which the desired result follows immediately. Suppose L is incident with at least two points of P. By the induction hypothesis the points of P on L are all in $\overline{\{a_0, \ldots, a_{k-1}\}}$. And then 4.4.2 and 4.4.3 show that $P \cap \langle L_k, L \rangle$ is in $\overline{\{a_0, \ldots, a_k\}}$. Now suppose that L is a tangent line whose points are not all in P. If $\langle L_k, L \rangle$ contains no point of P not on L_k, there is nothing more to show. So suppose p is a point of $P \cap \langle L_k, L \rangle$ but not on L_k. Consider the plane α generated by L and a secant line through a_0 in the space $\langle L_1, \ldots, L_{k-1} \rangle$ (such a line exists since $S(a_0) \not\supset \langle a_0, \ldots, a_{k-1} \rangle$). This plane is not in the tangent hyperplane $S(a_0)$, so L is the unique tangent line at a_0 in α . Hence there are two secant lines A, K in α and through a_0. Each of the planes $\langle L_k, A \rangle$, $\langle L_k, K \rangle$ is not in $S(a_0)$, and hence contains exactly one tangent line at a_0. Consider in $\langle L_k, A \rangle$ a secant line C ($C \neq L_k$) such that the plane $\langle C, p \rangle$ intersects $\langle L_k, K \rangle$ in a secant line D. (The line C exists because $\langle L_k, A \rangle$ has at least four lines through a_0). By the induction hypothesis the points of P on A and K belong to $\overline{\{a_0, \ldots, a_{k-1}\}}$. Hence by 4.4.2 and 4.4.3 the points of P on C and

D belong to $\{\overline{a_0,\ldots,a_k}\}$. But as $p \in \langle C,D\rangle$, again by 4.4.2 and 4.4.3 $p \in \{\overline{a_0,\ldots,a_k}\}$.

4.4.5. $S_x \neq P$.

Proof. Clearly we may suppose $x \notin P$, and there are two cases : $d = 3$ and $d = 4$. First suppose that $d = 3$ and that $S_x = P$. Each line through x intersecting P must be a tangent line, so the number of tangent lines through x is $|P| = (1+s)(1+st)$. As $t > 1$, there are at least $(1+s)^2 = 1+2s+s^2$ lines of PG(3,s) through x, of which there are only $1+s+s^2$. So we may suppose $d = 4$ and $S_x = P$. Let $p \in P$. If L is a line of S through p, the plane $\langle x,L\rangle$ intersects P in the points of L, because all points of P are points of S_x. Hence the 1+t lines of S through p together with x generate t+1 distinct planes. Since all these planes are contained in $S(p)$ and $\dim S(p) = 3$, we have $1+t \leqslant s+1$, an impossibility since $t/s = \ell > 1$. □

4.4.6. $\pi(x)$ *is a hyperplane.*

Proof. This result is known for $x \in P$, so suppose $x \notin P$. Consider the intersection $\pi(x) \cap P$. By 4.3.1 and 4.4.4 all points of $\pi(x) \cap P$ are in $S(x)$, implying $S_x = \pi(x) \cap P$. If $\pi(x)$ were not a hyperplane, then by 4.3.3 $\pi(x) = PG(d,s)$, implying $S_x = \pi(x) \cap P = P$, an impossibility by 4.4.5. □

This completes the proof that conditions (a), (b), (c) of Section 4.3 hold, so that π is a polarity. We show that P is the set of absolute points of π . Since B is the set of all lines of PG(d,s) which contain x and are contained in $\pi(x) \cap P$, where x runs over P, B must be the set of totally isotropic lines of π .

4.4.7. $x \in \pi(x)$ *iff* $x \in P$.

Proof. If $x \in P$, we know that $x \in \pi(x)$. We shall prove that if $x \in \pi(x)$, then $x \in P$. First suppose $d = 3$, so the number of lines through x in $\pi(x)$ is equal to s+1. Suppose $x \in \pi(x)-P$. If $p \in P \cap \pi(x)$, then $\langle p,x\rangle$ is a tangent . If $\pi(x)$ contains a line L of S, then all points of $\pi(x) \cap P$ are on L. Since every line of S contains a point of $\pi(x)$, all lines of S are concurrent with L, a contradiction. If $\pi(x)$ contains no line of S, every line of S meets $\pi(x)$ in exactly one point, and every point of $\pi(x) \cap P$ is on 1+t lines of S. Hence $|\pi(x) \cap P| = 1+st$, and there are at least 1+st lines through x in $\pi(x)$, an impossibility for $t > 1$. Finally, we may suppose $d > 3$

(i.e. $d = 4$) and let $x \in \pi(x)-P$. Let H be the hyperplane containing x and two lines L_1, L_2 of S through a point p, $p \notin \pi(x)$ (notice that $x \notin \langle L_1, L_2 \rangle$). The intersection $H \cap S$ is a subquadrangle, since otherwise H would be the tangent hyperplane $S(p)$, forcing p to be in $\pi(x)$. Clearly H is the ambient space of $H \cap S$. If $\pi'(x)$ is the polar of x with respect to $H \cap S$, then $\pi'(x) = \pi(x) \cap H$. Hence $x \in \pi'(x)$, a contradiction since $\dim H = 3$ and $x \notin P$. □

This completes the proof of the theorem of F. Buekenhout and C. Lefèvre:

4.4.8. *A projective* GQ $S = (P,B,I)$ *with ambient space* PG(d,s) *must be obtained in one of the following ways :*

(i) *There is a unitary or symplectic polarity* π *of* PG(d,s), $d = 3$ *or* 4, *such that* P *is the set of absolute points of* π *and* B *is the set of totally isotropic lines of* π .

(ii) *There is a nonsingular quadric* Q *of projective index* 1 *in* PG(d,s), $d = 3$, 4 *or* 5, *such that* P *is the set of points of* Q *and* B *is the set of lines on* Q .

Hence S must be one of the classical examples described in Chapter 3.

5 Combinatorial characterizations of the known generalized quadrangles

5.1. INTRODUCTION

In this chapter we review the most important purely combinatorial characterizations of the known GQ. Several of these theorems appeared to be very useful and were important tools in the proofs of certain results concerning strongly regular graphs with strongly regular subconstituents [34], coding theory [34], the classification of the antiflag transitive collineation groups of finite projective spaces [35], the Higman-Sims group [8], small classical groups (E.E. Shult, private communication), etc.

In the first part we shall give characterizations of the classical quadrangles W(q) and Q(4,q). The second part will contain all known characterizations of $T_3(O)$ and Q(5,q). Next an important characterization of $H(3,q^2)$ by G. Tallini [174] is given. Then we prove two characterization theorems of $H(4,q^2)$. In the final part conditions are given which characterize several GQ at the same time, and the chapter ends with a characterization by J.A. Thas [204] of all classical GQ and their duals.

5.2. CHARACTERIZATIONS OF W(q) AND Q(4,q)

Historically, this next result is probably the oldest combinatorial characterization of a class of GQ. A proof is essentially contained in R.R. Singleton [168] (although he erroneously thought he had proved a stronger result), but the first satisfactory treatment may have been given by C.T. Benson [10]. No doubt it was discovered independently by several authors (e.g. G. Tallini [174]).

5.2.1. *A GQ S of order* s (s > 1) *is isomorphic to* W(s) *iff all its points are regular.*

Proof. By 3.2.1 and 3.3.1 all points of W(s) are regular. Conversely, let us assume that S = (P,B,I) is a GQ of order s (s ≠ 1) for which all points are regular. Now we introduce the incidence structure S' = (P',B',I'), with P' = P, B' the set of spans of all point-pairs of P, and I' the natural incidence. Then S is isomorphic to the substructure of S' formed by all points and the spans of all pairs of points collinear in S. By 1.3.1 and using the fact that any triad of points is centric by 1.3.6, it follows that any three noncollinear points of S' generate a projective plane. Since $|P'| = s^3+s^2+s+1$, S' is the design of points and lines of PG(3,s). Clearly all spans (in S) of collinear point-pairs containing a given point x, form a flat pencil of lines of PG(3,s). Hence the set of all spans of collinear point-pairs is a linear complex of lines of PG(3,s) (cf. [159]), i.e. is the set of all totally isotropic lines for some symplectic polarity. Consequently $S \cong W(s)$. □

5.2.2. ([195]). *A GQ S of order* (s,t), $s \neq 1$, *is isomorphic to* W(s) *iff* $|\{x,y\}^{\perp\perp}| \geq s+1$ *for all* x,y *with* $x \neq y$.
Proof. For W(s) we have $|\{x,y\}^{\perp\perp}| = s+1$ for all points x,y with $x \neq y$. Conversely, suppose S has order (s,t), $s \neq 1$, and $|\{x,y\}^{\perp\perp}| \geq s+1$ for all x,y with $x \neq y$. By 1.4.2 (ii) we have $st \leq s^2$. Since $|\{x,y\}^{\perp\perp}| \leq t+1$ for $x \not\sim y$, there holds $t \geq s$. Hence $s = t$ and $|\{x,y\}^{\perp\perp}| = s+1$ for all x,y with $x \neq y$. Then $S \cong W(s)$ by 5.2.1. □

5.2.3. *Up to isomorphism there is only one GQ of order* 2.
Proof. Let S be a GQ of order 2. Consider two points x,y with $x \not\sim y$, and let $\{x,y\}^{\perp} = \{z_1,z_2,z_3\}$. If $\{z_1,z_2\}^{\perp} = \{x,y,u\}$, then by 1.3.4 (iv) we have $u \sim z_3$. Hence (x,y) is regular. So every point is regular and $S \cong W(2)$. □

5.2.4. (*J.A. Thas* [184]). *A GQ S = (P,B,I) of order* s, $s \neq 1$, *is isomorphic to* $W(2^h)$ *iff it has an ovoid* O *each triad of which is centric.*
Proof. The GQ $W(2^h)$ has an ovoid O by 3.4.1 (i) and each triad of O is centric by 1.3.6 (ii) and 3.3.1 (i). Conversely, suppose the GQ S of order s, $s \neq 1$, has an ovoid O each triad of which is centric. Consider a point $p \in P-O$. The s+1 lines incident with p are incident with s+1 points of O. Such a subset C of order s+1 of O is called a circle. The number of circles is at most $(s^2+1)(s+1)-|O| = s(s^2+1)$. Since every triad of O is centric, there are at least $(s^2+1)s^2(s^2-1)/(s+1)s(s-1) = s(s^2+1)$ circles. Consequently, there are exactly $s(s^2+1)$ circles, every three elements of O are contained in just one circle, and each circle is determined by exactly one point $p \notin O$. It follows that O together with the set of circles is a $3-(s^2+1,s+1,1)$ design, i.e. an inversive plane [49] of order s. This

inversive plane will be denoted by $I^*(O)$. The point $p \notin O$ defining the circle C will be called the nucleus of C.

Now consider two circles C and C' with respective nuclei p and p', where $p \sim p'$. If $p \mathrm{I} L \mathrm{I} p'$ and if x is the point of O which is incident with L, then $C \cap C' = \{x\}$. Hence the $s-1$ circles distinct from C which are tangent to C at x have as nuclei the $s-1$ points distinct from x and p, which are incident with L.

Consider a circle C, a point $x \in C$, and a point $y \in O-C$. Through y there passes a unique circle C' with $C \cap C' = \{x\}$. Now take a point $u \in C-\{x\}$, and consider the unique circle C'' with $u \in C''$ and $C' \cap C'' = \{y\}$. We shall prove that $|C \cap C''| = 2$. If not, then $C \cap C'' = \{u\}$. And the nucleus of C (resp., C', C'') is denoted by p (resp., p', p''). By the preceding paragraph there are distinct lines L, L', L'' such that $p' \mathrm{I} L \mathrm{I} p''$, $p'' \mathrm{I} L' \mathrm{I} p$, $p \mathrm{I} L'' \mathrm{I} p'$, giving a contradiction. Hence $|C \cap C''| = 2$. If u runs through $C-\{x\}$, then we obtain a partition of $C-\{x\}$ into pairs of distinct points. Hence $|C-\{x\}| = s$ is even. Since s is even, $I^*(O)$ is egglike by the celebrated theorem of P. Dembowski [49], and hence $s = 2^h$. Consequently there exists an ovoid O' in $PG(3,s)$ together with a bijection σ from O' onto O, such that for every plane π of $PG(3,s)$ with $|\pi \cap O'| > 1$, we have that $(\pi \cap O')^\sigma$ is a circle of $I^*(O)$. If $W(s)$ is the GQ arising from the symplectic polarity θ defined by O' [49], i.e. if $W(s)$ is the GQ formed by the points of $PG(3,s)$ together with the tangent lines of O', then we define as follows a bijection ϕ from the pointset and lineset of $W(s)$ onto the pointset and lineset of S : (i) $x^\phi = x^\sigma$ for $x \in O'$; (ii) for $x \notin O'$ the point x^ϕ is the nucleus of the circle $(x^\theta \cap O')^\sigma$ of $I^*(O)$; and (iii) if L is a line of $W(s)$ which is tangent to O' at x, L^ϕ is the line of S joining x^ϕ to the nucleus of the circle $(\pi \cap O')^\sigma$, where π is a plane of $PG(3,s)$ which contains L but is not tangent to O'. In one of the preceding paragraphs it was shown that L^ϕ is independent of the plane π. Now it is an easy exercise to show that ϕ is an isomorphism of $W(s)$ onto S . □

In view of 1.3.6 (ii), there is an immediate corollary.

5.2.5. *A GQ* S *of order* s, $s \neq 1$, *is isomorphic to* $W(2^h)$ *iff it has an ovoid* O *each point of which is regular.*

5.2.6. (*S.E. Payne and J.A. Thas* [143]). *A GQ* $S = (P,B,I)$ *of order* s, $s \neq 1$, *is isomorphic to* $W(2^h)$ *iff it has a regular pair* (L_1,L_2) *of nonconcurrent lines with the property that any triad of points lying on lines of* $\{L_1,L_2\}^\perp$ *is centric.*

Outline of proof. By 3.3.1 all lines and points of $W(2^h)$ are regular, and then by 1.3.6 (ii) all triads of points and lines are centric.

Conversely, suppose the GQ S of order s, $s \neq 1$, has a regular pair (L_1,L_2) of nonconcurrent lines with the property that any triad of points lying on lines of $\{L_1,L_2\}^\perp$ is centric. Let $\{L_1,L_2\}^\perp = \{M_1,\ldots,M_{s+1}\}$, $\{L_1,L_2\}^{\perp\perp} = \{L_1,\ldots,L_{s+1}\}$, and $L_i \mathrm{I} x_{ij} \mathrm{I} M_j$, $i,j = 1,\ldots,s+1$. Consider a point $p \in P-V$, with $V = \{x_{ij} \| i,j = 1,\ldots,s+1\}$. The s+1 lines incident with p are incident with s+1 points of V. Such a subset C of order s+1 of V is called a circle. By an argument similar to that used in the proof of 5.2.4 one proves that each triad of V is contained in exactly one circle and that any circle C is determined by exactly one point $p \in P-V$. The point p will be called the nucleus of C. Now we consider the incidence structure $M^* = (V,B',I')$, where $B' = \{L_1,L_2\}^\perp \cup \{L_1,L_2\}^{\perp\perp} \cup \{C \| C \text{ is a circle}\}$ and I' is defined in the obvious way. Then it is clear that M^* is a Minkowski plane of order s [68]. That s is even follows from an argument analogous to the corresponding one in 5.2.4. Now by a theorem proved independently by W. Heise [72] and N. Percsy [146], the Minkowski plane M^* is miquelian [68], i.e. is isomorphic to the classical Minkowski plane arising from the hyperbolic quadric H in PG(3,s). Hence $s = 2^h$. If W(s) is the GQ arising from the symplectic polarity θ defined by H [80], which means that W(s) is the GQ formed by the points of PG(3,s) together with the tangent lines of H, then in a manner analogous to that used in the preceding proof one shows that $W(s) \cong S$. □

5.2.7. (*F. Mazzocca* [102], *S.E. Payne and J.A. Thas* [143]). *Let S be a GQ of order* s, $s \neq 1$, *having an antiregular point* x . *Then S is isomorphic to* Q(4,s) *iff there is a point* $y, y \in x^\perp-\{x\}$, *for which the associated affine plane* $\pi(x,y)$ *is desarguesian.*

Proof. Since $S = (P,B,I)$ has an antiregular point x, s is odd by 1.5.1 (i). And for Q(4,s), s odd, it is clear that each associated affine plane $\pi(x,y)$ (see 1.3.2) is the desarguesian plane AG(2,s).

Conversely, suppose that $y, y \in x^\perp-\{x\}$, is a point for which the associated affine plane $\pi(x,y)$ is desarguesian. We consider the incidence structure $L^* = (x^\perp - \{x\}, B', I')$, where $B' = B_1 \cup B_2$ with $B_1 = \{M \in B \parallel x \mathrm{I} M\}$ and $B_2 = \{\{x,z\}^\perp \parallel z \not\sim x\}$ and where I' is defined in the obvious way. We shall prove that L^* is a Laguerre plane of order s [68], for which the elements of B_1 are the generators (or lines) and the elements of B_2 are the circles.

Clearly each point of L^* is incident with a unique element of B_1, and a generator and a circle intersect in exactly one point. Next, let x_1, x_2, x_3 be pairwise noncollinear points of L^*. Hence (x_1, x_2, x_3) is a triad of S with center x. By the antiregularity of x, the triad has exactly one center $z \neq x$ (see 1.3.6 (iii)). Hence x_1, x_2, x_3 lie on a unique circle C_z. Further, we remark that each circle has s+1 points and that there exist some $C \in B_2$ and some $d \in x^\perp - \{x\}$ such that $d \not\mathrm{I}' C$. Finally, we have to show that for each $C \in B_2$, $d \in C$, $u \in (x^\perp - \{x\}) - C$, $u \not\sim d$, there is a unique circle C_1 with $u \in C_1$ and $C \cap C_1 = \{d\}$. But this is an easy consequence from the preceding properties and $|x^\perp - \{x\}| = s^2 + s$, $|\{w \parallel w \mathrm{I}' M\}| = s$ for all $M \in B_1$, $|C| = s+1$ for all $C \in B_2$.

It is clear that the internal structure L^*_y [68] of the Laguerre plane L^* with respect to the point y is essentially the affine plane $\pi(x,y)$. Since $\pi(x,y)$ is desarguesian, then by a theorem proved by Y. Chen and G. Kaerlein [39] and independently by S.E. Payne and J.A. Thas [143], there is an isomorphism σ from the Laguerre plane L^* onto the classical Laguerre plane arising from the quadric cone C^* in PG(3,s).

Let C^* be embedded in the nonsingular quadric Q of PG(4,s). The vertex of C^* is denoted by x_∞. Now let $x^\phi = x_\infty$ and $w^\phi = w^\sigma$ for all $w \in x^\perp - \{x\}$. If $z \not\sim x$ and $C = \{x,z\}^\perp \in B_2$, then z^ϕ is the unique point of the GQ Q(4,s) for which $\{z^\phi, x_\infty\}^\perp = C^\sigma$. Evidently ϕ is a bijection from P onto Q. Moreover, it is easy to check that collinear (resp., noncollinear) points of S are mapped by ϕ onto collinear (resp., noncollinear) points of Q(4,s). It follows immediately that $S \cong Q(4,s)$. □

There is an easy corollary.

5.2.8. *Let* S *be a* GQ *of order* s, $s \neq 1$, *having an antiregular point* x. *If* $s \leq 8$, *i.e. if* $s \in \{3,5,7\}$, *then* S *is isomorphic to* Q(4,s).

Proof. Since each plane of order s, $s \leq 8$, is desarguesian [49], the result follows. □

From the proof of 5.2.7 it follows that with each GQ S of order s, $s \neq 1$, having an antiregular point there corresponds a Laguerre plane L^* of order s. In [143] it is also shown that, conversely, with each Laguerre plane L^* of odd order s there corresponds a GQ of order s with at least one antiregular point.

5.3. CHARACTERIZATIONS OF $T_3(O)$ AND $Q(5,q)$

The following characterization theorem will appear to be very important, not only for the theory of GQ but also for other domains in combinatorics : see e.g. L. Batten and F. Buekenhout [8], and P.J. Cameron, J.-M. Goethals, and J.J. Seidel [34].

5.3.1. (*J.A. Thas* [197]). *A* GQ *of order* (s,s^2), $s > 1$, *is isomorphic to* $T_3(O)$ *iff it has a 3-regular point* x_∞.

Proof. By 3.3.2 (ii) the point (∞) of $T_3(O)$ is 3-regular. Conversely, suppose that $S = (P,B,I)$ is a GQ of order (s,s^2), $s > 1$, for which the point x_∞ is 3-regular. The proof that S is isomorphic to $T_3(O)$ is arranged into a sequence of five rather substantial steps.

Step 1. The inversive plane $\pi(x_\infty)$.

Let $y \in P-x_\infty^\perp$. In 1.3.3 we noticed that the incidence structure $\pi(x_\infty,y)$ with pointset $\{x_\infty,y\}^\perp$, with lineset the set of elements $\{z,z',z''\}^{\perp\perp}$ where z,z', $z'' \in \{x_\infty,y\}^\perp$, and with the natural incidence, is an inversive plane [49] of order s. Let O_∞ be the set $\{L_1,\ldots,L_{s^2+1}\}$, where $L_1,\ldots,L_{s^2+1}$ are the s^2+1 lines which are incident with x_∞. If C is a circle of $\pi(x_\infty,y)$, then C_y is the subset of O_∞ consisting of the lines L_i for which x_∞ I L_i I x_i, with $x_i \in C$. The set of the elements C_y is denoted by B_y. It is clear that $\pi_y(x_\infty) = (O_\infty,B_y,\in)$ is an inversive plane of order s which is isomorphic to $\pi(x_\infty,y)$. The goal of Step 1 is to show that B_y is independent of the point y

Suppose that L_i,L_j,L_k are distinct lines through x_∞, and that x_1,x_2,x_3, x_3',x_∞ are distinct points with x_1 I L_i, x_2 I L_j, x_3 I L_k I x_3'. We prove that each line of O_∞ which is incident with a point of $\{x_1,x_2,x_3\}^{\perp\perp} = C$ is also incident with a point of $\{x_1,x_2,x_3'\}^{\perp\perp} = C'$. So let $L_\ell \in O_\infty$, $\ell \notin \{i,j,k\}$ be incident with a point x_4 of C, and assume L_ℓ is incident with no point of C'. Then by 1.4.2 (iii) x_4 is collinear with two points x_∞ and x_4'' of $\{x_1,x_2,x_3'\}^\perp$. But $x_4'' \in \{x_1,x_2,x_4\}^\perp = \{x_1,x_2,x_3\}^\perp$ implies x_4'',x_3,x_3' are the vertices of a triangle, a contradiction. Hence each line of O_∞ which is incident with a point of C is also incident with a point of C'.

Now consider two points $y,z \in P-x_\infty^\perp$. Let L_i,L_j,L_k be distinct elements of O_∞, and let x_1,x_2,x_3 (resp., x_1',x_2',x_3') be the points of $\pi(x_\infty,y)$ (resp., $\pi(x_\infty,z)$) which are incident with L_i,L_j,L_k, respectively. The sets $\{x_1,x_2,x_3\}^{\perp\perp}$ and $\{x_1',x_2',x_3'\}^{\perp\perp}$ are denoted by C and C', respectively. We have to consider four cases :

(1) If C = C', then each line of O_∞ which is incident with a point of C is also incident with a point of C'.

(2) If $|C \cap C'| = 2$, then by the preceding paragraph each line of O_∞ which is incident with a point of C is also incident with a point of C'.

(3) Let $|C \cap C'| = 1$, say $C \cap C' = \{x_4\}$, with $x_1 \neq x_4 \neq x_2$. Each line of O_∞ which is incident with a point of C is also incident with a point of $\{x_1',x_2,x_4\}^{\perp\perp}$ and hence also with a point of $\{x_1',x_2',x_4\}^{\perp\perp} = C'$.

(4) Let $C \cap C' = \phi$. Each line of O_∞ which is incident with a point of C is also incident with a point of $\{x_1,x_2,x_3'\}^{\perp\perp}$ and hence also with a point of C', by the preceding cases.

From (1)-(4) it follows that the circle $L_iL_jL_k$ of the inversive plane $\pi_y(x_\infty)$ coincides with the circle $L_iL_jL_k$ of the inversive plane $\pi_z(x_\infty)$. Hence $B_y = B_z$, i.e. $\pi_y(x_\infty) = \pi_z(x_\infty)$. The inversive plane $\pi_y(x_\infty)$, which is independent of the choice of the point y, will be denoted by $\pi(x_\infty)$.

Step 2. The inversive plane $\pi(x_\infty)$ is egglike.

Here we must prove that $\pi(x_\infty)$ arises from an ovoid in PG(3,s). Since there is a unique inversive plane of order s for s = 2 or 3 (cf. [49]), we may assume $s \geq 4$.

Let $z \sim x_\infty$, $z \neq x_\infty$, and define the following incidence structure $S_z = (P_z,B_z,I_z)$. The set P_z is just $x_\infty^\perp - z^\perp$. The elements of type (i) of B_z are the sets $L^* = \{u \in P_z \parallel u \mathrm{I} L\}$, with $x_\infty \mathrm{I} L$ and $z \not\mathrm{I} L$. The elements of type (ii) of B_z are the sets $\{z,u_1,u_2\}^{\perp\perp}$, with (z,u_1,u_2) a triad and $u_1,u_2 \in P_z$. I_z is the natural incidence. It is clear that S_z is a 2-$(s^3,s,1)$ design. We shall prove that S_z is the design of points and lines of AG(3,s).

By a theorem of F. Buekenhout [26] it is sufficient to prove that any three noncollinear points u_1,u_2,u_3 of S_z generate an affine plane. We consider three cases :

(a) Let $u_i,u_j, i \neq j$, be incident with an element of type (i) in B_z, and let $\{i,j,k\} = \{1,2,3\}$. From the proof of Step 1 it follows that the s^2 points of P_z which are collinear (in S) with a point of $\{z,u_i,u_k\}^{\perp\perp}$ form a 2-$(s^2,s,1)$ subdesign of S_z. This subdesign is an affine plane containing u_1,u_2,u_3. So the triangle with vertices u_1,u_2,u_3 of S_z generates an affine plane.

(b) Let (u_1,u_2,u_3) be a triad and suppose that the line $x_\infty z$ of S is incident with some point of $\{u_1,u_2,u_3\}^{\perp\perp}$. From the proof of Step 1 it follows that the s^2 points of P_z which are collinear (in S) with a point of $\{u_1,u_2,u_3\}^{\perp\perp}$ form a 2-$(s^2,s,1)$ subdesign of S_z. So the triangle $u_1u_2u_3$ of S_z generates an affine plane.

(c) Let (u_1,u_2,u_3) be a triad and suppose that the line $x_\infty z$ of S is incident with no point of $\{u_1,u_2,u_3\}^{\perp\perp}$. By 1.4.2 (iii) there is exactly one point x' for which $x' \in z^\perp \cap \{u_1,u_2,u_3\}^\perp$, $x' \neq x_\infty$. Now the internal (or residual) [49] structure of the inversive plane $\pi(x_\infty,x')$ at z is an affine plane of order s which is a substructure of S_z and contains the points u_1,u_2,u_3.

Hence S_z is the design of points and lines of $AG(3,s)$. We are now in a good position to prove that the inversive plane $\pi(x_\infty,y)$, $y \not\sim x_\infty$ and $y \not\sim z$, is egglike.

Let $PG(3,s)$ be the projective completion of $AG(3,s)$. All lines of type (i) of B_z are parallel lines of $AG(3,s)$, and thus define a point (∞) of $PG(3,s)$. If y' is the point defined by $y' \mathrm{I}\, x_\infty z$ and $y' \sim y$, then let $O'_y = (\{x_\infty,y\}^\perp - \{y'\}) \cup \{(\infty)\}$. It is easy to check that no three points of O'_y are collinear in $PG(3,s)$. Hence O'_y is an ovoid of $PG(3,s)$. If C is a circle of $\pi(x_\infty,y)$ which does not contain y', then $C \subset O'_y$, and by (c) C is a plane intersection of the ovoid O'_y of $PG(3,s)$, the plane being the projective completion of the internal structure of $\pi(x_\infty,x')$ at z where x' is the unique element of $C^\perp - \{x_\infty\}$ that is collinear with z. If C is a circle of $\pi(x_\infty,y)$ which contains y', then $(C-\{y'\}) \cup \{(\infty)\}$ is the intersection of O'_y with the projective completion of the affine subplane of S_z, having as points the s^2 points of P_z which are collinear (in S) with a point of $C-\{y'\}$. Hence $\pi(x_\infty,y)$ is isomorphic to the egglike inversive plane arising from the ovoid O'_y of $PG(3,s)$.

Since $\pi(x_\infty) \cong \pi(x_\infty,y)$, we conclude that the inversive plane $\pi(x_\infty)$ is egglike.

<u>Step 3. The point x_∞ is coregular</u>.
It is convenient to adopt just for the duration of this proof a notation

inconsistent with the standard labeling of lines of S through x_∞. Let L_0 be a line through x_∞ and let L_1 be a second line of S not concurrent with L_0. The proof amounts to showing that (L_0,L_1) is regular.

Let L_0' be the line through x_∞ meeting L_1, and let $L_1',\ldots,L_s'$ be the remaining lines in $\{L_0,L_1\}^\perp$. Similarly, let $L_0,L_1,\ldots,L_s$ be the lines in $\{L_0',L_1'\}^\perp$. Let $x_{i2},\ldots,x_{is}$ be the points of L_i not on L_0' or L_1', and let $x_{i2}',\ldots,x_{is}'$ be the points of L_i' not on L_0 or L_1, $i = 2,\ldots,s$. To show that (L_0,L_1) is regular, it will suffice to show that each x_{ij} lies on some L_r'.

Let y,z,u be the points defined by L_0 I y I L_1', L_1 I z I L_1', and L_1 I u I L_0'. Let $C_{ij} = \{x_\infty,z,x_{ij}\}^\perp$, $C_{ij}' = \{x_\infty,z,x_{ij}'\}^\perp$, $2 \leqslant i, j \leqslant s$. Then each C_{ij} and C_{ij}' are circles in the inversive plane $\pi(x_\infty,z)$. Moreover, each $\{C_{i2},\ldots,C_{is}, \{u\}, \{y\}\}$ and each $\{C_{i2}',\ldots,C_{is}', \{u\}, \{y\}\}$ are partitions of the pointset of $\pi(x_\infty,z)$, i.e. each $F_i = \{C_{ij} \parallel 2 \leqslant j \leqslant s\}$ and each $F_i' = \{C_{ij}' \parallel 2 \leqslant j \leqslant s\}$ are flocks [49, 58] of $\pi(x_\infty,z)$ with carriers [49, 58] u and y. Since $\pi(x_\infty,z)$ is egglike, the flocks F_i and F_i' are linear by theorems of W.F. Orr and J.A. Thas [58]. This means that the flocks F_i and F_i' are uniquely defined by their carriers. Since they all have the same carriers, we necessarily have $F_2 = F_3 = \ldots = F_s = F_2' = \ldots = F_s'$. Then, for example, $F_i = F_r'$ says that for each j, $2 \leqslant j \leqslant s$, there is a k, $2 \leqslant k \leqslant s$, such that $\{x_\infty,z,x_{ij}\}^\perp = \{x_\infty,z,x_{rk}'\}^\perp$. Hence $\{x_\infty,z,x_{ij}\}^{\perp\perp}$ has a point x_{rk}' on L_r'. Fixing i and j, we see that each of the s-1 lines $L_2',\ldots,L_s'$ contains a point of $\{x_\infty,z,x_{ij}\}^{\perp\perp}-\{x_\infty,z\}$. So x_{ij} must be on some L_r', and consequently (L_0,L_1) is regular.

Note : An additional consequence of interest is that each set of points of the form $\{x_\infty,z,x_{ij}\}^{\perp\perp}$ lies entirely in the set of points covered simultaneously by $\{L_0,L_1\}^\perp$ and by $\{L_0,L_1\}^{\perp\perp}$.

Step 4. The affine space $A = (P^*,B^*,\in)$.

Let $P^* = P-x_\infty^\perp$. If y and z are distinct points of P^* collinear in S, define the block yz of type (i) to be the set of points of P^* on the line of S through y and z. If y and z are noncollinear points of P^*, define the block yz of type (ii) to be the set $\{x_\infty,y,z\}^{\perp\perp}-\{x_\infty\}$. Let B^* be the set of blocks just defined. Then $A = (P^*,B^*,\in)$ is a 2-$(s^4,s,1)$ design.

In the set B^* of blocks we now define a parallelism. Two blocks of type (i) are parallel iff the corresponding lines of S are concurrent with a same element of O_∞ (recall that O_∞ consists of the lines of S incident with

x_∞). The blocks $\{x_\infty,y,z\}^{\perp\perp}-\{x_\infty\}$ and $\{x_\infty,y',z'\}^{\perp\perp}-\{x_\infty\}$ of type (ii) are parallel iff each line of O_∞ which is incident with a point of $\{x_\infty,y,z\}^\perp$ is also incident with a point of $\{x_\infty,y',z'\}^\perp$, i.e. iff they both determine the same circle of $\pi(x_\infty)$. A block of type (i) is never parallel to a block of type (ii). The parallelism defined in this manner will be denoted by $\|$.

By a well known theorem of H. Lenz [99] the design A is the design of points and lines of AG(4,s) iff the conditions (i) and (ii), or (i) and (ii)', are satisfied.

(i) Parallelism is an equivalence relation in the set B^*, and each class of parallel blocks is a partition of the set P^*.

(ii) Let $s \geqslant 3$ and let $L \| L'$, $L \neq L'$, $y \in L$, $y' \in L'$, $z' \in L'-\{y'\}$, $p \in yy'-\{y,y'\}$. Then $L \cap pz' \neq \phi$.

(ii)' Let $s = 2$ and let y,z,u be three distinct points of P^*. If L is the block defined by $y \in L$ and $L \| zu$, and M is the block defined by $z \in M$ and $M \| yu$, then $L \cap M \neq \phi$.

It is clear that parallelism is an equivalence relation in the set B^* and that each class of parallel blocks of type (i) is a partition of P^*. Since there are no triangles in S, any two distinct parallel blocks of type (ii) are disjoint. Now let $L = \{x_\infty,y,z\}^{\perp\perp}-\{x_\infty\}$ be a block of type (ii) and let $u \in P^*$. If we "project" $\{x_\infty,y,z\}^\perp$ from x_∞, then there arises a circle C of $\pi(x_\infty)$. By "intersection" of C and $\{x_\infty,u\}^\perp$, we obtain a circle C' of $\pi(x_\infty,u)$. Clearly $C'^\perp-\{x_\infty\}$ is the unique block which contains u and is parallel to L. Hence condition (i) is satisfied.

Now we assume $s \geqslant 3$ and prove that (ii) is satisfied. So let $L \| L'$, $L \neq L'$, $y \in L$, $y' \in L'$, $z' \in L'-\{y'\}$, $p \in yy'-\{y,y'\}$. It is clear that (ii) is satisfied if we show that the substructure of A generated by L and L' is an affine plane (of order s). Note that the substructure of A generated by L and L' has at least s^2 points. We have to consider several cases.

Let L and L' be of type (ii), say $L = \{x_\infty,y,z\}^{\perp\perp}-\{x_\infty\}$ and $L' = \{x_\infty,y',z'\}^{\perp\perp}-\{x_\infty\}$, and let $\{x_\infty,y,z\}^\perp \cap \{x_\infty,y',z'\}^\perp = \{x_1,x_2\}$. Then the substructure of A generated by L and L' is contained in $\{x_1,x_2\}^\perp-\{x_\infty\}$, and hence has at most s^2 points. Consequently that substructure is an affine plane.

Let L and L' be of type (ii), with notation as in the preceding case, but suppose $\{x_\infty,y,z\}^\perp \cap \{x_\infty,y',z'\}^\perp = \{x_1\}$. We first prove that for an arbitrary $u \in L$, the line ux_1 of S is incident with a point of L'. Suppose the contrary. Then by 1.4.2 (iii) u is collinear with two points x_1 and y_1 of $\{x_\infty,y',z'\}^\perp$. Since L ∥ L', the line $x_\infty y_1$ is incident with a point z_1 of $\{x_\infty,y,z\}^\perp$. So there arises a triangle uz_1y_1 in S, a contradiction. Consequently, for each point u of L, the line ux_1 is incident with a point of L'. The blocks of type (i) corresponding to the lines ux_1, $u \in L$, are denoted $M_1,\ldots,M_s$. By an argument just like that used in Step 1, one shows that each block which has a point in common with M_i,M_j, $i \neq j$, has a point in common with all s blocks $M_1,\ldots,M_s$. Clearly the substructure of A generated by M_1,M_2 has s^2 points and contains the blocks L and L'. Hence the blocks L and L' generate an affine plane.

Let L and L' be of type (i), and suppose that the corresponding lines of S have a point y in common ($y \sim x_\infty$). Further, let N be a block of type (ii) having a point in common with L and L'. The blocks of type (i) corresponding to the lines uy, $u \in N$, are denoted by $M_1 = L$, $M_2 = L',\ldots,M_s$. Just as in Step 1, one shows that each block that has a point in common with two of the M_i's has a point in common with each of the s blocks $M_1,\ldots,M_s$. Now it is clear that the blocks $M_1 = L$ and $M_2 = L'$ generate an affine plane.

Let L and L' be of type (i), and suppose that the corresponding lines of S are not concurrent. If the lines of S which correspond to L and L' are denoted N_1 and N_2, respectively, then O_∞ contains one line which is concurrent with N_1 and N_2. The set of all points of P^* which are incident with lines of $\{N_1,N_2\}^\perp$ (or $\{N_1,N_2\}^{\perp\perp}$) is denoted by V. We note that $|V| = s^2$. In the last paragraph of Step 3 we noted that each set of points of the form $\{x_\infty,y,z\}^{\perp\perp}-\{x_\infty\}$, with $y,z \in V$, $y \not\sim z$, lies entirely in V. Hence the substructure of A generated by L and L' has a pointset V of order s^2, and consequently is an affine plane of order s.

Finally, let L and L' be of type (ii), say $L = \{x_\infty,y,z\}^{\perp\perp}-\{x_\infty\}$ and $L' = \{x_\infty,y',z'\}^{\perp\perp}-\{x_\infty\}$, and let $\{x_\infty,y,z\}^\perp \cap \{x_\infty,y',z'\}^\perp = \phi$. First suppose that $\{x_\infty,y',z'\}^\perp$ contains a point z" which is collinear (in S) with z. By the hypothesis L ∥ L', the line $x_\infty z''$ is incident with some point u of $\{x_\infty,y,z\}^\perp$, and there arises a triangle zuz" in S, a contradiction. Hence by 1.4.2 (iii) the point z is collinear with two points u' and r' of L'. Let L_1 be the line of O_∞ which is concurrent with zu'. The set of all points of P^* which are incident with lines of $\{L_1,zr'\}^\perp$ (or $\{L_1,zr'\}^{\perp\perp}$) is denoted by V. Then $|V| = s^2$, and in the preceding paragraph we noticed that V is the pointset of an affine subplane of order s of A. Since clearly $L' \subset V$, it only remains to be shown that $L \subset V$. Let $M_1,\ldots,M_{s-1}$ be the blocks of type (ii) in V which contain z. One of these blocks is parallel to L', say M_1. Then M_i, $i \neq 1$, and L' have just one point in common, say v_i. It follows that there is no line $L_j \in O_\infty$ which is incident with a point of $M_i^\perp$, $i \neq 1$, and a point of $L'^\perp$, since otherwise there arises a triangle with vertex v_i and the other two vertices on L_j (keep in mind that by hypothesis $L^\perp$ and $L'^\perp$ are disjoint). Since each point of $M_i-\{z\}$ is collinear with two points of $M_j-\{z\}$, $i \neq j$, the sets $M_i^\perp$ and $M_j^\perp$ are disjoint. As $M_i \cap M_j = \{z\}$, $i \neq j$, there is no line of O_∞ which is incident with a point of $M_i^\perp$ and with a point of $M_j^\perp$. It now follows easily that the s+1 lines of O_∞ which are incident with a point of $L'^\perp$ coincide with the s+1 lines of O_∞ which are incident with a point of $M_1^\perp$. If these lines are denoted by $L_{i_0},\ldots,L_{i_s}$, then $M_1^\perp$ as well as $L^\perp$ consists of the points of $L_{i_0},\ldots,L_{i_s}$ which are collinear with z. Hence $M_1^\perp = L^\perp$, implying $M_1 = L$. It follows that $L \subset V$, and consequently L and L' generate an affine plane of order s.

It is now proved that for $s \neq 2$ the design A is the design of points and lines of AG(4,s). Finally, we assume that s = 2.

Let y,z,u be three distinct points of P^*. If L is the block defined by $y \in L$ and $L \parallel zu$, and M is the block defined by $z \in M$ and $M \parallel yu$, then we must prove that $L \cap M \neq \phi$. We have to consider several cases.

If uy and uz are blocks of type (i), then from the coregularity of x_∞ it follows immediately that L and M have a point in common.

Let uz be of type (i), uy of type (ii), and let $\{x_\infty,u,y\}^\perp \cap (M \cup \{x_\infty\})^\perp = \{r\}$. Just as in the case $s > 2$ one shows that u,z,r are collinear, that $y \sim r$, and that the line yr of S is incident with a point of M. Since L is the set of all points of P^* which are incident with yr, we have $L \cap M \neq \phi$.

Let uz be of type (i), uy of type (ii), and let $\{x_\infty,u,y\}^\perp \cap (M \cup \{x_\infty\})^\perp = \phi$ If $M = \{z,r\}$, then just as in the last part of the $s \neq 2$ case, one shows that $y \sim z$, $y \sim r$, and that the lines uz and yr of S are concurrent with a same element of O_∞. It follows immediately that L is of type (i) and that $L \cap M \neq \phi$.

Clearly the cases uy of type (i) and uz of type (ii) are analogous to the preceding two cases.

Let uz and uy be of type (ii), let $L = \{y,r\}$, and let ur be of type (i). In S the point u is collinear with exactly one point of L. If v is defined by v I ur and $v \in x_\infty^\perp$, then just as in the case $s \neq 2$ one sees that $z \sim v$ and that the line zv is incident with a point of L. Then clearly y I zv. Now from a preceding case it follows that the block {u,r} has a point in common with the block M. Hence $r \in M$, and $L \cap M \neq \phi$.

Finally, let uz and uy be of type (ii), let L = {y,r}, and let ur be of type (ii). If M = {z,v}, then by the preceding case uv is also of type (ii) (since otherwise ur would be of type (i)). As u is collinear with no point of L, it is collinear with two points x_1,x_2 of $(L \cup \{x_\infty\})^\perp$. Since the line $x_\infty x_i$ is incident with a point of $\{x_\infty,u,z\}^\perp$ and since S has no triangles, we have $x_1,x_2 \in \{x_\infty,u,z\}^\perp$. Hence r is collinear with the two common points x_1,x_2 of $\{x_\infty,u,z\}^\perp$, $\{x_\infty,u,y\}^\perp$, and $\{x_\infty,y,z\}^\perp$. Analogously, v is collinear with x_1 and x_2. Hence $\{x_1,x_2\}^\perp = \{x_\infty,u,y,z,r\} = \{x_\infty,u,y,z,v\}$, and it must be that r = v, implying $L \cap M \neq \phi$.

This completes the proof that also for s = 2 the design A is the design of points and lines of AG(4,s).

Step 5. The GQ $T(O_\infty)$.

The points of the hyperplane at infinity PG(3,s) of AG(4,s) can be identified in a natural way with the elements of O_∞, i.e. the points of $\pi(x_\infty)$, and with the circles of $\pi(x_\infty)$. Now we prove that O_∞ is an ovoid of the projective space PG(3,s).

Suppose $L_i,L_j,L_k \in O_\infty$ are collinear in PG(3,s). Projecting these three points L_i,L_j,L_k from a point $y \in P^*$ we obtain three blocks M_i,M_j,M_k of type (i) that must belong to an affine subplane of A of order s . If $y_i \in M_i-\{y\}$, $y_j \in M_j-\{y\}$, then the block y_iy_j (of type (ii)) has a point y_k ($\neq y$) in common with M_k (note that the blocks M_k and y_iy_j are not parallel since they are of different type). Consequently $y \in \{y_i,y_j,y_k\}^\perp$, implying $y \sim x_\infty$, a contradiction. Hence no three elements of O_∞ are collinear in PG(3,s). Since $|O_\infty| = 1+s^2$, O_∞ is an ovoid of PG(3,s), if $s \neq 2$ [49]. So we now assume that s = 2. Let $x_\infty \neq u$ I $L_i \in O_\infty$. Then it is easy to prove that $P' = \{y \in P^* \parallel y \sim u\}$ is the pointset of an affine subspace AG(3,2) of AG(4,2). Clearly L_i is the only point of infinity of AG(3,2) that belongs to O_∞. So the plane at infinity PG(2,2) of AG(3,2) has only the point L_i in common with O_∞. Consequently for each $L_i \in O_\infty$ there exists a plane of PG(3,2) which contains L_i and which has only the point L_i in common with O_∞. As $|O_\infty| = 5$, it follows immediately that O_∞ is an ovoid of PG(3,2).

Now we consider a point u I L_i, $u \neq x_\infty$. It is easy to show that

$P' = \{y \in P^* \parallel y \sim u\}$ is the pointset of an affine subspace AG(3,s) of AG(4,s). Clearly L_i is the only point at infinity of AG(3,s) that belongs to O_∞, so that the plane at infinity of AG(3,s) is the tangent plane $PG^{(i)}(2,s)$ of O_∞ at L_i. So with the s points u on L_i, $u \neq x_\infty$, there correspond the s three dimensional affine subspaces of AG(4,s) which have $PG^{(i)}(2,s)$ as plane at infinity.

At this point it is clear that S has the following description in terms of the ovoid O_∞. Points of S are (i) the points of AG(4,s), (ii) the three dimensional affine subspaces of AG(4,s) that possess a tangent plane of O_∞ as plane at infinity, (iii) one new symbol (∞). Lines of S are (a) the lines of AG(4,s) whose points at infinity belong to O_∞, and (b) the elements of O_∞. Points of type (i) are incident only with lines of type (a) and here the incidence is that of AG(4,s). A point AG(3,s) of type (ii) is incident with the lines of type (a) that are contained in AG(3,s) and with the unique point at infinity of AG(3,s) that belongs to O_∞. Finally, the unique point (∞) of type (iii) is incident with all lines of type (b) and with no line of type (a).

We conclude that S is isomorphic to the GQ $T_3(O_\infty)$ of J. Tits. □

There are some immediate corollaries.

5.3.2. (i) *If* S *is a* GQ *of order* (s,s^2), $s > 1$, *in which each point is 3-regular, then* $S \cong Q(5,s)$.

(ii) *Up to isomorphism there is only one* GQ *of order* (2,4).

(iii) *Up to isomorphism there is only one* GQ *of order* (3,9).

Proof. (i) By Step 3 of the preceding proof each line of S would be regular and a $T_3(O)$ with all lines regular is isomorphic to Q(5,q) by 3.3.3 (iii).

(ii) Let S be a GQ of order (2,4). If (x_1,x_2,x_3) is a triad of points, then clearly $\{x_1,x_2,x_3\}^{\perp\perp} = \{x_1,x_2,x_3\}$. Hence $|\{x_1,x_2,x_3\}^{\perp\perp}| = 1+s$, every point is 3-regular, and by part (i) we have $S \cong Q(5,2)$.

(iii) By 1.7.2 all points of any GQ of order (3,9) are 3-regular, so part (i) applies. □

The uniqueness of the GQ of order (2,4) was proved independently at least five times, by S. Dixmier and F. Zara [53], J.J. Seidel [164], E.E. Shult [166], J.A. Thas [188] and H. Freudenthal [63]. The uniqueness of a GQ of order (3,9) was proved independently by S. Dixmier and F. Zara [53] and by

P.J. Cameron [143] .

Using the same kind of argument and results from Sections 3.2 and 3.3 it is easy to conclude the following.

5.3.3. (i) *Let* S *be a* GQ *of order* (s,s^2), $s > 1$, *with* s *odd. Then* $S \cong Q(5,s)$ *iff* S *has a* 3*-regular point.*

(ii) *Let* S *be a* GQ *of order* (s,s^2) *with* s *even. Then* $S \cong Q(5,s)$ *iff one of the following holds :*

(a) *All points of* S *are* 3*-regular.*

(b) S *has at least one* 3*-regular point not incident with some regular line.*

Remark : Independently F. Mazzocca [103] proved the following result : A GQ S of order (s,s^2), $s \neq 1$ and s odd, is isomorphic to $Q(5,s)$ iff each point of S is 3-regular.

We now consider the role of subquadrangles in characterizing $T_3(O)$.

5.3.4. (*J.A. Thas* [197]). *Let* $S = (P,B,I)$ *be a* GQ *of order* (s,t), $s > 1$. *Then the following are equivalent :*

(i) S *contains a point* x_∞ *such that every centric triad of lines having a center incident with* x_∞ *is contained in a proper subquadrangle* S' *of order* (s,t').

(ii) $t > 1$ and S *contains a point* x_∞ *such that for every triad* (u,u',u'') *with distinct centers* x_∞ *and* x', *the points* u,u',u'',x_∞,x' *are contained in a proper subquadrangle of order* (s,t').

(iii) $s^2 = t$, S *contains a* 3*-regular point* x_∞, *and hence* $S \cong T_3(O)$.

Proof. By 3.5 (b) it is clear that (iii) implies (i) and (ii). Now we assume that (i) is satisfied. Clearly we have $t > 1$. Let K,L,M,N be lines for which x_∞ I N, $L \sim N$, $M \sim N$, $K \sim N$, $K \not\sim L$, $L \not\sim M$, $M \not\sim K$. Then K,L,M are contained in a proper subquadrangle $S' = (P',B',I')$ with order (s,t'). Suppose that K' is not a line of S' and that N,K,K' are concurrent. Then K',L,M are contained in a proper subquadrangle $S'' = (P'',B'',I'')$ with order (s,t''). Clearly $S' \neq S''$. By 2.3.1 $S''' = (P' \cap P'', B' \cap B'', I' \cap I'')$ is a proper subquadrangle of S'' of order (s,t''). Now by 2.2.2 (vi) $s^2 = t$, $t'' = s$ and $t''' = 1$. Since $t''' = 1$, the pair (L,M) is regular. It follows immediately that each line incident with x_∞ is regular, i.e. x_∞ is coregular.

Let us suppose that s is even. Consider a triad (x_∞,y,z) and suppose that $u,u' \in \{x_\infty,y,z\}^\perp$, $u \neq u'$. Let x', $x' \neq x_\infty$ and $x' \neq u$, be a point which is incident with the line $x_\infty u$. Further, let L be the line which is incident with x' and concurrent with yu'. Then the lines $x_\infty u'$, zu, and L are contained in a proper subquadrangle S' of order (s,t'). Clearly S' contains the lines $x_\infty u', zu, L, x_\infty u, yu', yu, zu'$, and the points x_∞, y, z, u, u'. Consequently $t' > 1$. By 2.2.2 we have $t' \leqslant s$ and since S' contains regular lines we have $t' \geqslant s$. Hence $s = t'$. Since s is even and x_∞ is a coregular point of S', the point x_∞ is regular for S' by 1.5.2 (iv). So each triad of points of S' containing x_∞ has exactly 1 or 1+s centers in S'. Since u and u' are centers of $\{x_\infty,y,z\}$, the triad (x_∞,y,z) has exactly 1+s centers $u_0 = u$, $u_1 = u'$, $u_2,\ldots,u_s$ in S'. Moreover, S' contains s+1 points $x_0 = x_\infty$, $x_1 = y$, $x_2 = z$, $x_3,\ldots,x_s$ which are collinear with each of the points $u_0,\ldots,u_s$. Hence (x_∞,y,z) is 3-regular in S . It follows that x_∞ is 3-regular and hence S is isomorphic to some $T_3(O)$.

Now suppose that s is odd. Let (x_∞,y,z) be a triad and suppose that $u,u' \in \{x_\infty,y,z\}^\perp$, $u \neq u'$. Just as in the preceding paragraph one shows that there is a subquadrangle $S' = (P',B',I')$ of S of order s which contains the points x_∞,y,z,u,u' and the lines $x_\infty u$, $yu, zu, x_\infty u', yu', zu'$. Since s is odd and x_∞ is coregular, the point x_∞ is antiregular for S' by 1.5.2 (v). Let $u'' \in \{x_\infty,y,z\}^\perp - \{u,u'\}$ and let $S'' = (P'',B'',I'')$ be a subquadrangle of S of order s containing the points x_∞,y,z,u,u''. If $S' = S''$, then in S' the triad (x_∞,y,z) has at least three centers, a contradiction by the antiregularity of x_∞. Hence $S' \neq S''$, $u' \notin P''$ and $u'' \notin P'$. Now we consider the incidence structure $S_1 = (P' \cap P'', B' \cap B'', I' \cap I'')$. We have $x_\infty,y,z,u \in P' \cap P''$ and $x_\infty u, yu, zu \in B' \cap B''$. By 2.3.1 one of the following occurs : (a) each point of $P' \cap P''$ is collinear with u and each line of $B' \cap B''$ is incident with u, and (b) S_1 is a proper subquadrangle of S' of order (s,t_1). If (b) occurs, then by 2.2.2 (vi) $t_1 = 1$, a contradiction since $B' \cap B''$ contains at least three lines through u. Hence we have (a). By 2.2.1 the point u' of S is collinear with the $1+s^2$ points of an ovoid of S''. Hence each line incident with u' has a point in common with S''. In particular, the 1+s lines of S' through u' are incident with a point of S''. It follows that $|B' \cap B''| = 1+s$ Now we consider a subquadrangle $S''' = (P''',B''',I''')$ of S of order s containing the points x_∞,y,z,u',u''. Then $S' \neq S''' \neq S''$. Then $(P' \cap P'' \cap P''' = P_2,\ B' \cap B'' \cap B''' = B_2,\ I' \cap I'' \cap I''' = I_2) =$

$((P' \cap P'') \cap P''', (B' \cap B'') \cap B''', (I' \cap I'') \cap I''')$ = (the set of the s+1 points of P''' which are collinear in S' (or in S'') with u,ϕ,ϕ). Analogously, we have P_2 = the set of the s+1 points of P' which are collinear in S'' (or S''') with u'' = the set of the s+1 points of P'' which are collinear in S''' (or S') with u'. Hence P_2 = trace of (u,u') in S' = trace of (u,u'') in S'' = trace of (u',u'') in S'''. It follows that each point of $\{x_\infty,y,z\}^\perp$ is collinear with each point of the trace of (u,u') in S'. Consequently (x_∞,y,z) is 3-regular in S . So x_∞ is 3-regular and S is isomorphic to a $T_3(O)$, i.e. to $Q(5,s)$.

Hence (i) implies (iii). Finally, we shall prove that (i) follows from (ii).

So assume that (ii) is satisfied. Consider a centric triad of lines (L,L',L'') with a center N which is incident with x_∞. Suppose that x_∞ is not incident with L'. Let $N' \in \{L,L'\}^\perp - \{N\}$ and L' I x' I N'. If $N' \not\sim L''$, then let $L''' = L''$; if $N' \sim L''$, then let L''' be a line for which $L''' \sim N$, $L''' \sim L''$, $L''' \notin \{N,L''\}$. Further, let N'' be the line which is incident with x' and concurrent with L''', let u be the point which is incident with N' and collinear with x_∞, let u' be the point which is incident with N'' and collinear with x_∞, and let N I u'' I L'. Then (u,u',u'') is a triad with centers x_∞ and x'. Hence u,u',u'',x_∞,x' are contained in a proper subquadrangle S' of order (s,t'). Clearly L,L',L'' are lines of S', so that (i) is satisfied. □

There is an easy corollary.

5.3.5. (i) *A GQ S of order* (s,t), $s > 1$, *is isomorphic to* $Q(5,s)$ *iff every centric triad of lines is contained in a proper subquadrangle of order* (s,t').

(ii) *A GQ S of order* (s,t), $s > 1$ *and* $t > 1$, *is isomorphic to* $Q(5,s)$ *iff for each triad* (u,u',u'') *with distinct centers* x,x' *the five points* u,u',u'', x,x' *are contained in a proper subquadrangle of order* (s,t').

Proof. (i) Let (L,L',L'') be a centric triad of lines of $Q(5,s)$. Then there is a $PG(4,s)$ which contains L,L',L''. If $Q \cap PG(4,s) = Q'$, then $Q'(4,s)$ is a proper subquadrangle of order s of $Q(5,s)$.

Conversely, suppose that $s > 1$ and that every centric triad of lines is contained in a proper subquadrangle of order (s,t'). Then from 5.3.4 it follows that $s^2 = t$ and that each point of S is 3-regular. By 5.3.2 we have $S \cong Q(5,s)$.

(ii) The proof is analogous and left to the reader . □

Let S be a GQ of order (s,t), and let (L_1,L_2,L_3) and (M_1,M_2,M_3) be two triads of lines for which $L_i \not\sim M_j$ iff $\{i,j\} = \{1,2\}$. Let x_i be the point defined by L_i I x_i I M_i, i = 1,2. This configuration of seven distinct points and six distinct lines is called a *broken grid with carriers* x_1 and x_2. First supppose S is classical and let M_4 be a line in $\{L_1,L_2\}^\perp$ not concurrent with any of M_1,M_2,M_3. There is a PG(4,s) containing the broken grid. Hence the threespace PG(3,s) defined by M_1 and M_2 has at least one point u in common with M_4. It is clear that there is a line L_4 which contains u and is concurrent with M_1,M_2, and M_4. Next suppose that S is the GQ $T_3(O)$ and assume that L_1 or M_1 contains the 3-regular point (∞). Then there is a $PG(3,s) \subset PG(4,s)$ for which $PG(3,s) \cap O$ is an oval O', and such that the corresponding subquadrangle $T_2(O')$ of $T_3(O)$ (see 3.5 (b)) contains the broken grid. If L_1 contains (∞), then the line L_1 is regular. Let $L_4 \in \{M_1,M_2\}^\perp$ (resp., $M_4 \in \{L_1,L_2\}^\perp$) with $L_4 \not\sim L_i$ (resp., $M_4 \not\sim M_i$) for i = 1,2,3. Then the pair (L_1,L_4) (resp., (M_1,M_4)) is regular. So there must be a line M_4 (resp., L_4) of $T_2(O')$ (by 1.3.6) which is concurrent with each of L_1,L_2,L_4 (resp., M_1,M_2,M_4). If M_1 contains (∞), we can proceed through the same discussion interchanging L_i and M_i. Similarly, the same argument holds if (∞) is on L_2 or M_2.

The preceding paragraph provides the motivation for the following definitions. Let Γ be a broken grid with carriers x_1 and x_2. Assume the same notation as above so that L_i and M_i are the lines of Γ incident with x_i, i = 1,2. We say that Γ satisfies *axiom* (D) with respect to the pair (L_1,L_2) provided the following holds : If $L_4 \in \{M_1,M_2\}^\perp$ with $L_4 \not\sim L_i$, i = 1,2,3, then (L_1,L_2,L_4) is centric. Interchanging L_i and M_i gives the definition of axiom (D) for Γ w.r.t. the pair (M_1,M_2). Further, Γ is said to satisfy *axiom* (D) provided it satisfies axiom (D) w.r.t. both pairs (L_1,L_2) and (M_1,M_2).

5.3.6. *Let* Γ *be a broken grid whose lines are those of the triads* (L_1,L_2,L_3) *and* (M_1,M_2,M_3) , *where* $M_i \not\sim L_j$ *iff* $\{i,j\} = \{1,2\}$. *If* Γ *satisfies axiom* (D) *w.r.t.* (L_1,L_2) (*or w.r.t.*(M_1,M_2)) *and if some line of* Γ *through one of its carriers* x_i (*here* L_i I x_i I M_i, i = 1,2) *is regular, then* Γ *satisfies axiom* (D).

Proof. Without loss of generality we may suppose that Γ satisfies axiom (D) w.r.t. (L_1,L_2). Let $L_j \in \{M_1,M_2\}^\perp$ with $L_1 \not\sim L_j \not\sim L_2$, $j \in J$ ($|J| = s$ if $x_1 \sim x_2$, and $|J| = s-1$ if $x_1 \not\sim x_2$). Then by hypothesis the triad (L_1,L_2,L_j) has a center M_j (clearly $M_1 \not\sim M_j \not\sim M_2$). Since $|\{M \in \{L_1,L_2\}^\perp \parallel L_1 \not\sim M \not\sim L_2\}| = |J|$, it is clear that Γ satisfies axiom (D) w.r.t. (M_1,M_2) if $L_j \neq L_k$ implies $M_j \neq M_k$, with $j,k \in J$. So suppose $M_j = M_k$ for distinct $j,k \in J$. Then for $i = 1$ or 2, (L_i,L_j,L_k) is a triad with two centers M_i and $M_j = M_k$. By hypothesis one of M_1,M_2,L_1,L_2 is regular. If either M_1 or M_2 is regular then the pair (L_j,L_k) is regular and hence the triad (L_i,L_j,L_k) must have 1+s centers, forcing $L_1 \sim M_2$, a contradiction. If L_i is regular, $i = 1$ or 2, the triad (L_i,L_j,L_k) also must have 1+s centers, giving a contradiction. □

Let x be any point of S. We say that S satisfies *axiom* $(D)'_x$ (respectively, *axiom* $(D)''_x$) provided the following holds : Let Γ be any broken grid whose lines are those of the triads (L_1,L_2,L_3) and (M_1,M_2,M_3), where $L_i \not\sim M_j$ iff $\{i,j\} = \{1,2\}$ and where x I L_1. Then Γ satisfies axiom (D) w.r.t. the pair (L_1,L_2) (respectively, w.r.t. the pair (M_1,M_2)). We say S satisfies *axiom* $(D)_x$ provided it satisfies both axiom $(D)'_x$ and axiom $(D)''_x$.

Then the following result is an immediate corollary of 5.3.6.

5.3.7. *Let S be a GQ of order* (s,t) *having a coregular point* x. *Then S satisfies* $(D)'_x$ *iff it satisfies* $(D)''_x$ *iff it satisfies* $(D)_x$.

5.3.8. (*J.A. Thas* [197]). *Let* $S = (P,B,I)$ *be a GQ of order* (s,t) *with* $s \neq t$, $s > 1$ *and* $t > 1$. *Then S is isomorphic to a* $T_3(O)$ *iff it has a coregular point* x_∞ *for which* $(D)'_{x_\infty}$ (*resp.*, $(D)''_{x_\infty}$) *is satisfied.*

Proof. We have already observed that $T_3(O)$ satisfies $(D)_{(\infty)}$. Conversely, let $S = (P,B,I)$ be a GQ of order (s,t) with $s \neq t$, $s > 1$ and $t > 1$, and suppose S has a coregular point x_∞ for which $(D)'_{x_\infty}$ (resp., $(D)''_{x_\infty}$) is satisfied. Then in fact S satisfies $(D)_{x_\infty}$. And since x_∞ is coregular we have $t > s$. If $s = 2$, then by 1.2.2 and 1.2.3 S is of order (2,4). So by 5.3.2 it must be that $S \cong Q(5,2) \cong T_3(O)$. We now assume $s > 2$.

Suppose that the triad of lines (L,L',L") has at least two centers N and N' where x_∞ I N. By the regularity of N, the lines L,L',L" are contained in a (proper) subquadrangle of order (s,1).

Now we consider the triad of lines (L,L',L") with a unique center N which is incident with x_∞. Let $\{L',L''\}^{\perp\perp} = \{L_0,L_1 = L', L_2 = L'',L_3,\ldots,L_s\}$, with $L_0 \sim L$, and for all $i \geqslant 1$ let $\{L,L_i\}^\perp = \{N_{i0} = N,N_{i1},\ldots,N_{is}\}$. Further, let $D(L,L',L") = \{y \in P \parallel y \text{ I } N_{ij},$ for some $i = 1,\ldots,s$ and some $j = 0,1,\ldots,s\}$. We have $|D(L,L',L")| = s^3+2s+1$.

Consider two points y_1, y_2 of $D(L,L',L")$ which are incident with a line V of S, and suppose that V does not contain the intersection z of L and N (in particular $L \neq V \neq N$). If $V \sim L$, then clearly V is some N_{ij}, and hence all points of V are contained in $D(L,L',L")$. If $V \sim N$ but $V \not\sim L$, then V is concurrent with some N_{ij}, $j \neq 0$, and belongs to $\{L,L_i\}^{\perp\perp}$. As all points on all lines of $\{L,L_i\}^{\perp\perp}$ belong to $D(L,L',L")$, all points of V are contained in $D(L,L',L")$. If $V \in \{L',L"\}^{\perp}$, then the s points of V not collinear with z are contained in $D(L,L',L")$. Now suppose that $V \not\sim L$, $V \not\sim N$, $V \notin \{L',L"\}^{\perp}$.

Evidently the point of V which is collinear with z is not contained in $D(L,L',L")$. Let $y_3 \text{ I } V$, $y_1 \neq y_3 \neq y_2$, $y_3 \not\sim z$. We shall prove that $y_3 \in D(L,L',L")$.

Let $y_1 \text{ I } N_{ik}$, $i \neq 0$, and $y_2 \text{ I } N_{j\ell}$, $j \neq 0$. Clearly $N_{ik} \neq N_{j\ell}$ and $L_i \neq L_j$. Now we have $N \sim L_i$, $N_{j\ell} \sim V$, $N_{j\ell} \sim L \sim N$, $V \sim N_{ik} \sim L_i$, $N_{ik} \sim L$, $N_{j\ell} \sim L_j \sim N$, and $x_\infty \text{ I } N \not\sim V$, $L_i \not\sim N_{j\ell}$. If $y_2 = N_{j\ell} \cap L_j$ or $y_1 = N_{ik} \cap L_i$, then trivially there is a unique line M which is concurrent with L_i, L_j and V. Suppose $y_2 \neq N_{j\ell} \cap L_j$ and $y_1 \neq N_{ik} \cap L_i$. Then (L_i,V,L) and $(N,N_{j\ell},N_{ik})$ are the two triads of a broken grid for which $(D)''_{x_\infty}$ guarantees that the triad (L_i,V,L_j) has a center M, which is unique because N is regular and $N \not\sim V$. Let N_3 be the line defined by $y_3 \text{ I } N_3 \sim L$. Since $V \notin \{L',L"\}^{\perp}$, we cannot have both $y_1 = N_{ik} \cap L_i$ and $y_2 = N_{j\ell} \cap L_j$. Without loss of generality we may suppose $y_2 \neq N_{j\ell} \cap L_j$. Then the triads $(N,M,N_{j\ell})$ and (L,V,L_j) give the lines of a broken grid with $N_3 \in \{L,V\}^{\perp}$ and N_3 not concurrent with any of $N,M,N_{j\ell}$. Hence by $(D)'_{x_\infty}$ there is a line W which is a center of (N,M,N_3). Clearly $z \not\text{I}\, W$. Since $y_3 \text{ I } N_3$, $N_3 \in \{L,W\}^{\perp}$ and $W \in \{N,M\}^{\perp} = \{L_i,L_j\}^{\perp\perp} = \{L',L"\}^{\perp\perp}$, we have $y_3 \in D(L,L',L")$. Hence V is incident with exactly s points of $D(L,L',L")$.

Let $P' = D(L,L',L") \cup D(L',L,L")$ and $P" = D(L,L',L") \cap D(L',L,L")$. We shall prove that $|P'| = s^3+s^2+s+1$ and $|P"| = s^3-s^2+3s+1$. Let z' be the point incident with N and L', and consider a point $y \in D(L,L',L")$ with $y \not\sim z'$. We show that $y \in D(L',L,L")$. Since the case $y \text{ I } L$ is trivial, we suppose that $y \not\text{I}\, L$. Let N_{ik} be the line through y meeting L. Then L_i is the line of $\{L',L"\}^{\perp\perp}$ which is concurrent with N_{ik}. If $L_i = L'$, then $N_{ik} \in \{L,L'\}^{\perp}$ and $y \in D(L',L,L")$ by definition. Now suppose that $L_i \neq L'$. Let $N_{ik} \text{ I } u \text{ I } L$, and let $u' \text{ I } L$ with $u' \notin \{z,u\}$. Further, let N' and V be defined by u' I N', N' ~ L', y I V, V ~ N'. If $V \in \{L',L''\}^{\perp}$, then y I L_i, and then clearly $y \in D(L',L,L'')$. So assume $V \notin \{L',L''\}^{\perp}$. Also $V \not\sim L$ and $V \not\sim N$. If we put $y_1 = y$, $y_2 = V \cap N'$, $L_j = L'$, then by the preceding paragraph $(L_i,V,L_j) = (L_i,V,L')$ has a unique center M. Note that $M \in \{L',L''\}^{\perp}$. Let $y_1' = V \cap M$ and $y_2' = y_2 = V \cap N'$. Clearly $y_1' = y_2'$ iff $L' \sim V$ iff $N' \cap L' \sim y$. Since $s > 2$, we may choose u' in such a way that $N' \cap L' \not\sim y$. Then we have $y_1' \neq y_2'$, $\{y_1', y_2'\} \subset D(L',L,L'')$, $V \not\sim L'$, $V \not\sim N$, $V \notin \{L,L''\}^{\perp}$, and $y \not\sim z'$. Now by the preceding paragraph $y \in D(L', L,L'')$. Since $D(L,L',L'')$ contains s^2-s points which are collinear with z' and not incident with L' or N, there holds $|P''| = s^3-s^2+3s+1$. It easily follows that $|P'| = s^3+s^2+s+1$.

Next let $p,p' \in P'$ with $p \sim p'$, $p \neq p'$. We shall prove that P' contains each element which is incident with the line pp'. There are several cases to be considered.

Let z or z' be incident with the line pp', say z I pp'. The case N = pp' is trivial. So suppose $N \neq pp'$. Since $p,p' \in D(L',L,L'')$ and z' $\not\mathrm{I}$ pp', it follows from a preceding paragraph that $D(L',L,L'')$ contains all elements incident with the line pp'.

Let z $\not\mathrm{I}$ pp', z' $\not\mathrm{I}$ pp', $pp' \sim N$. Since $p,p' \in D(L',L,L'')$ and z' $\not\mathrm{I}$ pp', the set $D(L',L,L'')$ contains each point incident with pp'.

Let $pp' \not\sim N$. Since $s > 2$, the line pp' contains points $w,w' (w \neq w')$ for which $z \not\sim w \not\sim z'$, $z \not\sim w' \not\sim z'$. Hence $w,w' \in P''$. If $pp' \sim L$ (resp., $pp' \sim L'$), then all points of pp' are contained in $D(L,L',L'')$ (resp., $D(L',L,L'')$). So assume $L \not\sim pp' \not\sim L'$. If w_1 is the point of pp' not contained in $D(L,L',L'')$, then $z \sim w_1$, so $z' \not\sim w_1$ and $w_1 \in D(L',L,L'')$. Hence P' contains each point incident with pp'.

Now from 2.3.1 it follows immediately that P' is the pointset of a subquadrangle S' of order (s,t'). Since $|P'| = (s+1)(s^2+1)$ we have $t' = s < t$, implying S' is proper. Consequently the lines L,L',L" are contained in a proper subquadrangle of order (s,t').

We have now proved that every centric triad of lines (L,L',L") having a center N incident with x_∞ is contained in a proper subquadrangle of order (s,t'). By 5.3.4 $s^2 = t$, the point x_∞ is 3-regular, and S is isomorphic to a $T_3(O)$. □

There is an easy corollary of 5.3.8, 3.3.3 and the note following 3.3.3 whose proof may be completed by the reader.

5.3.9. *Let* S *be a* GQ *of order* (s,t), *with* $s \neq t$, $s > 1$, $t > 1$.

(i) *If* s *is odd, then* $S \cong Q(5,s)$ *iff* S *contains a coregular point* x_∞ *for which* $(D)'_{x_\infty}$ *(resp.,* $(D)''_{x_\infty}$*) is satisfied.*

(ii) *If* s *is even, then* $S \cong Q(5,s)$ *iff all lines of* S *are regular and* S *contains a point* x_∞ *for which* $(D)'_{x_\infty}$ *(resp.,* $(D)''_{x_\infty}$*) is satisfied.*

In order to conclude this section dealing with characterizations of $T_3(O)$ and $Q(5,s)$, we introduce one more basic concept. Let $S = (P,B,I)$ be a GQ of order (s,t). If $B^{\perp\perp}$ is the set of all hyperbolic lines, i.e., the set of all spans $\{x,y\}^{\perp\perp}$ with $x \not\sim y$, then let $S^{\perp\perp} = (P,B^{\perp\perp},\in)$. For $x \in P$, we say S satisfies *property* $(A)_x$ if for any $M = \{y,z\}^{\perp\perp} \in B^{\perp\perp}$ with $x \in \{y,z\}^\perp$, and any $u \in cl(y,z) \cap (x^\perp - \{x\})$ with $u \notin M$, the substructure of $S^{\perp\perp}$ generated by M and u is a dual affine plane. The GQ S is said to satisfy *property* (A) if it satisfies $(A)_x$ for all $x \in P$. So the GQ S satisfies (A) if for any $M = \{y,z\}^{\perp\perp} \in B^{\perp\perp}$ and any $u \in cl(y,z) - (\{y,z\}^\perp \cup \{y,z\}^{\perp\perp})$, the substructure of $S^{\perp\perp}$ generated by M and u is a dual affine plane. The duals of $(A)_x$ and (A) are denoted by $(\hat{A})_L$ and $(\hat{A})$, respectively. If $(A)_x$ is satisfied for some regular point x, then the dual affine planes guaranteed to exist by $(A)_x$ are substructures of the dual net described in 1.3.1.

5.3.10. (*J.A. Thas* [204]). *Let* $S = (P,B,I)$ *be a* GQ *of order* (s,t), *with* $s \neq t$, $s > 1$ *and* $t > 1$. *Then* S *is isomorphic to a* $T_3(O)$ *iff* $(\hat{A})_L$ *is satisfied for all lines* L *incident with some coregular point* x_∞.

Proof. Let S be the GQ $T_3(O)$. Then it is easy to check that $(\hat{A})_L$ is satisfied for every line L incident with the coregular point (∞) of type (iii).

Now let S be a GQ of order (s,t), with $s \neq t$, $s \neq 1 \neq t$, and having a coregular point x_∞ such that $(\hat{A})_L$ is satisfied for all lines L incident with x_∞. We shall prove that $(D)''_{x_\infty}$ is satisfied. So suppose (L_1,L_2,L_3) and (M_1,M_2,M_3) are two triads of lines with x_∞ I L_1 and $L_i \not\sim M_j$ iff $\{i,j\} = \{1,2\}$. Let $M_4 \in \{L_1,L_2\}^\perp$ with $M_4 \not\sim M_i$, $i = 1,2,3$. We must show that the triad (M_1,M_2,M_4) is centric. Since L_1 is regular, any pair of noncon-current lines meeting L_1 is regular. Since (M_1,M_3) is regular, the line M_4 is an element of $cl(M_1,M_3)$. Because $M_1 \not\sim L_2$, we have $M_4 \notin \{M_1,M_3\}^{\perp\perp}$. Now consider the dual affine plane π generated by $\{M_1,M_3\}^{\perp\perp}$ and M_4 in the structure $\hat{S}^{\perp\perp} = (B,P^{\perp\perp},\in)$. Clearly $\{M_3,M_4\}^{\perp\perp}$ and $\{M_1,M_4\}^{\perp\perp}$ are lines of π. Since L_3 (resp., L_2) is an element of $\{M_1,M_3\}^\perp$ (resp., $\{M_3,M_4\}^\perp$), the point $L_3 \cap M_2$ (resp. $L_2 \cap M_2$) is incident with a line R (resp., R') of $\{M_1,M_3\}^{\perp\perp}$

(resp., $\{M_3,M_4\}^{\perp\perp}$). Then $\{R,R'\}^{\perp\perp}$ is a line of π. As any two lines of π intersect, the lines $\{M_1,M_4\}^{\perp\perp}$ and $\{R,R'\}^{\perp\perp}$ have an element R'' in common. Clearly $R'' \sim M_2$. If L_4 is the line which is incident with $M_2 \cap R''$ and concurrent with M_1, then by $R'' \in \{M_1,M_4\}^{\perp\perp}$, we have $L_4 \sim M_4$. Hence L_4 is a center of (M_1,M_2,M_4) and $(D)''_{x_\infty}$ is satisfied. Then by 5.3.8 S is isomorphic to a $T_3(O)$. □

There is an easy corollary.

5.3.11. *Let S be a GQ of order (s,t), $s \neq t$, $t > 1$.*

(i) *If $s > 1$, s odd, then S is isomorphic to $Q(5,s)$ iff $(\hat{A})_L$ is satisfied for all lines L incident with some coregular point x_∞.*

(ii) If s is even, then S is isomorphic to $Q(5,s)$ iff all lines of S are regular and $(\hat{A})_L$ is satisfied for all lines L incident with some point x_∞.

Proof. Left to the reader. □

We mention without proof one more result of interest which may turn out to be helpful in characterizing the GQ $Q(5,s)$.

5.3.12. (*J.A. Thas* [192]). *Suppose that the GQ $S = (P,B,I)$ of order (s,s^2), $s \neq 1$, has a subquadrangle $S' = (P',B',I')$ of order s with the property that every triad (x,y,z) of S' is 3-regular in S and $\{x,y,z\}^{\perp\perp} \subset P'$. Then $S' \cong Q(4,s)$ and S has an involution θ fixing P' pointwise.*

5.4. TALLINI'S CHARACTERIZATION OF H(3,s)

Let $S = (P,B,I)$ be a GQ of order (s,t), and let B^* be the set of all spans, i.e. let $B^* = \{\{x,y\}^{\perp\perp} \parallel x,y \in P,\ x \neq y\}$. Then $S^* = (P,B^*,\in)$ is a linear space in the sense of F. Buekenhout [27]. In order to have no confusion between collinearity in S and collinearity in S^*, points $x_1,x_2,\ldots$ of P which are on a line of S^* will be called S^*-collinear. A *linear variety* of S^* is a subset $P' \subset P$ such that $x,y \in P'$, $x \neq y$, implies $\{x,y\}^{\perp\perp} \subset P'$. If $P' \neq P$ and $|P'| > 1$, the linear variety is *proper*; if P' is generated by three points which are not S^*-collinear, P' is said to be a *plane* of S^*. Finally, if $L \in B$ with $x \mathrel{I} L \mathrel{I} y$ and $x \neq y$, then $\{x,y\}^{\perp\perp} \in B^*$ is denoted by L^*.

5.4.1. (*G. Tallini* [174]). *Let $S = (P,B,I)$ be a GQ of order (s,t), with $s \neq t$, $s > 1$ and $t > 1$. Then S is isomorphic to $H(3,s)$ iff*

(i) *all points of* S *are regular, and*

(ii) *if the lines* L *and* L' *of* B^* *are contained in a proper linear variety of* S^*, *then also the lines* $L^{\perp}$ *and* $L'^{\perp}$ *of* B^* *are contained in a proper linear variety of* S^*.

Proof. Let S be the classical GQ H(3,s). By 3.3.1 (ii) all points of S are regular, so (i) is satisfied. Let V be a proper linear variety of S^* containing at least three points x,y,z which are not S^*-collinear. Then x,y,z are noncollinear in PG(3,s) and V is contained in the plane π = xyz of PG(3,s). If V contains a line L of S, then it is clear that $|V| = s\sqrt{s}+s+1$ and that $V = \pi \cap H$. Now suppose that V does not contain a line of S. We shall show that V is an ovoid of S. Assume that the line M of S has no point in common with V. Since V is a linear variety of S^*, any point of M is collinear with 0 or 1 point of V. Hence $s+1 \geq |V|$. But V together with the lines of S^* contained in V is a 2-$(|V|, \sqrt{s}+1,1)$ design, which implies that $|V| \geq s+\sqrt{s}+1$. But then $s+1 \geq s+\sqrt{s}+1$, an impossibility. So V is an ovoid of S, and consequently $|V| = s\sqrt{s}+1$. It follows immediately that $V = \pi \cap H$. Now it is evident that the proper linear varieties of S^* which contain at least three points that are not S^*-collinear are exactly the plane intersections of the hermitian variety H. Clearly if the lines L and L' of S^* are contained in a plane of PG(3,s), then also $L^{\perp}$ and $L'^{\perp}$ are contained in a plane of PG(3,s). Hence (ii) is satisfied.

Now we consider the converse. Let S = (P,B,I) be a GQ of order (s,t) with $s \neq t$, $s > 1$ and $t > 1$. The proof is broken up into a sequence of steps, and to start with we assume only that S *satisfies* (i), *i.e. all points of* S *are regular.*

(a) Introduction and generalities.

By 1.3.6 (i) we have $s > t$. Let V be a proper linear variety of S^* that contains at least three points x,y,z which are not S^*-collinear. First, suppose that V contains L^* with $L \in B$, and assume $x \notin L^*$. If $u \sim x$ and u I L, then clearly V contains the proper linear variety $u^{\perp}$ of S^*. Suppose that $V \neq u^{\perp}$. Then V contains a subset M^* with $M \in B$ and $L^* \cap M^* = \phi$. The number of points on lines of S^* having a point in common with L^* and M^* equals $(st+1)(s+1) = |P|$. Hence V = P, a contradiction. So $u^{\perp} = V$ and $|V| = st+s+1$. Next, suppose that no two points of V are collinear in S. We shall show that V is an ovoid of S. Assume that the line M of S has no point in common with V. Since V is a linear variety of S^* and since every point of S

is regular, each point of M is collinear with 0 or 1 point of V. Hence $s+1 \geq |V|$. But V together with lines of S^* contained in V is a $2\text{-}(|V|,t+1,1)$ design, so that $|V| \geq t^2+t+1$. Hence $s \geq t^2+t$, an impossibility by Higman's inequality. So V is an ovoid of S, and consequently $|V| = st+1$.

Hence for a proper linear variety V of S^* which contains at least three non-S^*-collinear points, we have $|V| \in \{st+s+1, st+1\}$. If $|V| = st+s+1$, then $V = u^\perp$ for some $u \in V$; if $|V| = st+1$, then V is an ovoid of S.

Let V_1 and V_2 be two (distinct) proper linear varieties of S^* having at least three non-S^*-collinear points in common, and suppose that $|V_1| \leq |V_2|$. Since $V_1 \cap V_2$ is also a proper linear variety, we necessarily have $|V_1 \cap V_2| = st+1$, $|V_2| = st+s+1$. Hence the ovoid $V_1 \cap V_2$ is contained in some $u^\perp$, a patent impossibility. It follows that each three points which are not S^*-collinear are contained in at most one proper linear variety, and that each proper linear variety which contains at least three non-S^*-collinear points is a plane of S^*. If $|V| = st+s+1$, V will be referred to as an *absolute* plane; if $|V| = 1+st$, V will be referred to as a *nonabsolute* plane.

Now we introduce condition (ii)' : *every three non-S^*-collinear points are contained in a proper linear variety of S^**. If (ii)' is satisfied, then any hyperbolic line L of S is contained in 1+t absolute planes and s-t nonabsolute planes of S^*. This is easily seen by noticing that any plane containing L has exactly one point in common with M^*, with $M \in B$ and $L \cap M^* = \phi$.

Next we show that condition (ii) *implies condition* (ii)' . Suppose that (ii) is satisfied and that x,y,z are three non-S^*-collinear points. Clearly the lines $\{x,y\}^\perp$ and $\{x,z\}^\perp$ of S^* belong to the absolute plane $x^\perp$. By (ii) the lines $\{x,y\}^{\perp\perp}$ and $\{x,z\}^{\perp\perp}$ of S^* belong to a proper linear variety V of S^*, and hence $x,y,z \in V$. So S also satisfies (ii)'.

Condition (ii)' seems to be weaker than (ii), and we proceed as far as possible assuming only condition (ii)' (in addition to (i)).

(b) Let (ii)' be satisfied : the affine planes $\pi_x = (P_x, B_x)$, $x \in P$.

With (ii)' satisfied, let $x \in P$, let P_x be the set of hyperbolic lines of S containing x, and let B_x be the set of planes of S^* different from $x^\perp$ which contain x. If $L \in P_x$ and $V \in B_x$, then let $L\ I_x\ V$ iff $L \subset V$. It is clear that the incidence structure (P_x, B_x, I_x) (briefly (P_x, B_x) or π_x) is a $2\text{-}(s^2,s,1)$ design, i.e. an affine plane of order s. Let $x\ I\ M\ I\ y$, $x \neq y$. Then $y^\perp$ is a line of the affine plane π_x. So with M there corresponds lines of

π_x, and no two of them have a point of π_x in common. Hence these lines form a parallel class of π_x. The corresponding improper point of π_x is called *special*, and the lines of the parallel class are also called *special*. The special point defined by M, x I M, will also be denoted by M^*. We note that the special lines of π_x are exactly the absolute planes in B_x.

Let V be a nonabsolute plane of S^*, let $x \in V$, and let $y \notin V$ with $x \not\sim y$. If V' is a plane of S^* with $x,y \in V'$, then $V \cap V' = \{x\}$ iff V and V' are parallel lines of π_x. Since the point $\{x,y\}^{\perp\perp}$ of π_x belongs to just one line of π_x which is parallel to V, there is just one plane V' in S^* which contains x and y and is tangent to V at x. As we have $V \parallel V'$ in π_x and V is nonspecial, also V' is nonspecial, i.e. the plane V' is nonabsolute.

Further, let V be a nonabsolute plane with $x \notin V$. The points of V which are collinear (in S) with x are denoted $y_0,\ldots,y_t$. The hyperbolic lines containing x and a point of $V-\{y_0,\ldots,y_t\}$ are denoted by $M_1,\ldots,M_{st-t}$. The points $M_1,\ldots,M_{st-t}$ of π_x together with the t+1 special improper points of π_x form a set A of order st+1 of the projective completion $\bar{\pi}_x$ of π_x . Now it is an easy exercise to show that each line U of $\bar{\pi}_x$ intersects A in 0,1, or t+1 points (if U is the completion of a special line of π_x then it contains 1 or t+1 elements of A; if U is the completion of a nonspecial line of π_x then it contains 0,1 or t+1 elements of A). The lines U of $\bar{\pi}_x$ intersecting A in 1 point correspond to the absolute planes $y_0^\perp,\ldots,y_t^\perp$, and to the nonabsolute planes containing x and exactly one point of $V-\{y_0,\ldots,y_t\}$. By the preceding paragraph this number equals st+1. The number of lines U of $\bar{\pi}_x$ intersecting A in t+1 points equals $(st+1)st/t(t+1)$. Hence there are $s^2+s+1-(st+1)-(st+1)st/t(t+1) = s(s-t^2)/(t+1)$ lines in $\bar{\pi}_x$ having no point in common with A. Since this number is nonnegative and $s \leqslant t^2$, it must be that $s = t^2$ and every line U of $\bar{\pi}_x$ intersects A in 1 or t+1 points. Hence A is a unital [49] of $\bar{\pi}_x$.

Consequently, from (i) and (ii)' it follows that $s = t^2$, $|V| = t^3+t^2+1$ for an absolute plane, and $|V| = t^3+1$ for a nonabsolute plane. Finally, we shall show that two planes V and V' always intersect. If one of these planes is absolute, clearly $V \cap V' \neq \phi$. If V and V' are nonabsolute and $x \in V'-V$, then let A be the unital of $\bar{\pi}_x$ which corresponds to V. As the projective completion of the nonspecial line V' of π_x intersects A in 1 or t+1 (nonspecial) points, we have $|V \cap V'| = 1$ or t+1. We conclude that *any two planes of S intersect.*

(c) Let (ii)' be satisfied : bundles of planes .

If L is a line of S^*, then the set of all planes of S^* containing L is called the *bundle of planes with axis* L. That bundle is denoted by B_L, and $|B_L| = s+1$ (by one of the last paragraphs of (a)).

Let V_0 be a nonabsolute plane of S^*, and let $x \in V_0$. By considering the plane π_x we see that there are s-1 nonabsolute planes $V_1,\dots,V_{s-1}$ which are tangent to V_0 at x. The only absolute plane which is tangent to V_0 at x is $x^\perp = V_s$. Since $V_0,\dots,V_{s-1}$ are parallel lines of π_x, any two of the s+1 planes $V_0,\dots,V_s$ have only x in common. The set $\{V_0,\dots,V_s\}$ will be denoted by $\beta(V_0,x)$ and will be called a *bundle of mutually tangent planes*. Clearly $\beta(V_0,x) = \beta(V_i,x)$, $1 \leq i \leq s-1$. Further, two different bundles $\beta(V,x)$ and $\beta(V',x)$ of mutually tangent planes (at x) have only the plane $x^\perp$ in common.

Let β be a bundle, and let p be a point not belonging to two elements of β . If β has axis L with L a line of S^*, then p is contained in one element of β. If β is a bundle of mutually tangent planes $V_0,\dots,V_s$, then $|V_0 \cup \dots \cup V_s| = (t^2+1)(t^3+1) = |P|$, and consequently here too p is contained in just one element of β.

Finally, let us consider two planes V and V'. If $|V \cap V'| = t+1$, then the planes V and V' are contained in just one bundle, namely β_L with $L = V \cap V'$. Now let $V \cap V' = \{x\}$, and assume that V is nonabsolute. Then $\beta(V,x)$ is the only bundle containing V and V'.

(d) Let (ii) be satisfied : conjugacy .

Let L be a line of S^* . If $L^\perp$ is a subset of the plane V, we say that L is *conjugated to* V or that V is *conjugated to* L. The planes conjugated to L, with $L = M^*$ and $M \in B$, are the absolute planes containing L. The lines (of S^*) conjugated to the absolute plane $x^\perp$, are the lines (of S^*) which contain the point x. Let V be a plane and let $p \notin V$. Then the hyperbolic line $(p^\perp \cap V)^\perp$ is the only line of S^* which contains p and is conjugated to V. If V is nonabsolute we have $(p^\perp \cap V)^\perp \cap V = \phi$. Hence in this case the lines of S^* conjugated to V constitute a partition of P-V.

We say that the planes V and V' (not necessarily distinct) are *conjugated* if there is a line L in S^* for which $L \subset V$ and $L^\perp \subset V'$. It follows that a plane is conjugated to itself iff it is absolute. The set of planes conjugated to the plane V is called the *net of planes conjugated to* V, and it is denoted by $\tilde{V}$. If $V = x^\perp$, then $\tilde{V}$ consists of all planes through x, implying $|\tilde{V}| = t^4+t^2+1$. Conversely, if all elements of $\tilde{V}$ have a common point x, then

$V = x^{\perp}$. Since the absolute planes conjugated to the plane V are the planes $x^{\perp}$ with $x \in V$, we clearly have $\tilde{V} = \tilde{V}'$ iff $V = V'$.

Let V and V' be distinct conjugated planes, and let $p \in V'-V$. The unique line of S^* which contains p and is conjugated to V is denoted by N. We shall prove that $N \subset V'$. Since V and V' are conjugated, there is a line L in S^* such that $L \subset V$ and $L^{\perp} \subset V'$. If $N = L^{\perp}$, we have $N \subset V'$. So assume $N \neq L^{\perp}$. Since $N^{\perp}$ and L are contained in the plane V, by (ii) the lines (of S^*) N and $L^{\perp}$ are also contained in some plane V". Since $N \neq L^{\perp}$, clearly $p \notin L^{\perp}$. Then p and $L^{\perp}$ are contained in a unique plane, so that V' = V" and hence $N \subset V'$. This shows that if V is nonabsolute, the lines of S^* which are conjugated to V and contain at least one point of V' are all contained in V'-V and constitute a partition of V'-V.

Let V and V' be nonabsolute and conjugated. By the previous paragraph t+1 divides $|V'-V|$. Since $|V'-V| \in \{t^3, t^3-t\}$, it must be that $|V'-V| = t^3-t$, i.e. $V' \cap V$ is a hyperbolic line. It easily follows that for a nonabsolute plane V we have $|\tilde{V}| = t^2(t^2-t+1)(t^2-t)/(t^2-t)+t^2(t^2-t+1)(t+1)/t^2 = t^4+t^2+1$, where $t^2(t^2-t+1)$ is the number of hyperbolic lines conjugated to V; t^2-t (resp., t+1) is the number of nonabsolute (resp., absolute) planes containing an hyperbolic line; and t^2-t (resp., t^2) is the number of hyperbolic lines conjugated to V which are contained in a nonabsolute (resp., absolute) plane conjugated to V. Hence for any plane V of S^* we have $|\tilde{V}| = t^4+t^2+1$.

A plane V and a bundle β of planes are called *conjugated* iff V is conjugated to all elements of β . And we say that bundles β and β' are *conjugated* iff each element of β is conjugated to each element of β'.

Consider the bundle β_L with axis L. If the plane V is conjugated to β_L, then V contains all points x with $x^{\perp} \in \beta_L$. Consequently V contains $L^{\perp}$. Conversely, if V contains $L^{\perp}$, then it is evident that V is conjugated to β_L. It follows that there is just one bundle conjugated to β_L, namely $\beta_{L^{\perp}}$. Now consider a bundle $\beta(V,x)$ of mutually tangent planes. Let V' be a plane which is conjugated to the bundle $\beta(V,x)$. Then V' is conjugated to $x^{\perp}$, implying $x \in V'$. Now consider a plane V' which contains x and is conjugated to V. We shall show that V' is conjugated to $\beta(V,x)$.

If $V' = x^{\perp}$, then clearly V' is conjugated to $\beta(V,x)$. So assume that V' is nonabsolute. Let V" be a nonabsolute plane which contains x and is conjugated to V'. Suppose that $V \neq V"$ and $|V \cap V"| = t+1$. If $y \in V \cap V"$ and $y \notin V'$, then the hyperbolic line N containing y and conjugated to V' is a subset of V and V". Since both V and V" contain N and x, we have V = V", a contradiction. Hence $V \cap V'' \subset V'$. Let R be a hyperbolic line in one of the absolute planes containing $V \cap V''$, with $x \in R$ and $R \neq V \cap V''$. Then R is contained in just one plane V"' which is conjugated to V'. Clearly V"' is not absolute and does not contain $V \cap V''$. Since V and V" are not parallel in the affine plane π_x, at least one of these planes, say V, is not parallel to V"'. Hence $V \cap V'''$ is a hyperbolic line containing x. So V and V"' are distinct planes which contain the hyperbolic line $V \cap V''' \not\subset V'$ and are conjugated to V', a contradiction. Consequently, we have V = V" or $|V \cap V''| = 1$. Hence $V'' \in \beta(V,x)$. Since the number of nonabsolute planes containing x and conjugated to V' equals $(t^2(t^2-t+1)-t^2)/(t^2-t) = t^2 = |\beta(V,x)| - 1$, it is clear that V' is conjugated to $\beta(V,x)$.

Now consider the t^2+1 planes which contain x and are conjugated to V. By the preceding paragraph these planes are mutually tangent at x and hence form a bundle $\beta(V',x)$. So there is just one bundle which is conjugated to $\beta(V,x)$, namely the bundle $\beta(V',x)$, where V' is arbitrary nonabsolute plane which contains x and is conjugated to V.

If β is a bundle, then the unique bundle conjugated to β is denoted by $\tilde{\beta}$. We note that $\tilde{\beta}$ consists of all planes conjugated to β .

(e) Let (ii) be satisfied : some more properties of conjugacy.

Consider two planes V and V", with $|V \cap V''| = t+1$, which are conjugated to the nonabsolute plane V'. We shall prove that $V \cap V' \cap V'' = \phi$. The case where V or V" is absolute is easy. So assume that V and V" are nonabsolute. If $V \cap V' \cap V'' = \{x\}$, then there are at least two planes which contain $V \cap V'' \not\subset V'$ and are conjugated to V', a contradiction. If $V \cap V'' \subset V'$, then by one of the last paragraphs of (d) we obtain a contradiction. We conclude that always $V \cap V' \cap V'' = \phi$.

Let the plane V be conjugated to the planes V' and V", $V' \neq V''$. We shall prove that V is conjugated to the bundle β containing V' and V".

If $V' = x'^{\perp}$ and $V'' = x''^{\perp}$, then V contains x' and x", and consequently also $(V' \cap V'')^{\perp}$. By (d) V is conjugated to the bundle β defined by V' and V".

Now let $V' = x'^{\perp}$ and let V" be nonabsolute. If $x' \notin V''$, then V contains x' and the hyperbolic line L containing x' and conjugated to V". Since $L = (V' \cap V'')^{\perp}$, the plane V is conjugated to the bundle β . If $x' \in V''$, then by the last part of (d) V belongs to the bundle $\beta(W,x)$, where W is an arbitrary nonabsolute plane which contains x' and is conjugated to V". Since the bundles $\beta(V'',x')$ and $\beta(W,x')$ are conjugated, it is clear that V is conjugated to the bundle $\beta = \beta(V'',x')$ containing V" and V'.

Finally, let V' and V" be nonabsolute. If $V = x^\perp$, then $x \in V' \cap V''$, so V is conjugated to the bundle β. So assume V is nonabsolute. If $|V' \cap V''| = t+1$ and $x \in V' \cap V''$, then by the first paragraph of (e) we have $x \notin V$. The hyperbolic line L which contains x and is conjugated to V belongs to V' and V", hence must be the line $V' \cap V''$ of S^*. Consequently V contains $(V' \cap V'')^\perp$ which means that V is conjugated to the bundle β defined by V' and V". If $V' \cap V'' = \{x\}$, then $x \in V$, since otherwise V' and V" would contain the hyperbolic line containing x and conjugated to V. By the last part of (d) the plane V is conjugated to the bundle $\beta(V',x)$, i.e. to the bundle β containing V' and V".

This completes the proof that a plane V is conjugated to a bundle β iff it is conjugated to at least two planes of β. Now it is clear that for any two planes V and V', the set $\tilde{V} \cap \tilde{V}'$ is the unique bundle which is conjugated to the bundle defined by V and V'.

Let β be a bundle and let V be a plane which is not in $\tilde{\beta}$. We shall prove that $|\beta \cap \tilde{V}| = 1$. If we should have $|\beta \cap \tilde{V}| > 1$, then by the preceding paragraph β is conjugated to V, and hence $V \in \tilde{\beta}$, a contradiction. So we have only to show that $|\beta \cap \tilde{V}| \geq 1$. First suppose that there is a point x belonging to all elements of β and not contained in V. Then x is contained in just one hyperbolic line L conjugated to V. By (d) L is contained in an element V' of β. Clearly $V' \in \tilde{V} \cap \beta$. Next, we suppose that every element x common to all elements of β is also contained in V. If $\beta = \beta(V'',x)$, then $x^\perp \in \beta$ and $x^\perp \in \tilde{V}$. If β has axis M^*, $M \in B$, then $M^* \subset V$ and $V \in \tilde{\beta}$, a contradiction. So assume that β has axis N, with N a hyperbolic line. There is a plane contained in β and $\tilde{V}$ iff there is a plane conjugated to $\tilde{\beta}$ and V iff there is a plane conjugated to V, $y^\perp$ and $z^\perp$, with $y,z \in N$, $y \neq z$. If V is absolute, then it is clear that there is a plane conjugated to V, $y^\perp$ and $z^\perp$. If V is nonabsolute, then we have to show that $|\tilde{\beta}' \cap \tilde{z}^\perp| \geq 1$, with $\beta' = \beta(V,y)$. Since $z^\perp \notin \beta'$, this immediately follows from one of the preceding cases.

Finally, we give some easy corollaries : (1) if β is a bundle and V a plane not in β, then $|\tilde{V} \cap \tilde{\beta}| = 1$; (2) if β is a bundle and V is a plane such that $\beta \not\subset \tilde{V}$, then $|\beta \cap \tilde{V}| = 1$; and (3) if β is a bundle and V a plane not in β, then V and β are contained in exactly one net of planes.

(f) Let (ii) be satisfied : the space PG(3,s) and the final step.

Let $\mathcal{V}$ be the set of all planes, let $\mathcal{B}$ be the set of all bundles of planes, and let $\widetilde{\mathcal{V}}$ be the set of all nets of planes. An element $\widetilde{V}$ of $\widetilde{\mathcal{V}}$ is called "incident" with the element V' (resp., β) of $\mathcal{V}$ (resp., $\mathcal{B}$) iff $V' \in \widetilde{V}$ (resp., $\beta \in \widetilde{V}$); an element β of $\mathcal{B}$ is called "incident" with an element V of $\mathcal{V}$ iff $V \in \mathcal{B}$. We shall prove that for such an "incidence" the ordered triple $(\widetilde{\mathcal{V}},\mathcal{B},\mathcal{V})$ is the structure of points, lines and planes of the projective space PG(3,s). So we have to check that the following properties are satisfied :

(1) Every two nets are "incident" with exactly one bundle.

(2) For every two planes there is exactly one bundle which is "incident" with both of them.

(3) For every plane V and every bundle β which is not "incident" with V, there is exactly one net "incident" with V and β.

(4) Every bundle β and every net $\widetilde{V}$ which is not "incident" with β are "incident" with exactly one plane.

(5) There exist four nets which are not "incident" with a same plane, and for every bundle β there are exactly s+1 nets "incident" with β .

In (e) we have proved that for any two planes V and V', the set $\widetilde{V} \cap \widetilde{V}'$ is always a bundle, and hence (1) is satisfied. By the last paragraph of (c) also (2) is satisfied. By Corollary (3) in the last part of (e) Condition (3) is satisfied. Condition (4) is satisfied by Corollary (2) in the last part of (e). Let N and $N^{\perp}$ be hyperbolic lines and let $x,y \in N$, $x \neq y$, and $z, u \in N^{\perp}$, $z \neq u$. Then it is clear that the nets $\widetilde{x}^{\perp}, \widetilde{y}^{\perp}, \widetilde{z}^{\perp}, \widetilde{u}^{\perp}$ are not "incident" with a same plane. Finally, the number of nets "incident" with a bundle β equals the number of planes conjugated to β, hence equals $|\widetilde{\beta}| = s+1$ by (d). Hence $(\widetilde{\mathcal{V}},\mathcal{B},\mathcal{V})$ is the structure of points, lines and planes of PG(3,s).

Now we consider the following bijection $\theta : \widetilde{\mathcal{V}} \to \mathcal{V}$, $\widetilde{V} \mapsto V$. It is clear that the images of the s+1 nets "incident" with a bundle β are the planes which are "incident" with the bundle $\widetilde{\beta}$. Moreover, if $\widetilde{W}$ is "incident" with $\widetilde{V}^{\theta} = V$, then $\widetilde{V}$ is "incident" with $\widetilde{W}^{\theta} = W$. So θ defines a polarity of the projective space $(\widetilde{\mathcal{V}},\mathcal{B},\mathcal{V})$. The "absolute" [49] elements in $\widetilde{\mathcal{V}}$ and $\mathcal{V}$ for the polarity θ are the nets $\widetilde{x}^{\perp}$ and the planes $x^{\perp}$ (the absolute planes), $x \in P$. The "totally isotropic" [49] bundles are the bundles β for which $\beta = \widetilde{\beta}$, i.e. the bundles β_{M^*}, with $M \in B$. With respect to the "incidence" in $(\widetilde{\mathcal{V}},\mathcal{B},\mathcal{V})$, the "absolute" nets and "totally isotropic" bundles form a classical GQ $\bar{S}$.

Since $\tilde{x}^{\perp}$ is "incident" with β_{M^*}, $M \in B$, iff x I M in S, the classical GQ $\bar{S}$ is isomorphic to S. As there are $(s^3+1)(s+1)$ "absolute" nets, the polarity θ is unitary, implying that $\bar{S}$ is the GQ H(3,s). We conclude that $S \cong H(3,s)$. □

Remark : *Using* 5.3.5 *F. Mazzocca and D. Olanda* [106] *proved that a* GQ S *of order* (s^2,s), $s \neq 1$, *is isomorphic to* H(3,s) *iff the following conditions are satisfied :*

(i) *all points of* S *are regular,*

(ii)' *every three non-*S^**-collinear points are contained in a proper linear variety of* S^*, *and*

(iii) *for every point* x *and every triad* (y,z,u) *with center* x, *the affine plane* π_x *has an affine Baer subplane having only special improper points and containing the elements* $y^{\perp} \cap z^{\perp}, z^{\perp} \cap u^{\perp}, u^{\perp} \cap y^{\perp}$.

5.5. CHARACTERIZATIONS OF $H(4,q^2)$

5.5.1. (*J.A. Thas*[194]). *A* GQ S *of order* (s,t), $s^3 = t^2$ *and* $s \neq 1$, *is isomorphic to the classical* GQ H(4,s) *iff every hyperbolic line has at least* $\sqrt{s}+1$ *points.*

Proof. By 3.3.1 (iii) every hyperbolic line of H(4,s), $s = q^2$, has exactly q+1 points. Conversely, suppose that $S = (P,B,I)$ is a GQ of order (s,t), $s^3 = t^2$ and $s \neq 1$, for which $|\{x,y\}^{\perp\perp}| \geq \sqrt{s}+1$ for all $x,y \in P$ with $x \not\sim y$. To show that $S \cong H(4,q^2)$ will require a rather lengthy sequence of steps.

(a) Introduction and generalities .

By 1.4.2 (ii) we have $(|\{x,y\}^{\perp\perp}|-1)t \leq s^2$ if $x \not\sim y$. Hence $|\{x,y\}^{\perp\perp}| \leq \sqrt{s}+1$ if $x \not\sim y$. It follows that each hyperbolic line has exactly $\sqrt{s}+1$ points. Now, again by 1.4.2 (ii), every triad (x,y,z), $z \notin cl(x,y)$, has exactly $\sqrt{s}+1$ centers. Let u and v be two centers of the triad (x,y,z), $z \notin cl(x,y)$. Then $\{x,y\}^{\perp\perp} \cup \{z\} \subset \{u,v\}^{\perp}$, implying $\{u,v\}^{\perp\perp} \subset \{x,y,z\}^{\perp}$. As $|\{u,v\}^{\perp\perp}| = |\{x,y,z\}^{\perp}|$, we have $\{u,v\}^{\perp\perp} = \{x,y,z\}^{\perp}$. It follows that for any triad (x,y,z), $z \notin cl(x,y)$, the set $\{x,y,z\}^{\perp}$ is a hyperbolic line, and that {x,y,z} is contained in just one trace (of a pair of noncollinear points).

(b) The subquadrangles $S_{L,M}$.

Clearly each point of S is semiregular, so each point satisfies property (H). By 2.5.2, for any pair (L,M) of nonconcurrent lines the set $L^* \cup M^*$, with $L^* = \{x \in P \parallel x \text{ I } L\}$ and $M^* = \{x \in P \parallel x \text{ I } M\}$, is contained in a subquadrangle $S_{L,M}$ of order $(s,\sqrt{s})$. The pointset of this subquadrangle is the union of the sets $\{x,y\}^{\perp\perp}$ with x I L and y I M. Now we shall prove that the set $L^* \cup M^*$ is contained in just one proper subquadrangle of S.

Let $S_{L,M} = (P',B',I')$, and consider an arbitrary proper subquadrangle $S'' = (P'',B'',I'')$ of S for which $L^* \cup M^* \subset P''$. By 2.3.1 the structure $S''' = (P' \cap P'', B' \cap B'', I' \cap I'')$ is a subquadrangle of order (s,t''') of S. Since $t \neq s^2$, by 2.2.2 (vi) we have $S''' = S_{L,M}$ and $S''' = S''$. Hence $S'' = S_{L,M}$.

If $S_{L,M} = (P',B',I')$ and $x,y \in P'$, $x \neq y$, then we show that $\{x,y\}^{\perp\perp} \subset P'$. Clearly we have $\{x,y\}^{\perp\perp} \subset P'$ if $x,y \in P'$, $x \neq y$, and $x \sim y$. So assume $x \not\sim y$. Let x I U and y I V, with U and V nonconcurrent lines of $S_{L,M}$. The pointset $U^* \cup V^*$ is contained in the proper subquadrangles $S_{L,M}$ and $S_{U,V}$, implying that $S_{L,M} = S_{U,V}$ by the preceding paragraph. Since $\{x,y\}^{\perp\perp}$ is contained in the pointset of $S_{U,V}$, we also have $\{x,y\}^{\perp\perp} \subset P'$.

(c) $\underline{S_{L,M}} \cong H(3,s)$.

Next we shall prove that each subquadrangle $S_{L,M} = (P',B',I')$ is isomorphic to $H(3,s)$. The first step is to show that each point of $S_{L,M}$ is regular in $S_{L,M}$. Let y be a point of $S_{L,M}$ which is not collinear with x . Since every point of $\{x,y\}^{\perp\perp} \subset P'$ is collinear with every point of $\{x,y\}^{\perp} \cap P'$, it follows that the hyperbolic line of $S_{L,M}$ defined by x and y has at least $\sqrt{s}+1$ points. As the order of $S_{L,M}$ is $(s,\sqrt{s})$, the span of x and y in $S_{L,M}$ has exactly $\sqrt{s}+1$ points, implying (x,y) is regular. Consequently each point of $S_{L,M}$ is regular in $S_{L,M}$.

Let $S^*_{L,M} = (P',B'^*,\in)$ be the linear space introduced in 5.4. Notations and terminology of 5.4 will be used. In order to apply Tallini's theorem we must prove that if the lines L and L' of B'^* are contained in a proper linear variety of $S^*_{L,M}$, then also the lines $L^{\perp} \cap P'$ and $L'^{\perp} \cap P'$ of B'^* are contained in a proper linear variety of $S^*_{L,M}$. Consider three points x,y,z in P' which are not $S^*_{L,M}$-collinear. These points are contained in an absolute plane of $S^*_{L,M}$ iff $z \in cl(x,y)$ (cf. 5.4.1 (a)). If $z \notin cl(x,y)$, then $P' \cap T$, with T the unique trace of S containing x,y,z, is a proper linear variety of $S^*_{L,M}$ which contains x,y,z. Since by 5.4.1 (a) the nonabsolute plane $P' \cap T$ of $S^*_{L,M}$ contains $s\sqrt{s}+1$ points, we have $T \subset P'$. So condition (ii)' of 5.4.1 is satisfied, and moreover the absolute planes of $S^*_{L,M}$ are the traces of S which are contained in P', i.e. the traces of S

containing at least three non-$\mathcal{S}^*_{L,M}$-collinear points of P'. Now let L and L' be two lines of B'^* which are contained in a proper linear variety (i.e. a plane) of $\mathcal{S}^*_{L,M}$. There are four cases.

(i) If L and/or L' consists of the points incident with a line of $\mathcal{S}$, then clearly $L^\perp \cap P'$ and $L'^\perp \cap P'$ are contained in an absolute plane.

(ii) If L and L' are hyperbolic lines which are contained in the absolute plane $z^\perp \cap P'$, then $L^\perp \cap P'$ and $L'^\perp \cap P'$ contain z. Since (ii)' is satisfied the hyperbolic lines $L^\perp \cap P'$ and $L'^\perp \cap P'$ are contained in a plane of $\mathcal{S}^*_{L,M}$.

(iii) Now suppose that L and L' are hyperbolic lines which are contained in a nonabsolute plane of $\mathcal{S}^*_{L,M}$ and which have a nonvoid intersection. If $L \cap L' = \{z\}$, then clearly $L^\perp \cap P'$ and $L'^\perp \cap P'$ are contained in $z^\perp \cap P'$.

(iv) Finally, let L and L' be disjoint hyperbolic lines which are contained in a nonabsolute plane T of $\mathcal{S}^*_{L,M}$. If $T = \{x,y\}^\perp$, then it is easy to show that $\{x,y\}^{\perp\perp} \cap P' = \phi$ and $\{x,y\}^{\perp\perp} = L^\perp \cap L'^\perp$. Let d be a point of $L^\perp \cap P'$. Then d is collinear with no point of L', and consequently must be collinear with $\sqrt{s}+1$ points of $L'^\perp$. Let e be one of these points, and denote by V the line of $\mathcal{S}$ which is incident with d and e. Further, let R (resp., N) be the line of $\mathcal{S}$ which is incident with x (resp., y) and concurrent with V. If there is a point h with R I h I N, then h I V, implying h is collinear with d, e and all points of $\{x,y\}^{\perp\perp}$. Since h is collinear with at least $\sqrt{s}+2$ points of $L^\perp$ (resp., $L'^\perp$), we have $h \in L$ (resp., $h \in L'$). Hence $L \cap L' \neq \phi$, a contradiction. Hence R and N are not concurrent, and we may consider the subquadrangle $\mathcal{S}_{R,N} = (P'',B'',I'')$ of $\mathcal{S}$. Then we have $L^\perp \cup L'^\perp \subset P''$. Clearly $P' \cap P''$ is a linear variety of $\mathcal{S}^*_{L,M}$ which contains $L^\perp \cap P'$ and $L'^\perp \cap P'$. Since $|P'| = |P''|$ and $\{x,y\}^{\perp\perp} \cap P' = \phi$, we have $P' \cap P'' \neq P'$, implying $L^\perp \cap P'$ and $L'^\perp \cap P'$ are contained in a proper linear variety of $\mathcal{S}^*_{L,M}$.

Hence condition (ii) of 5.4.1 is satisfied, and by Tallini's theorem $\mathcal{S}_{L,M} \cong H(3,s)$.

(d) <u>Threespaces and bundles of threespaces.</u>

The sets $x^\perp$, $x \in P$, will be called *absolute threespaces*, and the pointsets of the subquadrangles $\mathcal{S}_{L,M}$ will be called *nonabsolute threespaces*. The traces of $\mathcal{S}$ will be called *nonabsolute planes*, the sets $cl(x,y) \cap z^\perp$, with $x \not\sim y$ and $z \in \{x,y\}^\perp$, will be called *absolute planes*, and the sets L^*, with $L \in B$ and $L^* = \{x \in P \parallel x \text{ I } L\}$ will be called *totally absolute planes*. We shall show that a plane T which is not totally absolute and a point x,

$x \notin T$, are contained in exactly one threespace. As usual, there are a few cases to consider.

(i) $T = \{y,z\}^{\perp}$, $y \not\sim z$, and $x \in cl(y,z)$. Let x be collinear with the point u of $\{y,z\}^{\perp\perp}$. Then $u^{\perp}$ is the unique absolute threespace which contains $\{x\} \cup T$. If there is a nonabsolute threespace E containing $\{x\} \cup T$, then E contains u and thus also all points of $u^{\perp}$. Hence the point u of the subquadrangle $S_{L,M}$ with pointset E is incident with t+1 lines of $S_{L,M}$, a contradiction.

(ii) $T = \{y,z\}^{\perp}$, $y \not\sim z$, and $x \notin cl(y,z)$. Then there is no absolute threespace containing $\{x\} \cup T$. The point x is collinear with $\sqrt{s}+1$ points $u_0,\ldots,u_{\sqrt{s}}$ of T. Let $w \in T-\{u_0,\ldots,u_{\sqrt{s}}\}$ and let L be the line incident with w and concurrent with the line V through x and u_0. If $x \; I \; M \; I \; u_1$, then the pointset E of $S_{L,M}$ contains x,w,u_0,u_1. Since $w \notin \{u_0,\ldots,u_{\sqrt{s}}\} = \{u_0,u_1\}^{\perp\perp}$, we have $T \subset E$ by (c). So $\{x\} \cup T \subset E$. Since any threespace through $\{x\} \cup T$ contains all points which are incident with L and M, by (b) there is just one threespace which contains x and T.

(iii) T is an absolute plane with $T \subset y^{\perp}$ and $x \sim y$. Then clearly $y^{\perp}$ is the only threespace through x and T.

(iv) T is an absolute plane with $T \subset y^{\perp}$ and $x \not\sim y$. Then $\{x\} \cup T$ is not contained in an absolute threespace. Let $T = L_0^* \cup \ldots \cup L_{\sqrt{s}}^*$ with $L_i \in B$ and $L_i^* = \{z \in P \parallel z \; I \; L_i\}$, and let M be the line which is incident with x and concurrent with L_0. Any threespace through $\{x\} \cup T$ contains L_1^* and all points incident with M. Hence the pointset of $S_{L_1,M}$ is the unique threespace which contains $\{x\} \cup T$.

Now we introduce bundles of threespaces.

A *nonabsolute bundle* is the set of all threespaces which contain a given nonabsolute plane T. From the first part of (d) it follows that a nonabsolute bundle contains $\sqrt{s}+1$ absolute threespaces and $s-\sqrt{s}$ nonabsolute threespaces. The $\sqrt{s}+1$ absolute threespaces are the threespaces $x^{\perp}$, with $x \in T^{\perp}$.

The set of all threespaces which contain a given absolute plane is called an *absolute bundle*. From the first part of (d) it follows that an absolute bundle contains one absolute threespace and s nonabsolute threespaces.

The set of all absolute threespaces which contain a given totally absolute plane is called a *totally absolute bundle*. A totally absolute bundle contains s+1 absolute threespaces.

Hence each bundle of threespaces contains exactly s+1 elements.

(e) <u>The incidence structure $\mathcal{D} = (\mathcal{E},\mathcal{B},\in)$.</u>

The set of all threespaces is denoted by $\mathcal{E}$ and the set of all bundles by $\mathcal{B}$.

We shall show that the incidence structure $\mathcal{D} = (\mathcal{E},\mathcal{B},\in)$ is a 2-$(s^4+s^3+s^2+s+1,s+1,1)$ design.

The number of absolute threespaces equals $(s+1)(s^2\sqrt{s}+1)$, and the number of nonabsolute threespaces equals $(s\sqrt{s}+1)(s^2\sqrt{s}+1)s^4/(\sqrt{s}+1)(s\sqrt{s}+1)s^2 = s^4-s^3\sqrt{s}+s^3-s^2\sqrt{s}+s^2$. Hence $|\mathcal{E}| = s^4+s^3+s^2+s+1$.

In (d) we noticed that each element of $\mathcal{B}$ contains $s+1$ elements of $\mathcal{E}$.

Let E be a nonabsolute threespace. The number of threespaces which intersect E in an absolute plane or a nonabsolute plane is equal to $s|\{$nonabsolute planes in E$\}| + s|\{$absolute planes in E$\}|$. By (c) and the first part of the proof of Tallini's theorem (5.4.1), this number of threespaces is equal to $s|\{$set of planes in PG(3,s)$\}| = s(s^3+s^2+s+1) = |\mathcal{E}|-1$. It follows that two given threespaces E and E', with E nonabsolute, are contained in exactly one bundle. Now consider two absolute threespaces $x^\perp$ and $y^\perp$. If $x \sim y$, then clearly $x^\perp$ and $y^\perp$ are contained in the totally absolute bundle defined by the totally absolute plane L^*, with x I L I y, and in no other bundle. If $x \not\sim y$, then $x^\perp$ and $y^\perp$ are contained in the nonabsolute bundle defined by the nonabsolute plane $\{x,y\}^\perp$, and in no other bundle. Hence any two threespaces are contained in a unique bundle.

We conclude that $\mathcal{D}$ is a 2-$(s^4+s^3+s^2+s+1,s+1,1)$ design.

(f) An interesting property of bundles.

Let β be the bundle defined by the plane T. Intersect β with a nonabsolute threespace E, with $T \not\subset E$ if T is not totally absolute (i.e. if T does not consist of all points incident with some line in β). By (c) E may be considered as a nonsingular hermitian variety of a PG(3,s). If $\beta = \{E_0,\ldots,E_s\}$, then by (e) the sets $E_0 \cap E,\ldots,E_s \cap E$ are plane intersections of the hermitian variety E. Consequently $(E_i \cap E) \cap (E_j \cap E) = T \cap E$ $(i \neq j)$ is a point, a hyperbolic line, or an L^* with $L \in B$.

If $T \cap E$ is not a point, then clearly the planes $E_0 \cap E,\ldots,E_s \cap E$ (no one of which is totally absolute) are exactly the intersections of the hermitian variety E and the s+1 planes of PG(3,s) through $T \cap E$. Hence with β there corresponds a bundle of planes in PG(3,s).

Next let $T \cap E = \{x\}$. Since $E_0 \cap E,\ldots,E_s \cap E$ are s+1 plane intersections of the hermitian variety E, having in pairs only the point x in common, their planes $\pi_0,\ldots,\pi_s$ in PG(3,s) have a tangent line of E (in PG(3,s)) in common. Consequently $\pi_0,\ldots,\pi_s$ constitute a bundle of planes in PG(3,s).

(g) <u>$\mathcal{D}$ is the design of points and lines of PG(4,s).</u>

Let E be a threespace common to the bundles β and β', $\beta \neq \beta'$. Let E_1 and E_2 be elements of β with E, E_1, E_2 distinct, and let E_1' and E_2' be elements of β' with E, E_1', E_2' distinct. The bundle containing E_1 and E_1' (resp., E_2 and E_2') is denoted by E_1E_1' (resp., E_2E_2'). We have to show [225] that $E_1E_1' \cap E_2E_2' \neq \phi$.

Suppose that β (resp., β') is defined by the plane T (resp., T'). Now we prove that $T \cap T' \neq \phi$. Evidently $T \cup T' \subset E$. If E is nonabsolute, then T and T' are plane intersections of a nonsingular hermitian variety of PG(3,s), and hence $T \cap T' \neq \phi$. Now let E be the threespace $x^{\perp}$, $x \in P$. If at least one of β, β' is absolute or totally absolute, then clearly $T \cap T' \neq \phi$. So we assume that β and β' are nonabsolute. Then we have $T = \{x,z\}^{\perp}$ and $T' = \{x,z'\}^{\perp}$ for some z and z'. If x,z,z' are contained in a nonabsolute plane $\{u,v\}^{\perp}$, then clearly $\{u,v\}^{\perp\perp} = T \cap T'$; if x,z,z' are contained in an absolute plane, then there is exactly one point w which is collinear with x,z,z', and $T \cap T' = \{w\}$. Hence in all cases $T \cap T' \neq \phi$.

Let $w \in T \cap T'$ and let E' be a nonabsolute threespace which does not contain w. By (c) E' may be considered as a nonsingular hermitian variety of a PG(3,s). The planes $E \cap E'$, $E_1 \cap E'$, $E_2 \cap E'$, $E_1' \cap E'$, $E_2' \cap E'$ are plane intersections of the hermitian variety E'. Let $\pi, \pi_1, \pi_2, \pi_1', \pi_2'$ be the respective planes of PG(3,s) in which these intersections are contained. By (f) π, π_1, π_2 (resp., π, π_1', π_2') are elements of a bundle γ (resp., γ') of planes in PG(3,s). We notice that π, π_1, π_2 (resp., π, π_1', π_2') are distinct. We shall now show that $\gamma \cap \gamma' = \{\pi\}$. If $\pi' \in \gamma \cap \gamma'$, then $\pi' \cap E'$ is the intersection of E' and an element R (resp., R') of β (resp., β'). Since R and R' both contain $\pi' \cap E'$ and w ($w \notin \pi' \cap E'$), we have R = R', implying R = R' = E. Hence $\pi' = \pi$, i.e. $\gamma \cap \gamma' = \{\pi\}$. Clearly the bundles of planes $\pi_1\pi_1'$ and $\pi_2\pi_2'$ have a plane π_3 in common. The plane intersection $\pi_3 \cap E'$ of the hermitian variety E' is the intersection of E' with an element E_3 (resp., E_3') of the bundle E_1E_1' (resp., E_2E_2'). Since w is a point of each element of the bundle E_1E_1' (resp., E_2E_2'), the threespace E_3 (resp., E_3') is the unique threespace containing $\pi_3 \cap E'$ and w. Consequently $E_3 = E_3'$, implying $E_1E_1' \cap E_2E_2' \neq \phi$.

This completes the proof that $\mathcal{D} = (\mathcal{E}, \mathcal{B}, \in)$ is the design of points and lines of a PG(n,s). Since $|\mathcal{E}| = s^4+s^3+s^2+s+1$ and $|\beta| = s+1$ for all $\beta \in \mathcal{B}$, it must be that n = 4.

(h) The final step.

Let $\hat{P}$ be the set of all absolute threespaces, and let $\hat{B}$ be the set of all totally absolute bundles. Then $\hat{S} = (\hat{P},\hat{B},\in)$ is a GQ of order $(s,s\sqrt{s})$ which is isomorphic to S. The elements of $\hat{P}$ are points of $PG(4,s)$ and the elements of $\hat{B}$ are lines of $PG(4,s)$. So by the theorem of F. Buekenhout and C. Lefèvre (cf. Chapter 4) $\hat{S}$ is a classical GQ. Since $t = s\sqrt{s}$, clearly $\hat{S} \cong H(4,s)$. This completes the proof that $S \cong H(4,s)$. □

We now turn to a characterization of $H(4,s)$ in terms of linear spaces. Let $S = (P,B,I)$ be a GQ of order (s,t), and let $S^* = (P,B^*,\in)$, with $B^* = \{\{x,y\}^{\perp\perp} \parallel x,y \in P \text{ and } x \neq y\}$, be the corresponding linear space. Recall (cf. 5.4) that points of P which are on a line of S^* are called S^*-collinear, and that any linear variety of S^* generated by three non-S^*-collinear points is called a plane of S^*.

5.5.2. (*S.E. Payne and J.A. Thas* [213]). *Let S have order* (s,t) *with* $1 < s^3 \leqslant t^2$. *Then S is isomorphic to* $H(4,s)$ *iff each trace* $\{x,y\}^\perp$, $x \not\sim y$, *is a plane (of S^*) which is generated by any three non-S^*-collinear points in it.*

Proof. Let S be the classical GQ $H(4,s)$, and consider a trace $\{x,y\}^\perp$, $x \not\sim y$. Let u,v,w be three non-S^*-collinear points in $\{x,y\}^\perp$ and call T the plane of S^* generated by u,v,w. Suppose that $T \neq \{x,y\}^\perp$ and let $z \in \{x,y\}^\perp - T$. Consider a line L through z which is incident with no point of $\{x,y\}^{\perp\perp}$. If $z' \mathrm{I} L$, $z' \neq z$, then z' is collinear with the $\sqrt{s}+1$ points of a span in $\{x,y\}^\perp$. Since T is a plane of S^* and $z \notin T$, z' is collinear with at most one point of T. Hence $s \geqslant |T|$, a contradiction since $|T| \geqslant s+\sqrt{s}+1$ (note that T is the pointset of a 2-$(|T|,\sqrt{s}+1,1)$ design). Consequently $\{x,y\}^\perp$ is the plane T of S^*.

Conversely, let $S = (P,B,I)$ be a GQ of order (s,t) with $1 < s^3 \leqslant t^2$, and suppose that each trace $\{x,y\}^\perp$, $x \not\sim y$, is a plane of S^* which is generated by any three non-S^*-collinear points of it. Let $u,v \in z^\perp$, $u \not\sim v$, and note that $|\{u,v\}^{\perp\perp}| < t+1$, since $s < t$ (cf. 1.3.6). The number of traces T for which $\{u,v\}^{\perp\perp} \subset T \subset z^\perp$ is denoted by α. Let M be a line of S incident with z that has no point in common with $\{u,v\}^{\perp\perp}$, and let w, $w \neq z$, be a point incident with M. If two traces T_1 and T_2 could contain $\{u,v\}^{\perp\perp}$ and w, then $T_1 \cap T_2$ $(\neq T_1)$ would contain the plane of S^* generated by u,v,w, a contradiction. Hence $\{u,v\}^{\perp\perp}$ and w are contained in at most one T, implying that $\alpha \leq s$. It follows that in $\{u,v\}^{\perp}$ there are at most s hyperbolic lines containing z, and by 1.4.2 (ii) each such hyperbolic line has at most $s^2/t \leq t^{1/3}$ points different from z. Hence $|\{u,v\}^{\perp}-\{z\}| = t \leq st^{1/3} \leq t$, implying $s^3 = t^2$ and each hyperbolic line in $\{u,v\}^{\perp}$ containing z has exactly $1+\sqrt{s}$ points. Now it is clear that each span has exactly $1+\sqrt{s}$ points. By 5.5.1 $S \cong H(4,s)$. □

5.6. ADDITIONAL CHARACTERIZATIONS

5.6.1. (*J.A. Thas* [195]). *Let* S *have order* (s,t) *with* $s \neq 1$. *Then* $|\{x,y\}^{\perp\perp}| \geq s^2/t+1$ *for all* x,y, *with* $x \not\sim y$, *iff one of the following occurs* :

(i) $t = s^2$,

(ii) $S \cong W(s)$,

(iii) $S \cong H(4,s)$.

Proof. If one of the three conditions holds, then clearly $|\{x,y\}^{\perp\perp}| \geq s^2/t+1$ for all x,y with $x \not\sim y$ (cf. 3.3.1).

Conversely, let $S = (P,B,I)$ be a GQ of order (s,t), $s \neq 1$, for which $|\{x,y\}^{\perp\perp}| \geq s^2/t+1$ for all x,y, with $x \not\sim y$. On the other hand, by 1.4.2 (ii) we have $|\{x,y\}^{\perp\perp}| \leq s^2/t+1$ for all $x,y,x \not\sim y$. Hence $|\{x,y\}^{\perp\perp}| = s^2/t+1$ for all x,y, with $x \not\sim y$. If $s = t$, then all points of S are regular and by 5.2.1 $S \cong W(s)$. From $|\{x,y\}^{\perp\perp}| \leq t+1$, $x \not\sim y$, it follows that $s \leq t$. So we now assume that $s < t$. By 1.4.2 (ii), each triad (x,y,z), $z \notin cl(x,y)$, has exactly $1+t/s$ centers. Hence each point of S is semiregular. By 2.5.2 we have $t = s^2$ or $s^3 = t^2$. In the latter case every hyperbolic line has exactly $1+\sqrt{s}$ points. By 5.5.1 we have $S \cong H(4,s)$, and the theorem is proved. □

5.6.2. (*J.A. Thas* [195], *J.A. Thas and S.E. Payne* [213]). *In the* GQ S *of order* (s,t) *each point has property* (H) *iff one of the following holds* :

(i) *each point is regular,*

(ii) *each hyperbolic line has exactly two points,*

(iii) $S \cong H(4,s)$.

Proof. If one of (i), (ii), (iii) holds, then clearly each point has property (H) (cf. 1.6.1 and 3.3.1).

Conversely, assume that each point of the GQ S has property (H). By 2.5.1 we must have one of the following : (i) each point is regular, (ii) all hyperbolic lines have exactly two points, or (iii)' $s^3 = t^2 \neq 1$ and each hyperbolic line has $1+\sqrt{s}$ points. By 5.5.1 (iii)' implies (iii). □

5.6.3. (*J.A. Thas* [195], *J.A. Thas and S.E. Payne* [213]). *Let* S *be a* GQ *of order* (s,t) *in which each point is semiregular. Then one of the following occurs :*

(i) $s > t$ *and each point is regular,*

(ii) $s = t$ *and* $S \cong W(s)$,

(iii) $s = t$ *and each point is antiregular,*

(iv) $s < t$ *and each hyperbolic line has exactly two points,*

(v) $S \cong H(4,s)$.

Proof. With the given hypotheses on S , by 2.5.2 we have one of the following : (i) $s > t$ and each point is regular, (ii)' $s = t$ and each point is regular, (iii) $s = t$ and each point is antiregular, (iv) $s < t$ and each hyperbolic line has exactly two points, or (v)' $s^3 = t^2 \neq 1$ and each hyperbolic line has $\sqrt{s}+1$ points. But (ii)' implies (ii) by 5.2.1 and (v)' implies (v) by 5.5.1 . □

5.6.4. (*J.A. Thas* [195]). *In a* GQ S *of order* (s,t) *all triads* (x,y,z) *with* $z \notin cl(x,y)$ *have a constant number of centers iff one of the following occurs :*

(i) *all points are regular,*

(ii) $s^2 = t$,

(iii) $S \cong H(4,s)$.

Proof. If we have one of (i), (ii), (iii), then all triads (x,y,z), $z \notin cl(x,y)$, have a constant number of centers (cf. 1.2.4, 1.4.2 and 3.3.1).

Conversely, suppose that all triads (x,y,z), $z \notin cl(x,y)$, have a constant number of centers. Also, assume that not all points are regular and that $s^2 \neq t$. Then there is an hyperbolic line $\{x,y\}^{\perp\perp}$ with p+1 points, $p < t$. By 1.4.2 (ii) we have $pt = s^2$ and the number of centers of the triad (x,y,z), $z \notin cl(x,y)$, equals $1+t/s$. From $pt = s^2$ and $p < t$ it follows that $s < t$. From $s^2 \neq t$, it follows that $p \neq 1$. Moreover, since $1+t/s > 1$, each point of S is semiregular. Now by 5.6.3 we conclude that $S \cong H(4,s)$. □

5.6.5. (*J.A. Thas* [195]). *The* GQ S *of order* (s,t), $s > 1$, *is isomorphic to one of* W(s) , Q(5,s) *or* H(4,s) *iff for each triad* (x,y,z) *with* $x \notin cl(y,z)$ *the set* $\{x\} \cup \{y,z\}^{\perp}$ *is contained in a proper subquadrangle of order* (s,t').

Proof. If $S \cong W(s)$, $S \cong Q(5,s)$, or $S \cong H(4,s)$, then it is easy to show that each set $\{x\} \cup \{y,z\}^\perp$, where (x,y,z) is a triad with $x \notin cl(y,z)$, is contained in a proper subquadrangle of order (s,t') (cf. Section 3.5). Note that in $W(s)$ there is no triad (x,y,z) with $x \notin cl(y,z)$.

Conversely, suppose that for each triad (x,y,z) with $x \notin cl(y,z)$ the set $\{x\} \cup \{y,z\}^\perp$ is contained in a proper subquadrangle of order (s,t'). If there is no triad (x,y,z) with $x \notin cl(y,z)$, then for each pair (y,z), $y \not\sim z$, all points of S belong to $cl(y,z)$, from which it follows easily that (y,z) is regular and $s = t$ or $s = 1$. By hypothesis $s \neq 1$, so from 5.2.1 $S \cong W(s)$.

Now assume that $S \not\cong W(s)$, so there is a triad (x,y,z) with $x \notin cl(y,z)$. Let $S' = (P',B',I')$ be a proper subquadrangle of order (s,t') for which $\{x\} \cup \{y,z\}^\perp \subset P'$. Since S' contains a set $\{y,z\}^\perp$ consisting of $t+1$ points, no two of which are collinear, we have $t \leqslant st'$ (cf. 1.8.1). Since S' is a proper subquadrangle of S we have $t \geqslant st'$ (2.2.1). Hence $st' = t$ and $\{y,z\}^\perp$ is an ovoid of S'. So x is collinear with exactly $t'+1 = 1+t/s$ points of $\{y,z\}^\perp$, implying $s \leqslant t$. It follows that each triad (x,y,z) with $x \notin cl(y,z)$ has exactly $1+t/s$ centers. By 5.6.4 all points are regular, or $s^2 = t$, or $S \cong H(4,s)$. If all points of S are regular, then $s = 1$ or $s \geqslant t$ (1.3.6). Thus $s = t$, and by 5.2.1 $S \cong W(s)$, a contradiction. Hence $t = s^2$ or $S \cong H(4,s)$.

Assume $t = s^2$ with $s > 2$, and consider a centric triad of lines (L,M,N) with center N'. Let $x \mathrel{I} N'$, $x \not\mathrel{I} L$, $x \not\mathrel{I} M$, $x \not\mathrel{I} N$, $y \mathrel{I} N$, $y \not\mathrel{I} N'$, $z \mathrel{I} M$, $z \not\mathrel{I} N'$, where (x,y,z) is a triad (since $s > 2$, the points x,y,z exist). Let $\{x,y,z\}^\perp = \{u_0,\ldots,u_s\}$. Then $u_i \not\mathrel{I} L$, $u_i \not\mathrel{I} M$, $u_i \not\mathrel{I} N$, and $u_i \not\mathrel{I} N'$, $i = 0,\ldots,s$. Moreover, no point u_i is collinear with the point u defined by $N' \mathrel{I} u \mathrel{I} L$. Hence there is at least one point u' which is incident with L and is collinear with at least two points u_i, u_j. A proper subquadrangle S' of order (s,t') which contains $\{u\} \cup \{u_i,u_j\}^\perp$ contains u,u',x,y,z. Hence S' contains L,M,N. By 5.3.5 we have $S \cong Q(5,s)$.

Finally, let $s = 2$ and $t = 4$. Then by 5.3.2, $S \cong Q(5,2)$. □

5.6.6. (*J.A. Thas* [195]). *Let S be a GQ of order (s,t) for which not all points are regular. Then S is isomorphic to $Q(4,s)$, with s odd, to $Q(5,s)$ or to $H(4,s)$ iff each set $\{x\} \cup \{y,z\}^\perp$, where (x,y,z) is a centric triad with $x \notin cl(y,z)$, is contained in a proper subquadrangle of order (s,t').*

Proof. If we have one of $S \cong Q(4,s)$ with s odd, $S \cong Q(5,s)$, or $S \cong H(4,s)$, then it is easy to show that each set $\{x\} \cup \{y,z\}^\perp$ with (x,y,z) a centric triad and $x \notin cl(y,z)$ is contained in a proper subquadrangle of order (s,t') (cf. Section 3.5).

Conversely, suppose that for each centric triad (x,y,z) with $x \notin cl(y,z)$ the set $\{x\} \cup \{y,z\}^{\perp}$ is contained in a proper subquadrangle of order (s,t'). By the proof of the preceding theorem we have $st' = t$, and x is collinear with exactly $1+t/s$ points of $\{y,z\}^{\perp}$. Hence each centric triad (x,y,z) with $x \notin cl(y,z)$ has exactly $1+t/s$ (> 1) centers. So all points of S are semi-regular. By 5.6.3 we have one of the possibilities (iii) $s = t$ and each point is antiregular, (iv) $s < t$ and each hyperbolic line has exactly two points, or (v) $S \cong H(4,s)$.

Suppose that we have one of the cases (iii) or (iv). Then each hyperbolic line has exactly two points. Let (x,y,z) be a centric triad (since not all points are regular, we have $t \neq 1$, so that such a triad exists), and let $S' = (P',B',I')$ be a proper subquadrangle of order $(s,t') = (s,t/s)$ containing x and $\{y,z\}^{\perp}$. The $1+t/s$ centers of (x,y,z) are denoted by $u_0,u_1,\ldots,u_{t/s}$. Consider a point $z' \in (\{u_0,u_1\}^{\perp}-\{x\}) \cap P'$. Notice that $z' \notin \{y,z\}$, since $y,z \notin P'$. Now let $S'' = (P'',B'',I'')$ denote a proper subquadrangle of order $(s,t/s)$ containing $\{x\} \cup \{z',z\}^{\perp}$. As $z' \notin P''$, we have $S' \neq S''$. By 2.3.1 the structure $S''' = (P' \cap P'', B' \cap B'', I' \cap I'')$ is a proper subquadrangle of order (s,t''') of S' (and S''), or all the lines of $B' \cap B''$ are incident with x and $P' \cap P''$ consists of all points incident with these lines. First assume that S''' is a proper subquadrangle of S'. By 2.2.2 (vi) we have $t = s^2$. In this case each triad is centric, so each set $\{u\} \cup \{v,w\}^{\perp}$, with (u,v,w) a triad, is contained in a proper subquadrangle of order $(s,t/s) = (s,s)$. Then by 5.6.5 $S \cong Q(5,s)$.

Next, assume that for each choice of z' all elements of $B' \cap B''$ are incident with x and $P' \cap P''$ consists of all points incident with these lines. By 2.2.1 the point z' is collinear with exactly $1+t's = 1+t$ points of S'', i.e. every line incident with z' contains a point of S''. It follows easily that $|B' \cap B''| = 1+t/s$. Consequently the lines of S' which are incident with x coincide with the lines of S'' which are incident with x. Let L be a line of S' which is incident with x. Since $\{y,z\}^{\perp}$ (resp., $\{z,z'\}^{\perp}$) is an ovoid of S' (resp., S''), the line L is incident with one point p (resp., p') of $\{y,z\}^{\perp}$ (resp., $\{z,z'\}^{\perp}$). But S has no triangles, so $p = p'$. Consequently z' is collinear with the $1+t/s$ centers $u_0,\ldots,u_{t/s}$ of (x,y,z). Suppose that $s \neq t$. Then x,y,z,z' are centers of the triad (u_0,u_1,u_2). Since we have t/s

choices for the point z', the triad (u_0,u_1,u_2) has at least $3+t/s$ centers, a contradiction since each centric triad has exactly $1+t/s$ centers. So we have $s = t$, and moreover each point is antiregular by 5.6.3. Hence s is odd (cf. 1.5.1). Now we consider two lines V and V', with $V \not\sim V'$. Let u I V, u' I V', $u \not\sim u'$, and let $w \in \{u,u'\}^{\perp}$ with $w \not{I}\ V$ and $w \not{I}\ V'$. Further, let N be a line concurrent with V and V' for which $u \not{I}\ N$ and $u' \not{I}\ N$. The point u" is defined by $w \sim u''$ I N. Since all points are antiregular, the triad (u,u',u") has exactly two centers w and w'. If N I z I V', then the set $\{z\} \cup \{w,w'\}^{\perp}$ is contained in a proper subquadrangle S' of order $(s,t/s) = (s,1)$. Clearly V, V', N are lines of this subquadrangle S' of order (s,1). Hence the pair (V,V') is regular. It follows that all lines of S are regular. From the dual of 5.2.1 it follows that the GQ S is isomorphic to Q(4,s). □

We now give a characterization due to F. Mazzocca and D. Olanda in terms of matroids.

A finite *matroid* [234] is an ordered pair (P,M) where P is a finite set, where elements are called points, and M is a closure operator which associates to each subset X of P a subset $\bar{X}$ (the *closure* of X) of P, such that the following conditions are satisfied :

(i) $\bar{\phi} = \phi$, and $\overline{\{x\}} = \{x\}$ for all $x \in P$.

(ii) $X \subset \bar{X}$ for all $X \subset P$.

(iii) $X \subset \bar{Y} \Rightarrow \bar{X} \subset \bar{Y}$ for all $X, Y \subset P$.

(iv) $y \in \overline{X \cup \{x\}}$, $y \notin \bar{X} \Rightarrow x \in \overline{X \cup \{y\}}$ for all $x,y \in P$ and $X \subset P$.

The sets $\bar{X}$ are called the *closed sets* of the matroid (P,M). It is easy to prove that the intersection of closed sets is always closed. A closed set C has *dimension* h if h+1 is the minimum number of points in any subset of C whose closure coincides with C. The closed sets of dimension one are the lines of the matroid.

5.6.7. (*F. Mazzocca and D. Olanda* [107]). *Suppose that* $S = (P,B,I)$ *is a* GQ *of order* (s,t), $s > 1$ *and* $t > 1$, *and that* P *is the pointset and* $B^* = \{\{x,y\}^{\perp\perp} \parallel x,y \in P$ *and* $x \neq y\}$ *is the lineset of some matroid* (P,M). *If moreover all sets* $x^{\perp}$, $x \in P$, *are closed sets of the matroid* (P,M), *then we have one of the following possibilities :* $S \cong W(s)$, $S \cong Q(4,s)$, $S \cong H(4,s)$, $S \cong Q(5,s)$, *or all points of* S *are regular,* $s = t^2$ *and* S *satisfies condition* (ii)' *introduced in the proof of Tallini's characterization* (5.4.1) *of* H(3,s).

Proof. First of all we prove that $\dim x^\perp = (\dim P)-1$ for all $x \in P$, and that $\dim \{x,y\}^\perp = (\dim P)-2$ for all $x,y \in P$ with $x \not\sim y$. Let $Y = \overline{x^\perp \cup \{z\}}$, with z a point of $P-x^\perp$. Clearly Y contains $x^\perp \cup z^\perp$ and $\{z,x\}^{\perp\perp}$. Choose a point u not contained in $x^\perp \cup z^\perp \cup \{x,z\}^{\perp\perp}$. Since $u \notin \{x,z\}^{\perp\perp}$, we have $\{x,z\}^\perp \not\subset u^\perp$. Hence there is a line V incident with u for which the points u' and u'' defined respectively by $x \sim u'$ I V and $z \sim u''$ I V are distinct. It follows that $\{u',u''\}^{\perp\perp} \subset Y$, implying $u \in Y$. Consequently $Y = P$, i.e. $\dim x^\perp = (\dim P)-1$. Now let $x,y \in P$ with $x \not\sim y$. Since $x^\perp$ and $y^\perp$ are closed, also the set $x^\perp \cap y^\perp = \{x,y\}^\perp$ is closed. Clearly we have $x^\perp = \overline{\{x,y\}^\perp \cup \{x\}}$, so that $\dim \{x,y\}^\perp = (\dim x^\perp)-1 = (\dim P)-2$. It is now immediate that $\dim P \geqslant 3$.

Suppose that not all points of S are regular, and consider a set $\overline{\{x\} \cup \{y,z\}^\perp}$, where (x,y,z) is a centric triad with $x \notin cl(y,z)$. By 2.3.1 the set $\overline{\{y,z\}^\perp \cup \{x\}}$ is the pointset of a subquadrangle S' of order (s,t') of S. Since $\dim (\{y,z\}^\perp \cup \{x\}) = (\dim P)-1$, it follows that $\overline{\{y,z\}^\perp \cup \{x\}} \neq P$. Hence S' is a proper subquadrangle of S. By 5.6.6 we have one of $S \cong Q(4,s)$ and s odd, $S \cong Q(5,s)$, or $S \cong H(4,s)$.

Now we suppose that all points of S are regular. If $s = t$, then by 5.2.1 we have $S \cong W(s)$ (which is equivalent to $S \cong Q(4,s)$ if s is even). So assume $s \neq t$. Let x,y,z be three points of S which are not on one line of the matroid (P,M). Then $\dim \overline{\{x,y,z\}} = 2 < \dim P$. Now it is clear that $\overline{\{x,y,z\}}$ is a proper linear variety of the linear space $S^* = (P,B^*,\in)$. Hence S satisfies condition (ii)' introduced in Tallini's characterization (5.4.1) of $H(3,s)$. Finally, by 5.4.1 (b) the parameters of S satisfy $s = t^2$. $\square$

Note : The first paragraph of this proof is due to F. Mazzocca and D. Olanda The remainder is due to the authors and represents a considerable shortening of the original proof.

We conclude this section and this chapter with a fundamental characterization of all classical and dual classical GQ with $s > 1$ and $t > 1$ due to J.A. Thas [204].

We remind the reader of properties (A) and ($\hat{A}$) introduced in the paragraph preceding 5.3.10. Let $B^{\perp\perp}$ be the set of all hyperbolic lines of the GQ $S = (P,B,I)$, and let $S^{\perp\perp} = (P,B^{\perp\perp},\in)$. We say that S satisfies *property* (A) if for any $M = \{y,z\}^{\perp\perp} \in B^{\perp\perp}$ and any $u \in cl(y,z)-(\{y,z\}^\perp \cup \{y,z\}^{\perp\perp})$ the substructure of $S^{\perp\perp}$ generated by M and u is a dual affine plane. The dual of (A) is denoted by $(\hat{A})$.

5.6.8. (*J.A. Thas* [204]). *Let* $S = (P,B,I)$ *be a* GQ *of order* (s,t), *with* $s > 1$ *and* $t > 1$. *Then* S *is a classical or a dual classical* GQ *iff it satisfies one of the conditions* (A) *or* $(\hat{A})$.

Proof. It is an exercise both interesting and not difficult to check that a classical or dual classical GQ with $s > 1$ and $t > 1$ satisfies one of the conditions (A) or $(\hat{A})$.

Conversely, assume that the GQ $S = (P,B,I)$ of order (s,t), $s > 1$ and $t > 1$, satisfies condition (A). We shall first prove that also property (H) is satisfied. To that end, consider a triad (u,y,z) for which $u \in cl(y,z)-\{y,z\}^{\perp\perp}$. Let π be the dual affine plane generated by $\{y,z\}^{\perp\perp}$ and u in $S^{\perp\perp}$. Evidently $\{z,u\}^{\perp\perp}$ is a line of π. In π the point y is not collinear with exactly one point of $\{z,u\}^{\perp\perp}$, i.e. in S the point y is collinear with exactly one point of $\{z,u\}^{\perp\perp}$. Hence $y \in cl(z,u)$, and (H) is satisfied. By 5.6.2 we have one of the following : (i) each point is regular, (ii) each hyperbolic line has exactly two points, or (iii) $S \cong H(4,s)$.

Now assume that $S \not\cong H(4,s)$. If $|\{y,z\}^{\perp\perp}| = 2$ for all $y,z \in P$ with $y \not\sim z$, then for any $M = \{y,z\}^{\perp\perp} \in B^{\perp\perp}$ and any $u \in cl(y,z)-\{y,z\}^{\perp}$ with $u \notin M$ (such a u exists since $s > 1$), the substructure of $S^{\perp\perp}$ generated by M and u has 3 points and consequently is not a dual affine plane, a contradiction. Hence all points of S are regular. If $s = t$, then by 5.2.1 $S \cong W(s)$. If $s \neq t$, then by dualizing 5.3.11 we obtain $S \cong H(3,s)$.

We have proved that if S satisfies (A), then S is isomorphic to one of $W(s)$, $H(3,s)$, $H(4,s)$. Hence if S (of order (s,t) with $s > 1$ and $t > 1$) satisfies one of the conditions (A) or $(\hat{A})$, then it is isomorphic to one of $H(4,s)$, the dual of $H(4,t)$, $W(s)$, $Q(4,s)$, $H(3,s)$, or $Q(5,s)$. □

6 Generalized quadrangles with small parameters

6.1. s = 2

Let $S = (P,B,I)$ be a GQ of order $(2,t)$, $2 \leqslant t$. By 1.2.2 and 1.2.3 we know $t = 2$ or $t = 4$. In either case, by 1.3.4 (iv) it is immediate that all lines are regular, and in case $t = 4$ all points are 3-regular. As was noted in 5.2.3 and 5.3.2 the GQ of orders (2,2) and (2,4) are unique up to isomorphism. Nevertheless it seems worthwhile to consider briefly an independent construction for these two examples, the first of which was apparently first discovered by J.J. Sylvester [171].

A *duad* is an unordered pair $ij = ji$ of distinct integers from among 1,2,...,6. A *syntheme* is a set $\{ij,k\ell,mn\}$ of three duads for which i,j,k,ℓ,m,n are distinct. It is routine to verify the following.

6.1.1. *Sylvester's syntheme-duad geometry with duads playing the role of points, synthemes playing the role of lines, and containment as the incidence relation, is the (unique up to isomorphism) GQ of order* (2,2) , *which is denoted* W(2).

It is also routine to check the following.

6.1.2. *For each integer* i, $1 \leqslant i \leqslant 6$, *the five duads* ij $(j \neq i)$ *form an ovoid of* W(2) . *These are all the ovoids of* W(2) *and any two have a unique point in common.*

The symmetric group S_6 acts naturally as a group of collineations of W(2). That S_6 is the full group of collineations also follows without too much effort. Since there is a unique GQ of order 2, it is clear that W(2) is self-dual. In fact it is self-polar. For example, it is easy to construct a polarity with the following absolute point-line pairs :
$1j \leftrightarrow \{1j,[j-1][j+1],[j-2][j+2]\}$, where $2 \leqslant j \leqslant 6$, and $[k]$ means k is to be reduced modulo 5 to one of 2,3,...,6. A complete description of the polarity may then be worked out using the following observation. Each point (resp., line)

is regular, and the set of absolute points (resp., lines) forms an ovoid (resp., spread). Hence each nonabsolute point (resp., line) is the unique center of a unique triad of absolute points (resp., lines). So if π is the polarity, and if u is the center of the triad (x,y,z) of absolute points, then u^π must be the unique center of the triad (x^π, y^π, z^π) of absolute lines. For example, the nonabsolute point 35 is the center of the triad (12,14,16) of absolute points, whose images under π are {12,63,54}, {14,35,26}, and {16,52,43}, respectively. This triad of absolute lines has the unique center {63,14,52}, implying that π : 35 $\leftrightarrow$ {63,14,52}.

Since W(2) $\cong$ Q(4,2) is a subquadrangle of Q(5,2), we may extend the above description of W(2) to obtain the unique GQ of order (2,4). In addition to the duads and synthemes given above for W(2), let 1,2,...,6 and 1',2',...,6' denote twelve additional points, and let {i,ij,j'}, $1 \leqslant i, j \leqslant 6$, $i \neq j$, denote thirty additional lines. It is easy to verify the following.

6.1.3. *The twenty-seven points and forty-five lines just constructed yield a representation of the unique* GQ *of order* (2,4).

H.Freudenthal [63] has written an interesting essay that contains an elementary account of many basic properties of these quadrangles, as well as references to their connections with classical objects such as the twenty-seven lines of a general cubic surface over an algebraically closed field.

6.2. s = 3

Applying 1.2.2 and 1.2.3 to those t with $3 \leqslant t \leqslant 9$, we find that $t \in \{3,5,6,9\}$. After some general considerations, each of these possibilities will be considered in turn.

Let x,y be fixed, noncollinear points of S, and let K_i be the set of points z for which (x,y,z) is a triad with exactly i centers, $0 \leqslant i \leqslant 1+t$. Put $N_i = |K_i|$, so that $N_t = 0$ by 1.3.4 (iv), and by 1.4.1 we have

$$N_i = 0 \text{ for } i \geqslant 6. \tag{1}$$

Equations (4)-(6) of 1.3 become, respectively,

$$N_0 = 6t-3t^2+(t^3+t)/2- \sum_{i=3}^{1+t} (i-1)(i-2)N_i/2, \tag{2}$$

$$N_1 = (t^2-1)(3-t)+ \sum_{i=3}^{1+t} (i^2-2i)N_i, \tag{3}$$

$$N_2 = (t^3-t)/2- \sum_{i=3}^{1+t} (i^2-i)N_i/2. \tag{4}$$

If $z \in K_i$, $0 \leqslant i \leqslant t-1$, then there are $t+1-i$ lines through z incident with no point of $\{x,y\}^\perp$, and since $s = 3$ each of these lines is incident with a unique point of $K_{t-1-i}-\{z\}$. This implies the following two observations of S. Dixmier and F. Zara [53].

$$N_i \neq 0 \Rightarrow N_{t-1-i} \geqslant t+1-i, \text{ for } 0 \leqslant i \leqslant t-1 \tag{5}$$

and

$$(t+1-i)N_i = (2+i)N_{t-1-i} \tag{6}$$

(count pairs (z,z'), $z \in K_i$, $z' \in K_{t-1-i}$, $z \sim z'$, $z \neq z'$ and zz' incident with no point of $\{x,y\}^\perp$).

The cases $t = 3,6,9$ are now easily handled.

6.2.1. *A GQ of order* (3,3) *is isomorphic to* W(3) *or to its dual* Q(4,3).
Proof. Equations (3) and (4) yield $N_1 = 8N_4$, $N_2 = 12-6N_4$, and (6) with $i = 0$ says $N_2 = 2N_0$. It is easy to check that $N_4 \neq 1$, hence either $N_4 = N_1 = 0$ and (x,y) is antiregular by 1.3.6 (iii), or $N_4 = 2$ so that $N_2 = N_0 = 0$ and (x,y) is regular. It follows that in any triad (x,y,z), each pair is regular or each pair is antiregular. From this it follows that each point is regular or antiregular. If some point is antiregular, *S* is isomorphic to Q(4,3) by 5.2.8. Otherwise *S* is isomorphic to W(3) by 5.2.1. □

6.2.2. (*S. Dixmier and F. Zara* [53]).* *There is no GQ of order* (3,6).
Proof. Equations (1)-(6) with $t = 6$ yield $N_0 = N_5 = 0$, $N_1 = 4$, $N_2 = 12$, $N_3 = 15$, $N_4 = 8$.

Let $z \in K_1$. The one line through z meeting a point of $\{x,y\}^\perp$ necessarily is incident with two points z_1,z_2 of K_4. Each of the other six lines through z must be incident with a unique point of K_4. Hence each of the four points

* We thank Jack van Lint for helping us to streamline the argument of [53].

of K_1 is collinear with each of the eight points of K_4. So if $z' \in K_1-\{z\}$, then z' is collinear with z_1 and z_2, giving a triangle with vertices z', z_1,z_2, a contradiction. □

6.2.3. (*P.J. Cameron* [143], *S. Dixmier and F. Zara* [53]) . *Any* GQ *of order* (3,9) *must be isomorphic to* Q(5,3).

Proof. This was proved, of course, in 5.3.2 (iii), using 1.7.2 to show that each point is 3-regular. We offer here an alternative proof relying on the equations just preceding 1.7.2 to show that each point is 3-regular. Let $T = (x,y,z)$ be a triad of points in S, and recall the notation M_i of 1.7, $0 \leq i \leq 4$, with $s = 3$, $t = 9$. Then multiply eq. (19) by 8, eq. (20) by -8, eq. (21) by 4, eq. (22) by -1, and sum to obtain $\sum_{i=0}^{4} (i-1)(i-2)(4-i)M_i = 0$. Since all terms on the left are nonnegative , in fact they must be zero, implying $M_0 = M_3 = 0$. Hence T is 3-regular. □

The remainder of this section is devoted to handling the final case $t = 5$, which requires several steps.

6.2.4. (*S. Dixmier and F. Zara* [53]). *Any* GQ *of order* (3,5) *must be isomorphic to the* GQ $T_2^*(O)$ *arising from a complete oval in* PG(2,4).

Proof. (a) From now on we assume $s = 3$ and $t = 5$. Then solving equations (2), (3), (4) and (6) simultaneously we have $N_1 = 6(2-N_0)$, $N_2 = 12N_0$, $N_3 = 10(2-N_0)$, $N_4 = 3N_0$. Moreover, by (5), if $N_0 \neq 0$, then $N_4 \geq 6$. So either $N_0 = 0$ or $N_0 = 2$. First suppose $N_0 = 0$, so that $N_4 = N_2 = 0$, $N_1 = 12$, $N_3 = 20$. This says each triad containing (x,y) has 1 or 3 centers. But consider a line L passing through some point of $\{x,y\}^\perp$ but not through x or y. For the three points w of L not in $\{x,y\}^\perp$ it is impossible to arrange all triads (x,y,w) having 1 or 3 centers. Hence we must have the following

$$N_0 = 2,\ N_2 = 24,\ N_4 = 6,\ N_1 = N_3 = 0 . \tag{7}$$

(b) Put $\{x,y\}^\perp = \{c_1,\ldots,c_6\}$. The line through x and c_i is denoted A_i, and a line through x is of *type* A. The line through y and c_i is denoted B_i and a line through y is of *type* B. If L is a line incident with no point of $\{x,y\} \cup \{x,y\}^\perp$, it is of *type* AB. The remaining lines are of *type* C.

A line of type C has two points of K_2 and one of K_4. A line of type AB has one point of K_0 and one of K_4, or it has two points of K_2. Now it is clear that the two points of K_0 are not collinear, and that each of the two points of K_0 is collinear with all six points of K_4. Hence $K_0^\perp = K_4$.

Let $K_0 = \{x',y'\}$ and $L_{x,y} = \{x,y,x',y'\}$. If z is a center of (y,x',y'), then $z \in K_0^\perp = K_4$, implying $z \not\sim y$, a contradiction. So $L_{x',y'} = L_{x,y}$. Now it is also clear that $L_{x',y} = L_{x,x'} = L_{x,y'} = L_{y,y'} = L_{x,y} = L_{x',y'}$. Let us define an *affine* line to be a line of S or a set $L_{x,y}$. Then the points of S together with the affine lines (and natural incidence) form a 2-(64,4,1) design.

(c) If (x,y,z) is a triad, then π_z is the permutation of $N_6 = \{1,2,3,4,5,6\}$ defined by : the line through z meeting A_i also meets $B_{\pi_z(i)}$. So $\pi_z(i) = i$ iff $z \sim c_i$. And if $z \in K_4$, then π_z is a transposition.

Put $\mathcal{D} = \bigcup_{i=0}^{4} K_i = K_0 \cup K_2 \cup K_4$.

For $i \neq j$, $1 \leqslant i, j \leqslant 6$, it is clear that there are precisely 4 points z of $\mathcal{D}$ such that $\pi_z(i) = j$.

For $z \in \mathcal{D}$, π_z interchanges i and j iff there is a line C_{ji} (resp., C_{ij}) which is incident with z and concurrent with A_j and B_i (resp., A_i and B_j). Then the lines $C_{ij}, C_{ji}, A_i, A_j, B_i, B_j$ define a 3×3 grid G (i.e. a grid consisting of 9 points and 6 lines). Let u_1,u_2,u_3,v_1,v_2,v_3 be the other points on the respective lines $C_{ji}, A_i, B_j, C_{ij}, A_j, B_i$. Since $s = 3$, we have $u_1 \sim u_2 \sim u_3 \sim u_1$ and $v_1 \sim v_2 \sim v_3 \sim v_1$. So u_1,u_2,u_3 are on a line L and v_1,v_2,v_3 are on a line M. Let u_4 (resp., v_4) be the fourth point on L (resp., M). Since $s = 3$, we have $u_3 \sim v_4 \sim u_2$, $u_1 \sim v_4$, implying $u_4 = v_4$. So we have shown that the grid G can be completed in a unique way to a grid with 8 lines and 16 points . The four points whose permutations map i to j (and j to i) are z,u_1,u_4,v_1. It also follows that if z and z' are distinct collinear points of $\mathcal{D}$ for which both π_z and $\pi_{z'}$ interchange i and j, the line through z and z' must be of type AB.

(d) Consider a 4×4 grid (i.e. a grid consisting of 8 lines and 16 points)

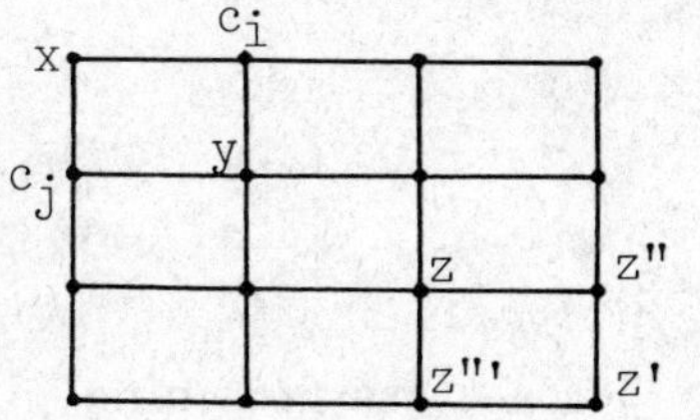

Fig. 6.1

containing x,y,c_i,c_j, and with points z,z',z'',z''' as indicated on the diagram. Then the lines $zz'',z''z',z'z''',z'''z$ are of type AB. Clearly z,z',z'',z''' are all in K_2, or $\{z,z'\} = K_0$ and $z'', z''' \in K_4$, or $\{z'',z'''\} = K_0$ and $z,z' \in K_4$. Assume we have the first case. Then each of the eight lines joining z,z',z'',z''' to a point of $\{x,y\}^{\perp}$ contains exactly one point of K_4. Since $N_4 = 6$, at least two of these lines, say L and M, contain a common point of K_4, say u. Clearly L and M are incident with z and z' or with z'' and z'''. Without loss of generality we may assume that z I L and z' I M. Let c_m (resp., c_ℓ) be the point of $\{x,y\}^{\perp}$ on M (resp., L). Since $c_m \sim x$, $c_m \sim y$, $c_m \not\sim z''$, we have $c_m \sim z$, giving a triangle $c_m zu$. Hence the first case does not arise, and there is no 3×3 grid containing x,y and a point $z \in K_2$. As a consequence we have : A 4×4 grid defines a linear subspace of the 2-(64,4,1) design, i.e. a 4×4 grid together with the affine lines on it is AG(2,4).

(e) Let $z \in K_4$, so π_z is a transposition, say interchanging i and j, and z is collinear with c_k,c_ℓ,c_m,c_n. Let z' be the other point of K_4 on the 4×4 grid containing x,y,z (and c_i and c_j). Clearly $\pi_z = \pi_{z'}$. If $K_0 = \{u,u'\}$, then u and u' are on the grid and both π_u and $\pi_{u'}$ interchange i and j. So we have proved that we may order i,j,k,ℓ,m,n and $z_1,\ldots,z_6$ in K_4 in such a way that $\pi_{z_1} = \pi_{z_2}$ interchanges i and j, that $\pi_{z_3} = \pi_{z_4}$ interchanges k and ℓ, that $\pi_{z_5} = \pi_{z_6}$ interchanges m and n, and that $\pi_u = \pi_{u'} = (ij)(k\ell)(mn)$.

Let L be a line and y a point not on L. Choose x,x I L, $x \not\sim y$. If,for example, $x \sim c_k \sim y$, then by the preceding paragraph the 4×4 grid containing x,y,z_3,z_4 is the unique 4×4 grid containing L and y.

(f) In the set of lines of S we define a *parallelism* in the following way : $L \parallel M$ iff L = M, or $L \not\sim M$ and L and M belong to a same 4×4 grid (i.e. $L \parallel M$ iff L = M or (L,M) is a regular pair of nonconcurrent lines). By (e) all lines parallel to a given line form a spread of S . Now we show that parallelism is an equivalence relation. Clearly the relation is reflexive and symmetric, and all that remains is to show that it is transitive.

Let $L \parallel M$ and $M \parallel N$, with L,M,N distinct. Let $\{L,M\}^{\perp\perp} = \{L,M,U,V\}$. If N contains a point of the 4×4 grid defined by L and M, then clearly $N \parallel L$.

So assume N contains no point of the grid. Let u I N, L I $u_1 \sim u$, M I $u_2 \sim u$, U I $u_3 \sim u$, V I $u_4 \sim u$, and let $R \in \{L,M\}^{\perp}$ and u_1 I R. Clearly $uu_i \nparallel R$ and $uu_i \nparallel L$. Hence the two lines through u and different from uu_i are parallel to L and R, respectively. So $N \parallel L$ or $N \parallel R$. Since R intersects M and $M \parallel N$, we have $N \parallel L$.

An equivalence class E contains 16 lines. If $L,M \in E$, $L \neq M$, then $\{L,M\}^{\perp\perp} \subset E$ and $\{L,M\}^{\perp}$ belongs to another equivalence class. Hence the elements of an equivalence class together with the line spans contained in it form a 2-(16,4,1) design, i.e. AG(2,4).

Note : If (L,M) is regular, $L \not\sim M$, and u is a point that does not belong to the grid defined by $\{L,M\}^{\perp}$, then u is on two lines having no point in common with the grid : one of these lines is parallel to all elements of $\{L,M\}^{\perp}$; the other line is parallel to all lines of $\{L,M\}^{\perp\perp}$.

(g) Choose a distinguished equivalence class E. Define a new incidence structure $S' = (P',B',I')$ as follows : $B' = (B-E) \cup \{E_1,\ldots,E_5\}$, with $E_1,\ldots,E_5$ the other equivalence classes. The elements of P' are of three types : (i) the elements of P, (ii) the traces $\{L,M\}^{\perp}$ with $L,M \in E$, $L \neq M$, (iii) (∞). Incidence is defined in the following manner : if $x \in P$, $L \in B-E$, then $x \mathrm{I}' L$ iff $x \mathrm{I} L$; if $x \in P$ and $L = E_i$, then $x \not\mathrm{I}' L$; if $x = \{L,M\}^{\perp}$, $L,M \in E$, $N \in B-E$, then $x \mathrm{I}' N$ iff $N \in \{L,M\}^{\perp}$; if $x = \{L,M\}^{\perp}$, $L,M \in E$, $N = E_i$, then $x \mathrm{I}' N$ iff $\{L,M\}^{\perp} \subset E_i$; $(\infty) \mathrm{I}' E_i$, $i = 1,\ldots,5$. It is now rather straightforward to check that S' is a GQ of order 4. There are the correct numbers of points and lines, each point is on five lines, each line is incident with five points, and there are no triangles. We leave the somewhat tedious details to the reader.

(h) We prove that $S' \cong W(4)$. Let $x,y \in P$, with x and y not collinear in S'. The lines of E incident (in S) with x and y are denoted by L and M, respectively. If $L \neq M$, then $\{L,M\}^{\perp}$ is a point of S' which is collinear with (∞), x,y in S'. If $L = M$ and $N \in E$, $N \neq L$, then $\{L,N\}^{\perp}$ is a center of $((\infty),x,y)$ in S'. Hence every triad containing (∞) is centric and (∞) is regular in S'. It follows from 1.3.6 (iv) and 5.2.1 that $S' \cong W(4)$ if all points z of S', $(\infty) \neq z \in (\infty)^{\perp'}$, are regular in S'. Since (∞) is regular, it is sufficient to prove that each triad (x,y,z), with x,y of type (i) and z of type (ii), is centric in S'. Let $z = \{L,M\}^{\perp}$, $L,M \in E$, so x and y are not on the 4×4 grid defined by L and M in S. The elements of E containing x and y are denoted by U and V, respectively. First suppose $U = V$. Let R and T be the lines containing x and y, respectively, and parallel to the elements of $\{L,M\}^{\perp}$. Then $\{R,T\}^{\perp\perp}$ is a center of (x,y,z) in S'. Now suppose $U \neq V$. By (f) $|\{U,V\}^{\perp\perp} \cap \{L,M\}^{\perp\perp}| \in \{0,1\}$. By the note in (f), if $\{U,V\}^{\perp\perp} \cap \{L,M\}^{\perp\perp} = \phi$, then the elements of $\{U,V\}^{\perp}$ are parallel to the elements of $\{L,M\}^{\perp}$. Hence $\{U,V\}^{\perp}$ is a center of (x,y,z). Finally, let $\{U,V\}^{\perp\perp} \cap \{L,M\}^{\perp\perp} = \{N\}$. Then, with respect to (x,y), N contains a point $u \in K_4$. The line of $\{L,M\}^{\perp}$ which contains u is denoted by H. The line H_x defined by $x \mathrm{I} H_x \sim H$ clearly does not belong to the 4×4 grid defined by U and V. Hence on H_x is a center of (x,y,u) in S. Since S does not contain triangles, this center is the intersection of H_x and H. So H contains a point n of $\{x,y\}^{\perp}$. Clearly n is a center of (x,y,z) in S'. We conclude that $S' \cong W(4)$.

(i) In S' the hyperbolic lines through (∞) are exactly the elements of E. Now it is clear that $S = P(S', (\infty))$. Since $S' \cong W(4)$ and $W(4)$ is homogeneous in its points, the GQ S is unique up to isomorphism. □

6.3. s = 4

Using 1.2.2 and 1.2.3 it is easy to check that $t \in \{4,6,8,11,12,16\}$. Nothing is known about $t = 11$ or $t = 12$. In the other cases unique examples are known, but the uniqueness question is settled only in the case $t = 4$.

Let $S = (P,B,I)$ be a GQ of order 4. The goal of this section is to prove that each pair of distinct lines (or points) is regular, so that S must be isomorphic to $W(4)$. The long proof is divided into a fairly large number of steps.

Since $s = t = 4$ is even, no pair of points (respectively, lines) may be antiregular by 1.5.1 (i). Hence each pair of noncollinear points (respectively, nonconcurrent lines) must belong to some triad with at least three (and thus by 1.3.4 (iv) with exactly three or five) centers. Let (x,y,z) and (u,v,w) be triads of points of S. We say that (x,y,z) is *orthogonal to* (u,v,w) (written $(x,y,z) \perp (u,v,w)$) provided the following two conditions hold : $\{x,y,z\}^{\perp} = \{u,v,w\}$ and $\{u,v,w\}^{\perp} = \{x,y,z\}$. Dually, the same terminology and notation are used for lines. Our characterization of S begins with a study of orthogonal pairs.

Until further notice let $L = (L_1,L_2,L_3)$ *and* $M = (M_1,M_2,M_3)$ *be fixed, orthogonal triads of lines of* S. *Let* x_{ij} *be the point at which* L_i *meets* M_j, $1 \leqslant i, j \leqslant 3$, *and put* $R = \{x_{ij} \parallel 1 \leqslant i, j \leqslant 3\}$. *Let* T *denote the set of points incident with some* L_i *or some* M_j, *but not both, and put* $V = R \cup T$, $P' = P - V$.

$$|P| = 85; \quad |V| = 21; \quad |P'| = 64. \tag{1}$$

An L_i or M_j will be called a *line of* R. A line incident with two points of T (but no point of R) will be called a *secant*. A line incident with precisely one point of V (respectively R, T) will be called *tangent to* V (respectively R,T). A line of S incident with no point of V will be called an *exterior line*. A point of P' collinear with three points of R will be called a *center of* R. Let B' denote the set of exterior lines. An easy count reveals the following :

There are 6 *lines of* R, 12 *secants,* 27 *tangents to* R, 24 *tangents to* T, 16 *exterior lines.* (2)

For a point $y \in P'$ *there are precisely the following possibilities :*

(i) y *is collinear with three points of* R (*i.e.* y *is a center of* R), *with no point of* T, *and is on two exterior lines; or*

(ii) y *is collinear with two points of* R, *with two points of* T, *is on two tangents to* T *and is on one exterior line; or*

(iii) y is *collinear with one point of* R, *with four points of* T, *and is on zero, one or two exterior lines, zero, one or two secants, and four, two or zero tangents to* T, *respectively; or*

(iv) y *is collinear with no point of* R, *with six points of* T, *and is on zero or one exterior lines, one or two secants, and four or two tangents to* T, *respectively.* (3)

Let n_i be the number of points of P' on i exterior lines, i = 0,1,2. Let k_i be the number of points of P' collinear with i points of R, i = 0,1,2,3.

$$|P'| = 64 = \sum_{i=0}^{3} k_i = \sum_{i=0}^{2} n_i . \tag{4}$$

Count the pairs (x,y) with $x \in R$, $y \in P'$ and $x \sim y$, to obtain the following :

$$108 = \sum_{i=0}^{3} ik_i . \tag{5}$$

Similarly, count the ordered triples (x,y,z), with $x,y \in R$, $x \neq y$, $z \in P'$, and $x \sim z \sim y$:

$$108 = 2k_2 + 6k_3. \tag{6}$$

Solving (4), (5) and (6) for k_i, $0 \leqslant i \leqslant 2$, we have

$$\begin{aligned} k_0 &= 10-k_3 \geqslant 0, \\ k_1 &= 3k_3, \\ k_2 &= 54-3k_3. \end{aligned} \tag{7}$$

Count pairs (x,L) with $x \in P'$, $L \in B'$, $x\ I\ L$:

$$80 = n_1+2n_2. \tag{8}$$

Using (4) and (8), solve for n_0 :

$$n_0 = n_2-16 \geqslant 0. \tag{9}$$

A point of P' is called *special* provided it lies on two secants. In general there are two possibilities.

Case (a). No secant is incident with two special points.

Case (b). Some secant is incident with two special points.

We say that the orthogonal pair (L,M) of triads of lines is of *type* (a) or of *type* (b) according as case (a) or case (b) occurs.

If y_1 *and* y_2 *are distinct special points incident with a secant* N, *and if the other secant through* y_i *is* K_i, $i = 1,2$, *then* K_1 *and* K_2 *do not meet the same two lines of* R . (10)

Proof. We may suppose that the two special points y_1 and y_2 lie on a secant $N \in \{M_2,M_3\}^{\perp}$. Let K_i be the other secant through y_i, $i = 1,2$, and suppose that both K_1 and K_2 are in $\{L_2,L_3\}^{\perp}$. As M_2,M_3,K_1,K_2 are all centers of the triad (L_2,L_3,N), this triad must have five centers, so that $M_1 \sim N$. But then (M_1,M_2,M_3) has four centers, contradicting the hypothesis that $L \perp M$. □

The secants meeting L_1,L_2,L_3 are naturally divided into two *opposite* families : two such secants are in the same family iff they do not meet the same two L_i's and do not meet each other. Similarly, there are two opposite families of secants meeting M_1,M_2,M_3.

If (L,M) *is an orthogonal pair of triads of lines, then* $k_3 = 10$, $k_2 = 24$, $k_1 = 30$ *and* $k_0 = 0$, *so that each point of* P' *is collinear with some point of* R, *and some triad of points of* R *has three centers. If* (L,M) *is of type* (a), *then* $n_2 = 16$, $n_1 = 48$ *and* $n_0 = 0$. (11)

Proof. Suppose $k_0 > 0$, so there is some point $y \in P'$ collinear with no point of R. By (3) (iv) y must lie on some secant; say y is on $N \in \{M_2,M_3\}^{\perp}$. Then the secants meeting M_1 and belonging to the family opposite to that containing N make it impossible for y to be collinear with a point of M_1 lying in T. Hence y must be collinear with some point of R, implying $k_0 = 0$, $k_3 = 10$, $k_1 = 30$, $k_2 = 24$. Now assume that the triad (x_1,x_2,x_3) of points of R has centers y_1 and y_2. If x_i I N_i, i = 1,2,3, with $N_i \notin \{x_iy_1, x_iy_2\}$ and N_i not a line of R, then clearly $N_1 \sim N_2 \sim N_3 \sim N_1$. Hence there is a point y_3 incident with N_i, i = 1,2,3, so that (x_1,x_2,x_3) has three centers. Since there are ten centers of R and six triads consisting of points of R, some triad of R must have three centers.

Suppose (L,M) is of type (a). Since there are six secants concurrent with a pair of L_i's and any special point must lie on such a secant, there are at most six special points. So $n_2 \leqslant 6+k_3 = 16$, and by (9) $n_2 \geqslant 16$. Hence $n_2 = 16$, $n_0 = 0$ and $n_1 = 48$. □

If a secant passes through two special points, it must be incident with three special points. The other secants through these special points must be the secants of one family. (12)

Proof. Let N be a secant incident with two special points y_1 and y_2. We may suppose $N \in \{M_2,M_3\}^{\perp}$, and that if K_i is the other secant through y_i, i = 1,2, then $K_1 \in \{L_2,L_3\}^{\perp}$ and $K_2 \in \{L_1,L_3\}^{\perp}$. Clearly K_1 and K_2 must belong to the same family. By considering which points of N are collinear with which points of L_1, L_2 and L_3 we see easily that the third point y_3 on N and on no M_j must lie on the third secant of the family containing K_1 and K_2. □

Let N_1 *be a secant incident with two special points* y_1 *and* y_2, *and let* K_1 *be the other secant through* y_1. *If* (N_1,N_2,N_3) *is the family of secants containing* N_1 *and* (K_1,K_2,K_3) *is the family of secants containing* K_1, *then* $(K_1,K_2,K_3) \perp (N_1,N_2,N_3)$. *Moreover, the nine intersection points* $N_i \cap K_j$ *are all special points.* (13)

Proof. By (12) there must be a third special point y_3 on N_1. Let K_i be the other secant on y_i, i = 1,2,3, and suppose that the K_j's are incident with no special points other than y_1,y_2,y_3. We may suppose $N_1 \in \{M_2,M_3\}^{\perp}$. Let a,b,c be the points of M_1 and M_2 as indicated in Fig. 6.2. As there are only

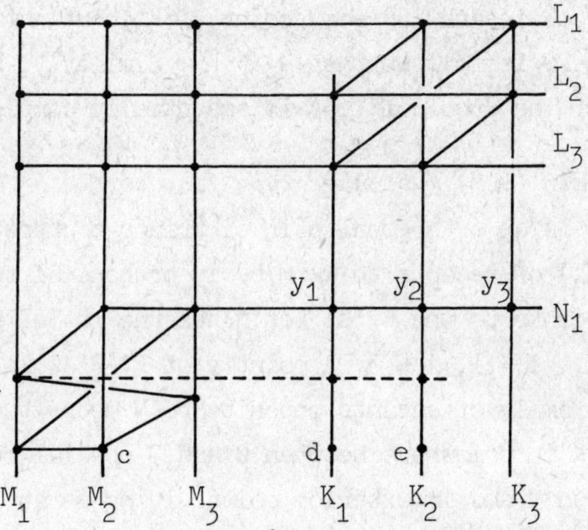

Fig. 6.2

two available lines through the point a to meet K_1,K_2, and K_3, one of them must hit two of the K_i's, say K_1 and K_2. Let d and e be the remaining points of K_1 and K_2 as indicated. The point b must be collinear with some point of K_1 and some point of K_2. It follows readily that d ~ b ~ e. Similarly, c must be collinear with some point of K_1 and some point of K_2. But the only available points are those at which the line through a meets K_1 and K_2, respectively. Of course, c cannot be collinear with both of these. Hence at least one of K_1,K_2,K_3 must pass through some additional special point. For example, if K_1 has an additional special point, then by (12) K_1 must have three special points. Moreover, by relabeling we may assume that the points and lines are related as in Fig. 6.3. But now the three points of N_2 on M_1,M_3,

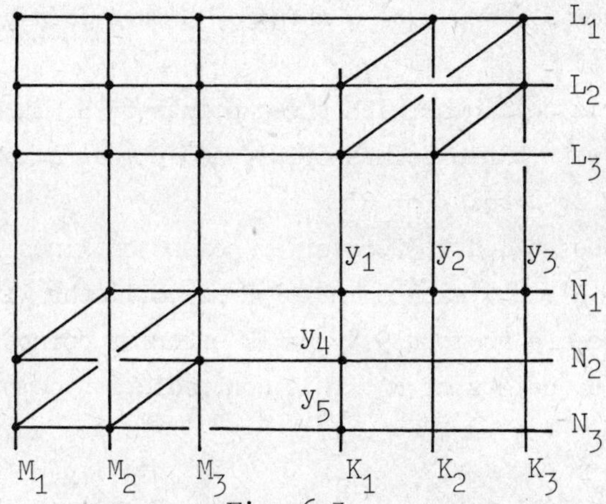

Fig. 6.3

and K_1 must each be collinear with some point of K_2, but not with any point of K_2 on L_1,L_3, or N_1 . It follows that $N_2 \sim K_2$. Similarly, $N_2 \sim K_3$, $N_3 \sim K_2$, and $N_3 \sim K_3$. The proof of (13) is essentially completed. □

Each orthogonal pair (L,M) *must have type* (a). (14)

Proof. Suppose (L,M) is an orthogonal pair of triads of type (b), so that a family $N = (N_1,N_2,N_3)$ of secants to the M_j's is orthogonal to a family $K = (K_1,K_2,K_3)$ of secants to the L_i's. Let R' be the set of points at which some N_i meets some K_j, $1 \leq i, j \leq 3$. A point y of P'-R' will be called an *exterior* point. The family of secants opposite to N meets the family of secants opposite to K in somewhere between 0 and 9 special points, implying that there are between 9 and 18 exterior points lying on at least one secant. As there are 55 exterior points, there must be at least 37 exterior points lying on no secant. Let y be an exterior point lying on no secant. The argument used to prove (11) may now be used to show that y must be collinear with some point of R' (alternatively, by (3)(iv) it is immediate that y is collinear with some point of R').

Case 1. The point y is collinear with one point of R and lies on four tangents to T (since by assumption y is on no secant). It follows that y is collinear with one point of R', *necessarily on the same line joining it to a point of* R.

Case 2. The point y is collinear with two points of R and lies on two tangents to T, one meeting some L_i and one meeting some M_j. It follows readily that y cannot be collinear with one or three points of R'. Hence y is collinear with two points of R'. As y is on five lines, including two tangents to T, *one of the lines joining* y *to a point of* R *must join* y *to a point of* R'.

Case 3. The point y is collinear with three points of R. It follows readily that y is collinear with three points of R', and y *must be on some line joining a point of* R *to a point of* R'.

Hence there must be at least 37 exterior points on lines joining a point of R with a point of R'. But each point of R is collinear with a unique point of R', so there are at most $9 \times 3 = 27$ exterior points lying on lines joining points of R to points of R'. This contradiction completes the proof of (14). □

This completes our preliminary study of orthogonal pairs, with (11) and (14) being the main results, and we drop the notation used so far.

Until further notice let S have a regular pair (L_0,L_1) *of nonconcurrent lines.* Let $\{L_0,L_1\}^{\perp} = \{M_0,\dots,M_4\}$, $\{L_0,L_1\}^{\perp\perp} = \{L_0,\dots,L_4\}$. Let x_{ij} be the point at which L_i and M_j meet, and put $R = \{x_{ij} \parallel 0 \leqslant i, j \leqslant 4\}$.

Each line of S is in $\{L_0,L_1\}^{\perp} \cup \{L_0,L_1\}^{\perp\perp}$ *or meets* R *in a unique point.* (15)
Proof. This accounts for all 85 lines. □

Either each triad of R *has a unique center, so* $S \cong W(4)$ *by* 5.2.6*, or each triad of* R *has exactly* 0 *or* 3 *centers.* (16)
Proof. Clearly no triad of R could have four or five centers. Suppose some triad, say (x_{00},x_{11},x_{22}) has two centers y_0 and y_1. Let N_{ij} be the line through y_i and x_{jj}, $i = 0,1$, $j = 0,1,2$. Then L_j,M_j,N_{0j},N_{1j} are four of the five lines through x_{jj}, $j = 0,1,2$. Moreover, for $0 \leqslant j < k \leqslant 2$, each one of L_j,M_j,N_{0j},N_{1j} meets one of L_k,M_k,N_{0k},N_{1k}. Hence the fifth lines through x_{00},x_{11}, and x_{22} all meet at some point y_2, showing that no triad of R has exactly two centers. It follows that either each of (x_{00},x_{11},x_{22}), (x_{00},x_{11},x_{32}), (x_{00},x_{11},x_{42}) has a unique center, or one of them has three centers and the other two have no center. It is easy to move around the grid R to complete the proof of (16). □

As mentioned in (16), if each triad of R has a unique center, then $S \cong W(4)$ by 5.2.6. *Hence until further notice we assume that each triad of* R *has exactly* 0 *or* 3 *centers.*

If a triad (y_0,y_1,y_2) *has three centers in* R *, it must have five centers in* R . (17)
Proof. Suppose (x_{00},x_{11},x_{22}) has three centers y_0,y_1,y_2. Then for $0 \leqslant j \leqslant 2$, y_j is collinear with both x_{33} and x_{44} or y_j is collinear with both x_{34} and x_{43}. By relabeling we may suppose that y_0 and y_1 are both collinear with x_{33} and x_{44}. If y_2 were collinear with both x_{34} and x_{43}, then the two lines y_2x_{43} and y_2x_{34} must meet the lines y_jx_{33} and y_jx_{44} in some order, $j = 0,1$. Any such possibility quickly yields a triangle. Hence y_2 must also be collinear with x_{33} and x_{44}. This shows that if a triad has three centers in R, it must have five centers in R. □

It follows that each pair of noncollinear points of R belongs to a unique

5-tuple of noncollinear points of R having three centers y_0,y_1,y_2. Such a 5-tuple will be called a *circle* of R with centers y_0,y_1,y_2. For each $y \in P-R$, the points of R collinear with y form a circle denoted C_y. Moreover, given $y \in P-R$, there are two other points $y',y'' \in P-R$ for which $C_y = C_{y'} = C_{y''}$.

There are 25 points x_{ij} of R with each x_{ij} lying on L_i and on M_j and on 4 circles. Two distinct points of R lie on a unique one of the ten lines L_i, M_j, or on a unique circle. It follows readily that the points of R together with the lines and circles of R are the points and lines, respectively, of the affine plane AG(2,5). The line of AG(2,5) defined by distinct points x,y of R will be denoted (xy).

Our goal, of course, is to obtain a contradiction under the present hypotheses. At this point in the published "proof" [132] the argument is incomplete, and the authors thank J. Tits for providing the argument given here to finish off this case.

We continue to consider R as the pointset of the affine plane AG(2,5) in which the two families $L_0,\dots,L_4$ and $M_0,\dots,M_4$ of lines are two distinguished sets of parallel lines called *horizontal* and *vertical*, respectively. A *path* is a sequence xyz... of points of R for which $x \not\sim y \not\sim z \not\sim \dots$ Let P denote the set of all paths. Each $x \in R$ is incident in S with three tangents to R, which are labeled $[x,i]$, i = 1,2,3, in a fixed but arbitrary manner. To each $xy \in P$ we associate a permutation ϕ_{xy} of the elements of {1,2,3} as follows : $i^{\phi_{xy}} = j$ iff $[x,i] \sim [y,j]$ in S. For any path $x_1x_2 \dots x_n$ we denote by $\phi_{x_1\dots x_n}$ the composition $\phi_{x_1x_2}\cdot\phi_{x_2x_3}\cdot \dots \cdot\phi_{x_{n-1}x_n}$. If $x_1 = x_n$ and if $\phi_{x_1\dots x_n}$ is the identity permutation, we write $x_1x_2\dots x_n \sim 0$.

By our construction, the following condition is seen to hold for all paths of the form xyzx.

For $xyzx \in P$, *either* x,y,z *are collinear in* AG(2,5) *and* $xyzx \sim 0$, *or they are not collinear in* AG(2,5) *and* ϕ_{xyzx} *is fixed-point free (i.e. is a 3-cycle).* (18)

If $xyztx \in P$, *if* (xy) *and* (zt) *are parallel in* AG(2,5), *and if the ratio of the slopes of the lines* (yz) *and* (tx)*(w.r.t. horizontals and verticals) is different from* ±1, *then* $xyztx \sim 0$. (19)

Proof. Let a be the intersection of the lines (xt) and (yz). Note : a,x,y,z, t must all be distinct. The points of the lines (axt) and (ayz) can be labeled, respectively, a, b_0, b_1, b_2, b_3 and a, c_0, c_1, c_2, c_3 in such a way that the lines $(b_i c_i)$ are all parallel and neither horizontal nor vertical, and similarly for the lines $(b_i c_{i+1})$, where subscripts run over the integers modulo 4. (For example, by exchanging horizontals and verticals, if necessary, one may assume that the slopes of (axt) and (ayz) are 1 and 2 , respectively, and for a coordinate system centered at a take $b_i = (2^i, 2^i)$, $c_i = (2^{i+1}, 2^{i+2})$. Here the coordinates for AG(2,5) are taken from Z_5.) Now

$$\phi_{ab_1c_1b_0a} \cdot \phi_{ac_1b_1a} \cdot \phi_{ab_0c_1a} = \phi_{ab_1c_1a} \cdot \phi_{ac_1b_0a} \cdot \phi_{ac_1b_1a} \cdot \phi_{ab_0c_1a} = \mathrm{id}, \quad (20)$$

since 3-cycles on {1,2,3} commute and $\phi_{axya} = \phi^{-1}_{ayxa}$.

As $\phi_{ab_1c_1b_0a} = \phi_{ab_1} \cdot \phi_{b_1c_1b_0b_1} \cdot \phi^{-1}_{ab_1}$ is a 3-cycle also, all three factors of the original product must be equal. In particular, $\phi_{ab_0c_1a} = \phi_{ac_1b_1a}$. Repeating the argument we find that $\phi_{ab_0c_1a} = \phi_{ac_1b_1a} = \phi_{ab_1c_2a} = \phi_{ac_2b_2a} = \dots$. (To derive the second equality, in (20) replace b_0, c_1, b_1 by c_1, b_1, c_2, respectively.) Then from $\phi_{ab_ic_ia} = \phi_{ab_jc_ja}$ with $j \neq i$ we have $\mathrm{id} = \phi_{ab_ic_ia} \cdot \phi^{-1}_{ab_jc_ja} = \phi_{ab_ic_ia} \cdot \phi_{ac_jb_ja} = \phi_{ab_i} \cdot \phi_{b_ic_ic_jb_jb_i} \cdot \phi_{b_ia}$, from which it follows that $b_ic_ic_jb_jb_i \sim 0$. Similarly, starting with $\phi_{ab_ic_{i+1}a} = \phi_{ab_jc_{j+1}a}$, $j \neq i$, we find $b_ic_{i+1}c_{j+1}b_jb_i \sim 0$. The relation xyztx ~ 0 must be one of these two, since the lines (b_ic_i) and (b_ic_{i+1}) are the only nonhorizontal and nonvertical lines connecting points b_j and c_k. □

We are now ready to obtain the desired contradiction.

If S has even one regular pair of nonconcurrent lines (respectively, points), then $S \cong W(4)$. (21)

Proof. Continuing with the assumptions and notations adopted just preceding (17), consider five distinct points x,y,z,t,u such that u,t and z are collinear in AG(2,5), (xy) is parallel to (utz), the lines (xy), (yz), (xt), (xu) represent the four nonhorizontal and nonvertical directions, and the lines (xy) and (yz) have opposite slopes. (For example, take x = (0,0), y = (1,1), z = (0,2), t = (1,3), u = (2,4).) By (19) xyztx ~ 0 and xyzux $\sim$ 0. Combining these we obtain $\phi_{zt}\cdot\phi_{tx} = \phi_{zu}\cdot\phi_{ux} = (\phi_{zt}\cdot\phi_{tu})\cdot\phi_{ux}$, and finally, id $= \phi_{tu}\cdot\phi_{ux}\cdot\phi_{xt}$. But this says tuxt $\sim$ 0, which is impossible by (18). □

If S is a GQ *of order* 4 *not isomorphic to* W(4), *then any triad of points or lines having three centers must have exactly three centers.* (22)

Proof. Let *S* be a GQ of order 4. Then by 1.5.1 (i) each pair (L_1,L_2) of non-concurrent lines must belong to some triad $L = (L_1,L_2,L_3)$ with at least three centers $(M_1,M_2,M_3) = M$. If both L and M have five centers, then (L_1,L_2) is regular. (For suppose $L^\perp = \{M_1,\ldots,M_5\}$ and $M^\perp = \{L_1,\ldots,L_5\}$. Let $j,k \in \{4,5\}$ and consider which points of L_j are collinear with which points of M_k. It follows readily that $\{L_1,\ldots,L_5\}^\perp = \{M_1,\ldots,M_5\}$.) Hence $S \cong$ W(4) by (21). If L has five centers but M has only three, it easily follows that the ten points on the centers of L but on no line of L may be split into two sets of five, with one set being the perp (or trace) of the other. This would force *S* to have a regular pair of points, contradicting (21). □

For the remainder of this section we assume that S is a GQ *of order* 4, $S \not\cong$ W(4), *and let* $L = (L_1,L_2,L_3)$ *and* $M = (M_1,M_2,M_3)$ *denote an orthogonal pair of triads of lines, necessarily of type* (a). The notation and terminology of the beginning of this section, up through the proof of (11), will also be used throughout the rest of this section. From the proof of (11), $n_2 = 16$ and $k_3 = 10$. By (3) this leaves exactly 6 special points, proving the following :

Each secant is incident with a unique special point. (23)

Let a and b denote distinct special points of the pair (L,M). Let N_a and K_a be the secants through a meeting lines of M and L, respectively. Similarly, N_b and K_b denote the secants through b meeting lines of M and L, respectively. The pair (a,b) of special points is said to be *homologous* provided N_a and N_b belong to the same family of secants and K_a and K_b belong to the same family of secants.

If (a,b) *is an homologous pair of special points, then* a $\sim$ b. (24)

Proof. With no loss in generality we may suppose that (a,b) is a homologous pair of special points with a I $N_1 \in \{M_2,M_3\}^\perp$ and a I $K_1 \in \{L_2,L_3\}^\perp$, and with b I $N_2 \in \{M_1,M_3\}^\perp$ and b I $K_2 \in \{L_1,L_3\}^\perp$. Then a $\sim x_{11} = L_1 \cap M_1$, and

$b \sim x_{22} = L_2 \cap M_2$. Let N_3 and K_3 be secants for which $N = (N_1,N_2,N_3)$ is one of the two families of secants meeting lines of M and $K = (K_1,K_2,K_3)$ is one of the two families of secants meeting lines of L. Let $a_{ij} = L_i \cap K_j$, $i \neq j$, $1 \leqslant i, j \leqslant 3$, and let $b_{ij} = M_i \cap N_j$, $i \neq j$, $1 \leqslant i, j \leqslant 3$. Let c and d be the two remaining points of K_1, e and f the two remaining points of K_2. Suppose c and d are labeled so that $b_{13} \sim c$ and $b_{12} \sim d$. Then the "projection" from M_1 onto K_1 is complete. Consider the projection from M_2 onto K_1. Clearly $x_{12} = L_1 \cap M_2$ and b_{23} must be collinear, in some order, with c and d. It follows easily that $b_{23} \not\sim c$ (since the secant K_1 may not pass through

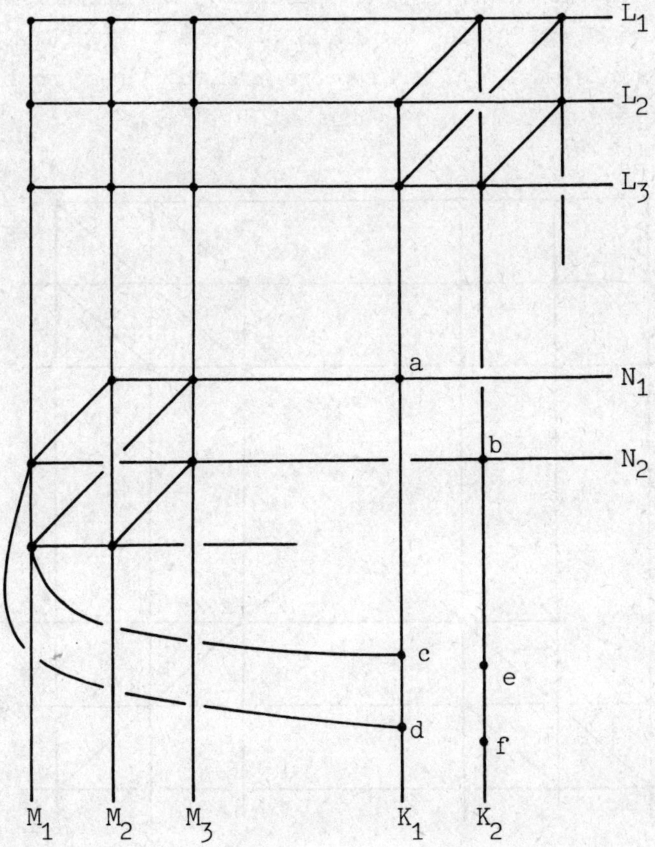

Fig. 6.4

two special points), so $b_{23} \sim d$ and $x_{12} \sim c$. Projecting from L_1 onto K_1 we find that $x_{13} \sim d$. Projecting from M_3 onto K_1, we find $b_{32} \sim c$. In projecting from K_2 onto K_1, it is clear that b must be collinear with one of a,c,d.

But $b_{12} \sim d$ precludes $b \sim d$, and $b_{32} \sim c$ precludes $b \sim c$, as no secant may have two special points. Hence $b \sim a$. □

An orthogonal pair (L,M) is called *rigid* provided that three special points lying on one family of secants of (L,M) are pairwise homologous.

No orthogonal pair is rigid. (25)

Proof. Suppose (L, M) is a rigid orthogonal pair. Hence the six special points are divided into two sets of three, say $S = \{a,b,c\}$ and $S' = \{a',b',c'\}$, with each pair of points in one set being homologous. By (24) and since S has no triangles, the points of S (respectively, S') lie on some exterior line L (respectively, L'). Let $N = (N_1,N_2,N_3)$ and $K = (K_1,K_2,K_3)$ be the two families of secants on a,b,c, and suppose that the lines are labeled so that the incidences are as described in part by Fig. 6.5.

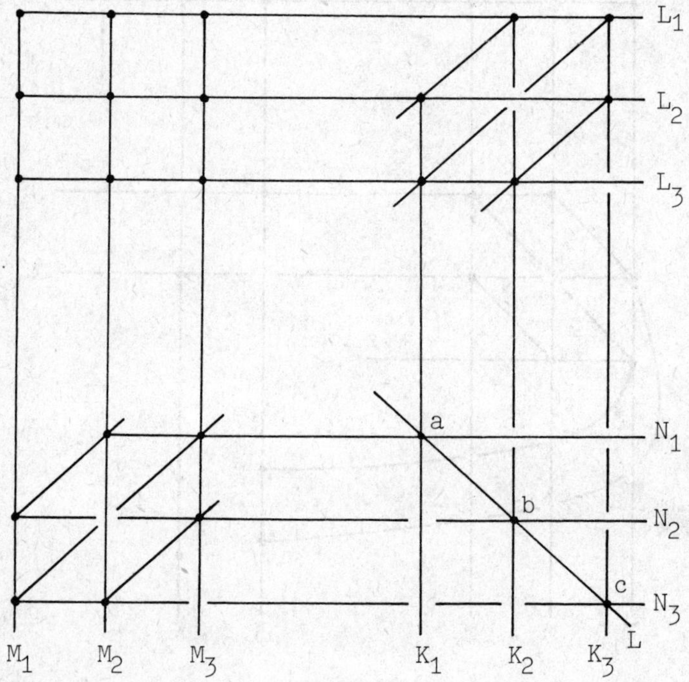

Fig. 6.5

Let $N' = (N'_1,N'_2,N'_3)$ (resp. $K' = (K'_1,K'_2,K'_3)$) be the family of secants opposite to N (resp., to K) with $M_i \not\sim N'_i$ (resp., $L_i \not\sim K'_i$), $i = 1,2,3$. Finally, let $a_{ij} = L_i \cap K_j$, $b_{ij} = M_i \cap N_j$, $i \neq j$, $1 \leqslant i, j \leqslant 3$. It is easy to check that $B_1 = (b_{12},b_{23},b_{31})$ and $B_2 = (b_{21},b_{32},b_{13})$ are orthogonal triads of points. The nine lines joining them are M_i,N_i,N_i', $1 \leq i \leq 3$. The six triads formed by these lines are M,N,N', and (M_i,N_i,N_i'), $i = 1,2,3$. By the dual of (11) there must be ten lines that are centers of these six triads. Of course M has three centers, and we claim that neither N nor N' can have three centers. The two cases are entirely similar, so consider N. N has the center L. Suppose there were two other centers K_4 and K_5 of N. For $i \neq j$, $1 \leq i, j \leq 3$, the point a_{ij} is collinear with $K_j \cap L$ on N_j, but must be collinear with a point of N_k lying on K_4 or K_5 if $k \neq j$, $1 \leq k \leq 3$. Since no secant of the family K' opposite to K can meet a member of N, it is easy to reach a contradiction by considering which points a_{ij} are collinear with which points of $K_t \cap N_k$, $i \neq j$, $k \neq j$, $1 \leq i,j,k \leq 3$, $t = 4,5$. It is also clear that if N has a second center K_4, it also has a third center K_5. It follows that the unique center of N is L, the unique center of N' is L', and one of (M_i,N_i,N_i'), $i = 1,2,3$, must have three centers while the other two each just have one center (by the proof of (11) (M_i,N_i,N_i') cannot have exactly two centers). By relabeling we may suppose (M_1,N_1,N_1') has three centers. Let d,e,f be the special points (w.r.t. (L, M)) lying on N_3',N_2',N_1', respectively ($\{d,e,f\} = \{a',b',c'\}$). The remainder of the proof of (25) is divided into three cases according as f is collinear with x_{11},x_{21}, or x_{31}.

Case 1. $f \sim x_{11}$ (cf. Fig. 6.6).

As (M_1,N_1,N_1') is assumed to have three centers and $a \sim x_{11}$, it must be that x_{11},f, a all lie on one line. Let p and q be the points of N_1 collinear with a_{12} and a_{13}, respectively. Then $a_{23} \sim p$ and $a_{32} \sim q$, so that $p \sim x_{31}$ and $q \sim x_{21}$. Let $v = N_1' \cap x_{31}p$ and $w = N_1' \cap x_{21}q$. Let the points r,s of N_2 and t,u of N_3 be labeled so that $p \sim r \sim t \sim q \sim s \sim u \sim p$. Of the three lines L_1,K_3' and K_2 through a_{12}, none can meet N_1',N_1 or N_3. Moreover, a line through a_{12} cannot meet both N_1' and one of N_1,N_3. Hence the line through a_{12} and p must be the line pu, and $a_{12} \sim w$ on the fifth line through a_{12}. A similar argument shows that a_{13},q,s lie on a line. This implies $a_{23} \not\sim s$, so $a_{23} \sim r$ and $a_{21} \sim s$. Again, a similar argument shows that a_{21},s,u lie on a line. Then $a_{31} \not\sim u$, so $a_{31} \sim t$. And $a_{31} \not\sim s$ implies $a_{31} \sim r$, so a_{31}, r,t lie on a line. This implies $a_{32} \sim u$. But as a_{12},p,u are on a line and $a_{12} \sim a_{32}$, a contradiction has been reached.

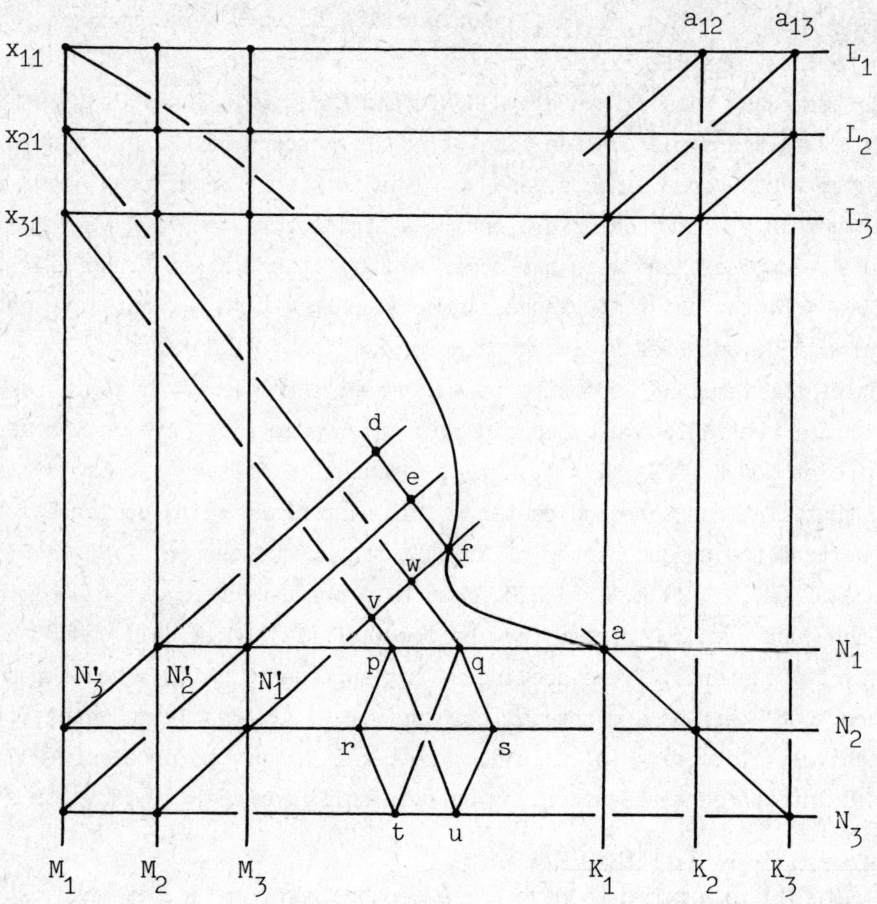

Fig. 6.6

Case 2. $f \sim x_{21}$.

The three secants K_1', K_2', K_3' pass through the points d,e,f in some order, and in this case it is clear that K_2' must pass through f. We then easily obtain a contradiction by considering the points of N_1' collinear with $x_{11}, x_{12}, x_{21}, x_{13}, x_{31}$.

Case 3. $f \sim x_{31}$.

In this case K_3' must pass through f, and again we obtain a contradiction by considering the points of N_1' collinear with $x_{11}, x_{12}, x_{21}, x_{13}, x_{31}$. This completes the proof of (25). □

From now on we may suppose that each orthogonal pair is *flexible*, i.e., it is not rigid. Let (L,M) be a (flexible) orthogonal pair. Let N and N' be the two opposite families of secants meeting lines of M, and let K and K'

be the two opposite families of secants meeting lines of $\mathcal{L}$. Then each of $\mathcal{N}$, $\mathcal{N}'$ is *paired with* just one of $\mathcal{K}$, $\mathcal{K}'$, in the following sense : $\mathcal{N}$ is paired with $\mathcal{K}$ provided that two of the secants of $\mathcal{N}$ meet two of the secants of $\mathcal{K}$. If $\mathcal{N}$ is paired with $\mathcal{K}$ and if $N \in \mathcal{N}$, $K' \in \mathcal{K}'$, with $N \sim K'$, we say N is the *odd* member of the family $\mathcal{N}$. (Also in this case K' must be the odd member of the family $\mathcal{K}'$, since $\mathcal{K}'$ is paired with $\mathcal{N}'$.) We may choose notation so that $\mathcal{N} = (N_1,N_2,N_3)$ is paired with $\mathcal{K} = (K_1,K_2,K_3)$, with $N_1 \sim K_1$, $N_3 \sim K_3$. If the odd member N_2 of $\mathcal{N}$ meets the secant of $\mathcal{K}'$ that belongs to $\{K_1,K_3\}^\perp$ and the odd member K_2 of $\mathcal{K}$ meets the secant of $\mathcal{N}'$ that belongs to $\{N_1,N_3\}^\perp$, then the pairing $\mathcal{N} \leftrightarrow \mathcal{K}$ is *strong* and the pair $(\mathcal{L},\mathcal{M})$ is *strongly flexible*. Clearly then also the pairing $\mathcal{N}' \leftrightarrow \mathcal{K}'$ is strong.

Every orthogonal pair $(\mathcal{L},\mathcal{M})$ *is strongly flexible.* (26)

Proof. Let $(\mathcal{L},\mathcal{M})$ be an orthogonal pair that fails to be strongly flexible. By labeling appropriately we may suppose that $\mathcal{N}$ and $\mathcal{K}$ are paired as in the preceding paragraph with the odd member N_2 of $\mathcal{N}$ meeting the secant K_1' of $\mathcal{K}'$ that belongs to $\{K_2,K_3\}^\perp$ (cf. Fig. 6.7). Put $a = N_1 \cap K_1$ and $b = N_3 \cap K_3$, so $a \sim b$ by (24) and ab is an exterior line. The unique point of R collinear

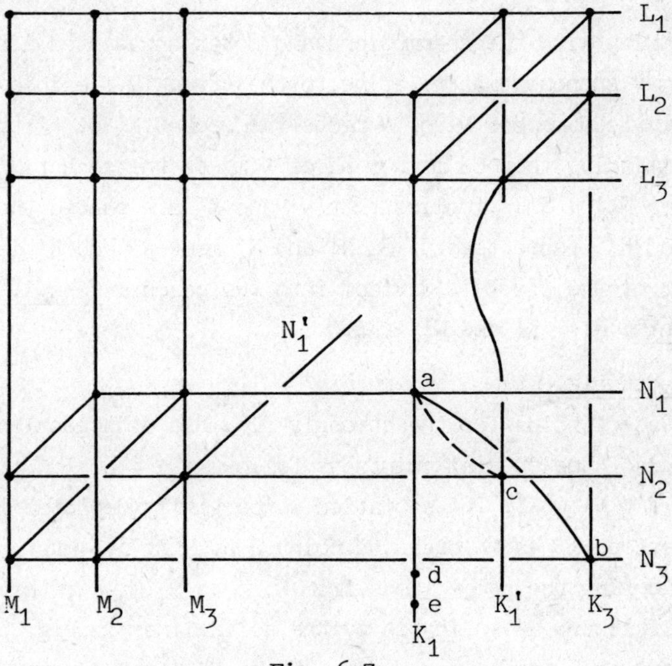

Fig. 6.7

with a is x_{11}, and the unique point of R collinear with b is x_{33}. Let $c = N_2 \cap K_1'$. Put $b_{ij} = M_i \cap N_j$, $i \neq j$, $1 \leq i, j \leq 3$. Let d and e be the remaining two points of K_1, say with $b_{13} \sim d$ and $b_{12} \sim e$. Then considering the projection from M_2 onto K_1, it follows that $x_{12} \sim d$ and $b_{23} \sim e$. Projecting M_3 onto K_1, we find that $x_{13} \sim e$ and $b_{32} \sim d$. As $b_{12} \sim e$ and $b_{32} \sim d$ clearly $d \not\sim c \not\sim e$. Projecting K_1' onto K_1, we find $c \sim a$. At this point we know that ab_{21},ad,ac,ab, and ax_{11} are the five distinct lines through a. One of these lines must be the line through a meeting the secant N_1' through b_{32} and b_{23}. The only possibility is the line ax_{11}. Say $N_1' \cap ax_{11} = y$. Now y is collinear only with the point x_{11} of R, but it must be collinear with some point of L_2 and some point of L_3. It is collinear with the point a of K_1, hence must be collinear with both $a_{23} = K_1' \cap L_2$ and $a_{32} = K_1' \cap L_3$. This forces y to lie on K_1'. Clearly $y \neq c$, so K_1' contains the two special points y and c. This completes the proof of (26). □

We are now nearing the end of the proof of the main result of this section.

6.3.1. (*S.E. Payne* [132,133]). *A GQ S of order* 4 *must be isomorphic to* W(4).

Proof. Continuing with the assumptions and notation adopted after the proof of (22), we may suppose that relative to the strongly flexible orthogonal pair (L,M) the odd member N_2 of N meets the secant K_2' of K' that belongs to $\{K_1,K_3\}^\perp$; similarly, the odd member K_2 of K meets the secant N_2' of N' that belongs to $\{N_1,N_3\}^\perp$. This implies that N' and K' are paired and have odd members N_2' and K_2', respectively. So N_3' and N_1' meet K_1' and K_3' in some order. The remainder of the proof is divided into two cases : <u>Case 1</u>. $N_1' \sim K_3'$ and $N_3' \sim K_1'$. <u>Case 2</u>. $N_1' \sim K_1'$ and $N_3' \sim K_3'$.

<u>Case 1 is impossible.</u> (27)

Assume that Case 1 holds for the strongly flexible orthogonal pair (L,M), and label four of the special points as follows : $a = K_1 \cap N_1$; $b = K_3 \cap N_3$; $c = N_1' \cap K_3'$; $d = N_3' \cap K_1'$. The situation is partially depicted in Fig. 6.8. Note that the four lines through the point d of Fig. 6.8 must be distinct. Then by considering the projections from K_1 onto M_1,M_2,M_3, the diagram may be filled in further, as indicated by the solid lines in Fig. 6.9 . Moreover the line from d to K_1 must be new and must hit K_1 at the point of the figure

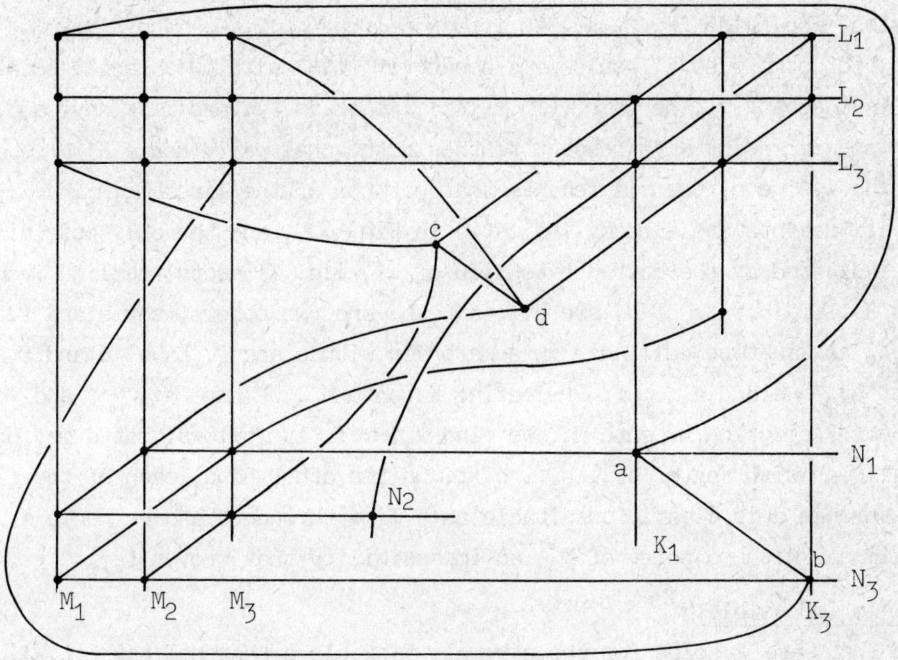

Fig. 6.8

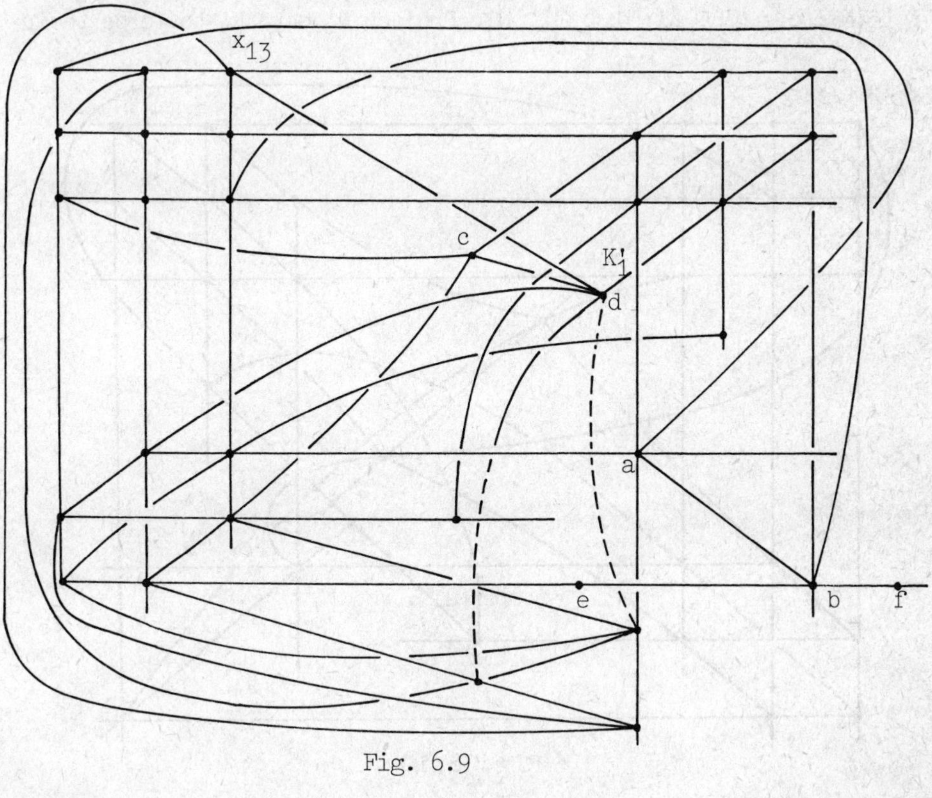

Fig. 6.9

indicated. The triad (M_2,M_3,K_1) is orthogonal to (L_2,L_3,N_1). And K_1' cannot hit N_1, for otherwise K_1' would be a secant of (L,M) with two special points. So K_1' is a secant of the pair $((M_2,M_3,K_1), (L_2,L_3,N_1))$, and must have a unique special point with respect to this orthogonal pair. Hence K_1' must meet exactly one of the six secants that hit two of the lines M_2,M_3,K_1. These six secants are already indicated in Fig. 6.9, and the only possibility is indicated by the dotted extension of K_1', i.e. K_1' meets the line from b_{23} to K_1. The points of N_3 are b_{13}, b_{23}, b, and two others, say e and f. And a_{12}, a_{32} must be collinear in some order with e and f. Label e and f so that $a_{12} \sim e$ and $a_{32} \sim f$. Projecting K_1 onto N_3, we have $a_{21} \sim f$ and $a_{31} \sim e$. Projecting L_1 onto N_3, we find $x_{13} \sim f$. It follows that d may not be collinear with any of b_{13}, b_{23}, b, f. On the other hand, each of the five lines through d is clearly unsuitable as a line through d and e. Hence d is collinear with no point of N_3, an impossibility that proves (27).

<u>Case 2 is impossible</u>. (28)

Assume that Case 2 holds for the strongly flexible orthogonal pair (L,M), and label the special points as indicated in Fig. 6.10 : $a = K_1 \cap N_1$; $b = K_3 \cap N_3$; $c = N_3' \cap K_3'$; $d = K_1' \cap N_1'$. Project N_2' onto K_2' to force their

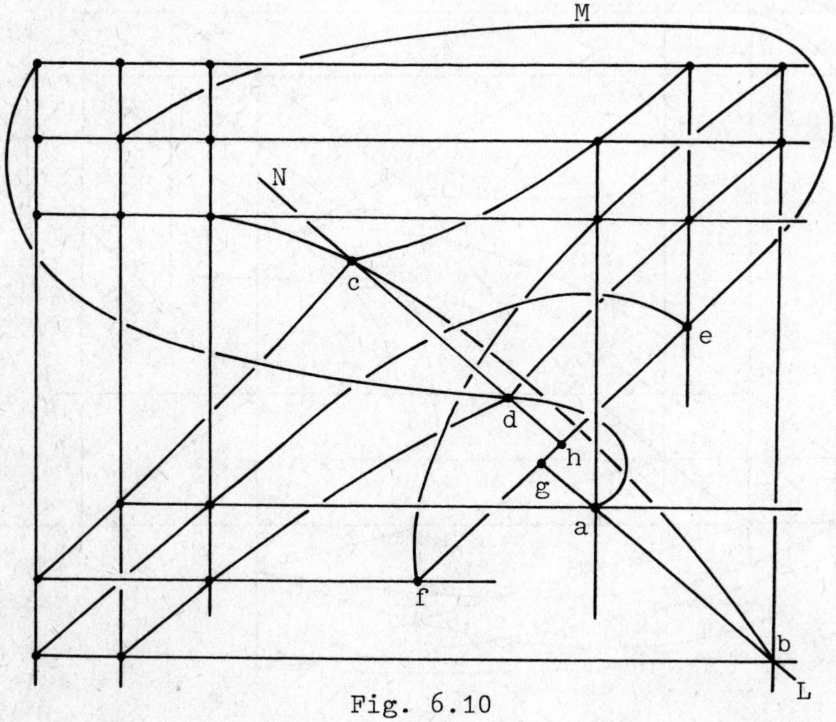

Fig. 6.10

special points e and f to be collinear on a line M through x_{22}. Project N_3' onto K_3 to find that c is collinear with b on a line through x_{33}. Similarly, project N_1' onto K_1 to find that d and a lie on a line through x_{11}.

Let p,q,g be the other three points on the line L through a and b. Notice that ab is an exterior line and that each secant concurrent with L is one of N_2, N_2', K_2, K_2'. If $N_2' \sim ab$ (resp., $K_2' \sim ab$) we have case (iii) in (3) and hence K_2', N_2' and ab are concurrent, a contradiction. Hence p,q,g are incident with no secant. It follows that p,q,g are each collinear with two or three points of R. One of p,q,g, say p, is collinear with a_{12} and with two points of R. One of g,q, say q, is collinear with a_{32} and with two points of R. By (ii) of (3) p and q, in some order, are collinear, respectively, with b_{12} and b_{32}. As neither a nor b is collinear with the special points $e = K_2 \cap N_2'$ and $f = K_2' \cap N_2$, it must be that $g \sim e$ and $f \sim g$. So $g = L \cap M$.

Let N = cd. Let N play the role of L in the above paragraph to find that N meets M at a point h. So f,g,h,e,x_{22} are the five distinct points of M. The points $x_{11}, x_{13}, x_{31}, x_{33}$ of R must each be collinear with a point of M. It follows that x_{11} and x_{33} are collinear with one of g,h, and x_{13} and x_{31} are collinear with the other. But it is also easy to see that x_{11} (resp., x_{33}) may not be collinear with g (resp., h). □

6.3.2. (*J.A. Thas* [209]). *If a* GQ $S = (P,B,I)$ *of order* (4,16) *contains a 3-regular triad, then it is isomorphic to* Q(5,4).

Proof. Let (x,y,z) be a 3-regular triad of the GQ S of order (4,16). Then by 2.6.2 $\{x,y,z\}^{\perp} \cup \{x,y,z\}^{\perp\perp}$ is contained in a subquadrangle $S' = (P',B',I')$ of order 4. By 6.3.1 S' may be identified with Q(4,4) ($\cong$ W(4)).

In Q(4,4) all points are regular. It follows immediately that any three distinct points of an hyperbolic line of Q(4,4) form a 3-regular triad of S.

Let u be a point of P-Q. The 17 points of Q which are collinear with u form an ovoid of Q(4,4). It is well known that each ovoid of Q(4,4) belongs to a hyperplane PG(3,4) of the space PG(4,4) containing Q (this easily follows from the uniqueness of the projective plane of order 4). So the number of ovoids of Q(4,4) equals 120. Since for any triad (u_1,u_2,u_3) of S we have $|\{u_1,u_2,u_3\}^{\perp}| = 5$, clearly any ovoid of Q(4,4) corresponds to at most two points of P-Q. Since $|P-Q| = 240$, any ovoid of Q(4,4) corresponds to exactly two points of P-Q.

Consider a triad (v_1,v_2,v_3) of S, with $v_i \in Q$, $i = 1,2,3$. We shall prove that (v_1,v_2,v_3) is 3-regular. We already noticed that this is the case if v_1,v_2,v_3 are points of an hyperbolic line of Q(4,4). So assume that v_1, v_2,v_3 do not belong to a common hyperbolic line. Since each point of Q(4,4) is regular, we have $\{v_1,v_2,v_3\}^{\perp'} = \{w\}$ in Q(4,4) (cf. 1.3.6 (ii)). Let C be the conic $Q \cap \pi$, where π is the plane $v_1v_2v_3$. Clearly w is collinear with each point of C. In Q(4,4) there are two ovoids O,O' which contain C. The points of P-Q which correspond to O,O' are denoted by u_1,u_2, u_1',u_2'. Since u_1,u_2,u_1',u_2' are collinear with all points of C, we have $\{v_1,v_2,v_3\}^{\perp} = \{w,u_1,u_2,u_1',u_2'\}$ and $\{v_1,v_2,v_3\}^{\perp\perp} = C$. Hence (v_1,v_2,v_3) is 3-regular in S.

Now we shall show that any point v of Q is 3-regular. If (v,v',v") is a triad consisting of points of Q, then we have already shown that (v,v',v") is 3-regular. Next, let (v,v',v") be a triad with $v' \in Q$, $v'' \in P-Q$. Let w be a point of Q which is collinear with v and v'. If (v_1,v_2,v_3) is a triad with $v_i \in w^{\perp'} \subset Q$, $i = 1,2,3$, then by the preceding paragraph $\{v_1,v_2,v_3\}^{\perp\perp}$ is contained in a subquadrangle S_1 of order 4. If $\{v_1,v_2,v_3\}^{\perp\perp}$ is not an hyperbolic line of Q(4,4), then $S_1 \neq Q(4,4)$. If the intersection S'' of S_1 and Q(4,4) contains a point which is not in $w^{\perp'}$, then by 2.3.1 S'' is a subquadrangle of order 4 of Q(4,4), i.e. $Q(4,4) = S_1$, a contradiction. Hence the intersection of the pointsets of S_1 and Q(4,4) is $w^{\perp'}$. Next, if (v_1',v_2',v_3') is another triad in $w^{\perp'}$ and if the corresponding subquadrangle S_1' is distinct from S_1, then clearly $w^{\perp'}$ is the intersection of the pointsets of S_1 and S_1'. The number of subquadrangles arising from triads in $w^{\perp'}$ is equal to the quotient of the number of irreducible conics in $w^{\perp'}$ and the number of hyperbolic lines in $w^{\perp'}$ of a given S_1, hence is equal to $64/16 = 4$. The total number of points of these 4 quadrangles is 277. Clearly no one of these quadrangles contains points of $w^{\perp}-w^{\perp'}$. Since $|w^{\perp}-w^{\perp'}| = 4$ and $|P| = 325$, the union of the 4 subquadrangles and $w^{\perp}-w^{\perp'}$ is exactly P. Now suppose that each point $w \in \{v,v'\}^{\perp'}$ is collinear with v". If w_1,w_2,w_3 are distinct points of $\{v,v'\}^{\perp'}$, then $v'' \in \{w_1,w_2,w_3\}^{\perp}$. But $\{w_1,w_2,w_3\}^{\perp} = \{v,v'\}^{\perp'\perp'}$, and so $v'' \in \{v,v'\}^{\perp'\perp'} \subset Q$, a contradiction. So we may assume that $w \not\sim v''$. Then one of the 4 subquadrangles corresponding to w contains v", say S_1. Interchanging the roles of Q(4,4) and S_1, we see that each triad in S_1 is 3-regular. Hence (v,v',v") is 3-regular. Finally, let (v,v',v") be a triad with $v',v'' \in P-Q$. Let v"' be a point of Q(4,4) which is not collinear with v or v'. Let S_1 be a subquadrangle of the type described above containing v,v',v"'. Now, interchanging the roles of S_1 and Q(4,4), we know by the preceding cases that (v,v',v") is 3-regular. We conclude that v is 3-regular.

Next, let $u \in P-Q$. Choose a triad (u,u',u") with $u', u'' \in Q$. Then there is a subquadrangle S_1 of order 4 containing u,u',u". Interchanging roles of S_1 and Q(4,4), we see that u is 3-regular.

Since all points of S are 3-regular, $S \cong Q(5,4)$ by 5.3.3. □

7 Generalized quadrangles in finite affine spaces

7.1. INTRODUCTION

By the beautiful theorem of F. Buekenhout and C. Lefèvre (cf. Chapter 4) we know that if a pointset of PG(d,s) together with a lineset of PG(d,s) form a GQ S of order (s,t), then S is a classical GQ. So all GQ of order (s,t) embedded in PG(d,s) are known.

In this chapter we solve the following analogous problem for affine spaces : find all GQ of order (s,t) whose points are points of the affine space AG(d,s+1), whose lines are lines of AG(d,s+1), and where the incidence is that of AG(d,s+1). In other words, we determine all GQ whose lines are lines of a finite space AG(d,q), whose points are all the points of AG(d,q) on these lines, and where the incidence is the natural one (here q = s+1). Such GQ are said to be *embedded* in AG(d,q). This embedding problem was completely solved in J.A. Thas [196]. The theorem on the embedding in AG(3,q) was proved independently by A. Bichara [12].

Finally, we note that in contrast with the projective case, there arise five nontrivial "sporadic" cases in the finite affine case.

7.2. EMBEDDING IN AG(2,s+1)

7.2.1. *If the* GQ S *of order* (s,t) *is embedded in* AG(2,s+1), *then the lineset of* S *is the union of two parallel classes of the plane and the pointset of* S *is the pointset of the plane.*
Proof. Easy exercise. □

7.3. EMBEDDING IN AG(3,s+1)

7.3.1. *Suppose that the* GQ S = (P,B,I) *of order* (s,t) *is embedded in* AG(3,s+1),*and that* P *is not contained in a plane of* AG(3,s+1). *Then one of the following cases must occur :*

(i) s = 1, t = 2 *(trivial case);*

(ii) t = 1 *and the elements of* S *are the affine points and affine lines of an hyperbolic quadric of* PG(3,s+1), *the projective completion of*

AG(3,s+1), *which is tangent to the plane at infinity of* AG(3,s+1);

(iii) P *is the pointset of* AG(3,s+1) *and* B *is the set of all lines of* AG(3,s+1) *whose points at infinity are the points of a complete oval* O *of the plane at infinity of* AG(3,s+1), *i.e.* $S = T_2^*(O)$ (*here* $s+1 = 2^h$ *and* $t = s+2$);

(iv) P *is the pointset of* AG(3,s+1) *and* $B = B_1 \cup B_2$, *where* B_1 *is the set of all affine totally isotropic lines with respect to a symplectic polarity* θ *of the projective completion* PG(3,s+1) *of* AG(3,s+1) *and where* B_2 *is the class of parallel lines defined by the pole* x (*the image with respect to* θ) *of the plane at infinity of* AG(3,s+1), *i.e.* $S = P(W(s+1),x)$ (*here* $t = s+2$);

(v) $s = t = 2$ (*an embedding of the* GQ *with* 15 *points and* 15 *lines in* AG(3,3)).

Proof. Suppose that $x \in P$, $L \in B$ and $x \not{I}\, L$. Then a substructure $S_\omega = (P_\omega, B_\omega, I_\omega)$ is induced in the plane $xL = \omega$. By 2.3.1 B_ω is the union of two parallel classes of lines in ω or B_ω is a set of lines with common point y, and in both cases P_ω is the set of all points on the lines of B_ω.

Assume that B_ω is a set of lines with common point y, and that there exists a line M in B which is incident with y and which is not contained in ω (hence $t > 1$). Let z I M, $z \neq y$. The lines of B through z are necessarily the line M and t lines in a plane ω' parallel to ω. We claim that $B_{\omega'}$ is a set of t lines with common point z. For otherwise $B_{\omega'}$ would consist of two parallel classes of lines in ω'. Then $t = 2$, and the number of lines of B which are incident with y and have a point in common with $P_{\omega'}$ equals $s + 1$. So there are at least $(s+1)+2 > 3$ lines of B which are incident with y, a contradiction which proves our claim. Analogously (interchange y and z) B_ω is a set of t lines with common point y.

It follows that if ω is a plane containing at least two lines of B, there are three possibilities for S_ω : If S_ω is a net, we say ω is of *type* I; if B_ω is a set of t lines having a common point y, we say ω is of *type* II (if M is the line defined by y I M, $M \in B - B_\omega$, and if z I M, then the t+1 lines of B incident with z are M and t lines in a plane ω' parallel to ω and also of type II); if B_ω is a set of t+1 lines having a common point y, we say ω is of *type* III.

The remainder of the proof is divided into three cases that depend on the value of t, beginning with the most general case.

(a) $\underline{t > 2}$.

Assume that ω is a plane which contains exactly one line L of B. Let L I y I M I x, with $M \in B-\{L\}$, $x \neq y$. The lines of B which are incident with x are M and t lines in a plane ω' parallel to ω. Since $t > 2$, the plane ω' is of type II. Consequently the lines of B which are incident with y are M and t lines in the plane ω, a contradiction. So any plane ω contains no line of B or at least two lines of B.

Now suppose that ω is a plane of type III, and let L be a line of B_ω. The common point of the t+1 lines of B_ω is denoted by y. Assume that each plane through L is of type II or III. As there are s+2 planes through L and only s+1 points on L, there is some point z on L which is incident with at least 2t-1 lines of B, a contradiction. So there must be a plane ω' through L which is of type I. In $S_{\omega'}$, there are two lines L,N which are incident with y, forcing y to be incident with at least t+2 lines of B, a contradiction. It follows that there are no planes of type III.

Next assume that there is at least one plane ω of type II. The common point of the lines of B_ω is denoted by y_0, and M denotes the line of S which is incident with y_0 but not contained in ω. Suppose that $y_0,y_1,\ldots,y_s$ are the points of M, and that $L_{i1},L_{i2},\ldots,L_{it},M$ are the t+1 lines of B incident with y_i, $i = 0,1,\ldots,s$. Each plane ω' which contains L_{ij} but not M, and which is not parallel to ω, is of type II, since otherwise y_i would be incident with at least t+2 lines of B. Next let ω'' be a plane which contains M, and suppose that ω'' is of type II. If y_i is the common point of the lines of $B_{\omega''}$, then y_i is incident with the t lines of $B_{\omega''}$ and also with the t lines $L_{i1},\ldots,L_{it}$, an impossibility. Hence any plane ω'' through M is of type I. It follows that for any $i \in \{0,1,\ldots,s\}$ there is a unique one of the lines L_{ij} which is contained in $B_{\omega''}$. So the number of planes ω'' through M is equal to $|\{L_{i1},\ldots,L_{it}\}| = t$. Consequently $t = s+2$ and $v = (s+1)^3$, i.e. P is the pointset of AG(3,s+1). From the preceding there also follows that any line of AG(3,s+1) which is parallel to M is an element of B. It is now also clear that any plane parallel to M is of type I, and that any plane not parallel to M contains a line L_{ij} and consequently is of type II. Also it is easy to see that the same conclusions hold if we replace M by any line parallel to M.

The plane at infinity of AG(3,s+1) is denoted by π_∞, and the point at infinity of M is denoted by y_∞. Let y_i' be a point of M', where M' is parallel to M, and let $L'_{i1},\dots,L'_{it},M'$ be the lines of B which are incident with y'_i. The lines $L'_{i1},\dots,L'_{it}$ are contained in a plane ω'_i, and the line at infinity M'_∞ of ω'_i is independent of the choice of the point y_i on M'. We notice that y_∞ is not on M'_∞. If the lines M' and M", M' ≠ M", are both parallel to M, then we show that $M'_\infty \neq M''_\infty$. Suppose the contrary. Then any plane with line at infinity M'_∞ contains at least 2t-1 lines of B, a contradiction. Hence $M'_\infty \neq M''_\infty$. So with the $(s+1)^2$ lines parallel to M there correspond the $(s+1)^2$ lines of π_∞ which do not contain y_∞. Now consider a line N_∞ of π_∞ through y_∞.

A plane ω" with line at infinity N_∞ is of type I, and the lines of B in ω" define two points at infinity, y_∞ and z_∞, on N_∞. Consequently with the s+1 lines of ω" which are parallel to M, there correspond the s+1 lines of π_∞ which contain z_∞ but not y_∞.

Now we define as follows an incidence structure $S' = (P',B',I')$: $P' = P \cup P_\infty$ with P_∞ the pointset of π_∞; $B' = (B-B_M) \cup B_\infty$, where B_M is the set of all lines parallel to M and where B_∞ is the set of all lines of π_∞ which contain y_∞; I' is the natural incidence relation. From the considerations in the preceding paragraph it follows readily that S' is a GQ of order s+1, which is embedded in the projective completion PG(3,s+1) of AG(3,s+1). By the theorem of F. Buekenhout and C. Lefèvre (cf. Chapter 4) B' is the set of all totally isotropic lines with respect to a symplectic polarity θ of PG(3,s+1). Hence $B = B_1 \cup B_2$, where B_1 is the set of all affine totally isotropic lines with respect to θ and B_2 is the class of parallel lines defined by y_∞, the pole of π_∞ with respect to θ. And with the notation of 3.1.4 we have $S = P(W(s+1),y_\infty)$. So in this case we have the situation described in part (iv) of 7.3.1.

Finally, we assume that there are no planes of type II. Let L be a line of B, and let ω be a plane containing L. Clearly ω is of type I. Consequently any point of ω is in P, and any line of ω parallel to L belongs to B. Since ω is an arbitrary plane containing L, P is the pointset of AG(3,s+1) and B contains all lines parallel to L. Let π_∞ be the plane at infinity of AG(3,s+1) and consider the points at infinity of the lines of B. The set of these points intersects any line of π_∞ in 2 points or in none at all. Consequently this set is a complete oval O of π_∞. So with the notation of 3.1.3 we have $S = T_2^*(O)$, i.e. we have case (iii) of 7.3.1.

(b) t = 1.

Suppose that $B = \{L_0,\ldots,L_s, M_0,\ldots,M_s\}$, $L_i \sim M_j$, and consider the projective completion PG(3,s+1) of AG(3,s+1). Since P is not contained in an AG(2,s+1), the projective lines M_i and M_j (resp., L_i and L_j), $i \neq j$, are not concurrent in PG(3,s+1). If $s \geq 2$, then the s+2 lines of PG(3,s+1) which are concurrent with the projective lines M_0,M_1,M_2 constitute a regulus R, i.e. a family of generating lines of an hyperbolic quadric Q. Consequently $L_0,L_1,\ldots,L_s$ are elements of R, and $M_0,M_1,\ldots,M_s$ are elements of the complementary regulus R' of Q. It follows that Q contains two lines at infinity. Hence we have case (ii) of 7.3.1. If s = 1 it is easy to see that case (ii) also arises.

(c) t = 2.

First of all we assume that there is a plane ω of type I. If x is a point of $P-P_\omega$, then the number of lines of B which are incident with x and a point of S_ω equals s+1. Hence $s+1 \leq t+1 = 3$, or $s \in \{1,2\}$.

Now we suppose that there is no plane of type I. Let $L \in B$ and assume that there is a plane ω which contains only the line L of B. If x is a point of P which is not in ω, then the lines of B which are incident with x are the line M defined by x I M I y I L, and two lines in a plane ω' parallel to ω. Clearly ω' is of type II. Consequently, the lines of B which are incident with y are M and two lines in ω, a contradiction. It follows that each plane containing L is of type II or III. Suppose that each plane ω through L is of type III. Since there are s+2 planes through L and only s+1 points on L, there is a point on L which is incident with at least five lines of B, a contradiction. Consequently, there is a plane ω of type II. Let ω be of type II and suppose that $L_1,L_2 \in B_\omega$, L_1 I x I L_2, and x I M with $M \in B-B_\omega$. If y I M, then the lines of B which are incident with y are M and two lines in a plane ω' parallel to ω. If a plane ω'' through M is of type III, then there is a point on M which is incident with at least four lines of B, a contradiction. Hence each plane ω'' through M is of type II. It follows that the number of lines of B having exactly one point in common with M is s+2. This number also equals $(s+1)t = 2(s+1)$, a contradiction.

So there is at least one plane of type I and $s \in \{1,2\}$. Consequently we have s = t = 2 or the trivial case s = 1, t = 2, i.e. we have cases (i) or (v) of 7.3.1. □

In the following theorem the "sporadic" case s = t = 2 is considered in detail.

7.3.2. *Up to a collineation of the space* AG(3,3) *there is just one embedding of a* GQ *of order* 2 *in* AG(3,3).

Before proceeding to the proof we describe the embedding as follows. Let ω be a plane of AG(3,3) and let $\{L_0,L_1,L_2\}$ and $\{M_x,M_y,M_z\}$ be two classes of parallel lines of ω. Suppose that $\{x_i\} = M_x \cap L_i$, $\{y_i\} = M_y \cap L_i$, and $\{z_i\} = M_z \cap L_i$, $i = 0,1,2$. Further, let N_x,N_y,N_z be three lines containing x_0,y_0,z_0, respectively, such that $N_x \notin \{M_x,L_0\}$, $N_y \notin \{M_y,L_0\}$, $N_z \notin \{M_z,L_0\}$, such that the planes N_xM_x, N_yM_y, N_zM_z are parallel, and such that the planes ω, L_0N_x, L_0N_y, L_0N_z are distinct. The points of N_x are x_0,x_3,x_4; the points of N_y are y_0,y_3,y_4; and the points of N_z are z_0,z_3,z_4; where notation is chosen in such a way that x_3,y_3,z_3 (resp., x_4,y_4,z_4) are collinear. Then the points of the GQ are $x_0,\dots,x_4,y_0,\dots,y_4,z_0,\dots,z_4$, and the lines are L_0,L_1,L_2,M_x, $M_y,M_z,N_x,N_y,N_z,x_3y_4,x_4y_3,x_3z_4,x_4z_3,y_3z_4,y_4z_3$.

Proof. Let $S = (P,B,I)$ be a GQ of order 2 which is embedded in AG(3,3). By the final part of the proof of the preceding theorem there is at least one plane ω of type I. Let $B_\omega = \{L_0,L_1,L_2,M_x,M_y,M_z\}$, $P_\omega = \{x_0,y_0,z_0,x_1,y_1,z_1,$ $x_2,y_2,z_2\}$, with $x_i \mathrel{I} M_x$, $y_i \mathrel{I} M_y$, $z_i \mathrel{I} M_z$, $x_i \mathrel{I} L_i$, $y_i \mathrel{I} L_i$, $z_i \mathrel{I} L_i$. Suppose that $x_0 \mathrel{I} N_x$, $y_0 \mathrel{I} N_y$, $z_0 \mathrel{I} N_z$, with $N_x \notin \{M_x,L_0\}$, $N_y \notin \{M_y,L_0\}$, $N_z \notin \{M_z,L_0\}$, that x_0,x_3,x_4 are the points of N_x, that y_0,y_3,y_4 are the points of N_y, and that z_0,z_3,z_4 are the points of N_z. Then $P = \{x_i,y_i,z_i \parallel i = 0,1,2,3,4\}$. Clearly the plane N_xM_x is of type I or II. If N_xM_x is of type I, then the fifteen points of S are contained in the planes N_xM_x and ω. Hence the points x_3,x_4,y_3,y_4,z_3,z_4 are in N_xM_x, so the points x_0,y_0,z_0 are in N_xM_x. Consequently, $N_xM_x = \omega$, a contradiction. It follows that N_xM_x is of type II, and also that N_xM_x, N_yM_y, N_zM_z are parallel planes of type II. Now assume that the planes L_0N_x, L_0N_y, L_0N_z are not distinct, e.g. $L_0N_x = L_0N_y$. Then the plane L_0N_x is of type I, and by a preceding argument ω is of type II, a contradiction. Hence the planes ω, L_0N_x, L_0N_y, L_0N_z are exactly the four planes that contain L_0. Now it is clear that the lines N_x,N_y,N_z, together with the line at infinity V_∞ of ω, form a regulus. Consequently, notation may be chosen in such a way that x_3,y_3,z_3 (resp., x_4,y_4,z_4) are on a line which is parallel to ω. As any line of B is incident with a point of P_ω, the lines $x_3y_4,x_4y_3,x_3z_4,x_4z_3,y_3z_4,y_4z_3$

are the remaining six lines of B.

From the preceding paragraph it easily follows that up to a collineation of AG(3,3) there is at most one GQ of order 2 which is embedded in AG(3,3) : If in PG(3,3), the projective completion of AG(3,3), the coordinate system is chosen in such a way that $x_0(0,0,0,1)$, $m(0,1,0,0)$ with m the point at infinity of the lines M_x, M_y, M_z, $\ell(0,0,1,0)$ with ℓ the point at infinity of the lines L_0, L_1, L_2, $z(1,0,0,0)$ with z the point at infinity of the line N_z, and $y_3(1,1,1,1)$, then the affine coordinates of the points of the GQ are given by $x_0(0,0,0)$, $x_1(0,1,0)$, $x_2(0,-1,0)$, $x_3(1,-1,0)$, $x_4(-1,1,0)$, $y_0(0,0,1)$, $y_1(0,1,1)$, $y_2(0,-1,1)$, $y_3(1,1,1)$, $y_4(-1,-1,1)$, $z_0(0,0,-1)$, $z_1(0,1,-1)$, $z_2(0,-1,-1)$, $z_3(1,0,-1)$, $z_4(-1,0,-1)$. And now it may be checked that the fifteen points with these coordinates together with the lines x_0x_1, y_0y_1, z_0z_1, x_0y_0, x_1y_1, x_2y_2, x_3x_4, y_3y_4, z_3z_4, x_3y_4, x_4y_3, x_3z_4, x_4z_3, y_3z_4, y_4z_3 form indeed a GQ. □

Remark : The existence of a GQ of order 2 which is embedded in AG(3,3) is also showed as follows. Consider the GQ described in Part (iv) of 7.3.1 in the case where s = 2. There arises a GQ of order (2,4) embedded in AG(3,3). Up to an isomorphism this GQ is unique (cf. 5.3.2 (ii)). Hence it must have a subquadrangle of order 2 (cf. 3.5), which is embedded in AG(3,3).

7.4. EMBEDDING IN AG(4,s+1)

7.4.1. *Suppose that the* GQ $S = (P,B,I)$ *of order* (s,t) *is embedded in* AG(4,s+1) *and that* P *is not contained in an* AG(3,s+1). *Then one of the following cases must occur :*

(i) s = 1, t ∈ {2,3,4,5,6,7} (*trivial case*);

(ii) s = t = 2, *i.e. an embedding of the* GQ *with* 15 *points and* 15 *lines in* AG(4,3) . *Moreover, up to a collineation of the space* AG(4,3) *there is just one embedding of a* GQ *of order* 2 *in* AG(4,3)(*so that the* GQ *is not contained in any subspace* AG(3,3)). This GQ may be described as follows . Let PG(3,3) be the hyperplane at infinity of AG(4,3); let ω_∞ be a plane of PG(3,3), and let ℓ be a point of PG(3,3)-ω_∞. In ω_∞ choose points m_{01}, m_{02}, m_{11}, m_{12}, m_{21}, m_{22}, in such a way that m_{01}, m_{21}, m_{11} are collinear, that m_{11}, m_{02}, m_{22} are collinear, that m_{21}, m_{02}, m_{12} are collinear, and that m_{01}, m_{22}, m_{12} are collinear. Let L be an affine line containing ℓ, and let the affine points of L be denoted by p_0, p_1, p_2 . The points of the GQ are the affine points of the lines p_0m_{01}, p_0m_{02}, p_1m_{11}, p_1m_{12}, p_2m_{21}, p_2m_{22}. The lines of the GQ are the affine lines of the (2-dimensional) hyperbolic quadric containing p_0m_{01}, p_1m_{11}, p_2m_{21}, resp. p_0m_{02}, p_1m_{11}, p_2m_{22}, resp. p_0m_{02}, p_1m_{12}, p_2m_{21}, and resp. p_0m_{01}, p_1m_{12}, p_2m_{22}.

(iii) $s = t = 3$ *and* S *is isomorphic to the* GQ $Q(4,3)$. *Moreover, up to a collineation (whose companion automorphism is the identity) of the space* $AG(4,4)$ *there is just one embedding of a* GQ *of order* 3 *in* $AG(4,4)$. This GQ may be described as follows. Let $PG(3,4)$ be the hyperplane at infinity of $AG(4,4)$, let ω_∞ be a plane of $PG(3,4)$, let H be a hermitian curve [80] of ω_∞, and let ℓ be a point of $PG(3,4)-\omega_\infty$. In ω_∞ there are exactly four triangles $m_{i1}m_{i2}m_{i3}$, $i = 0,1,2,3$, whose vertices are exterior points of H and whose sides are secants (non-tangents) of H [80]. Any line $m_{0a}m_{1b}$, $a, b \in \{1,2,3\}$, contains exactly one vertex m_{2c} of $m_{21}m_{22}m_{23}$ and one vertex m_{3d} of $m_{31}m_{32}m_{33}$, and the cross-ratio [80]$\{m_{0a},m_{1b}; m_{2c},m_{3d}\}$ is independent of the choice of $a, b \in \{1,2,3\}$. Let L be an affine line through ℓ, and let p_0,p_1,p_2,p_3 be the affine points of L, where notation is chosen in such a way that $\{p_0,p_1; p_2,p_3\} = \{m_{0a},m_{1b}; m_{2c},m_{3d}\}$. The points of the GQ are the 40 affine points of the lines p_im_{ij}, $i = 0,1,2,3$, $j = 1,2,3$. The lines of the GQ are the affine lines of the (2-dimensional) hyperbolic quadric containing p_0m_{0a}, p_1m_{1b}, p_2m_{2c}, p_3m_{3d}, $a, b = 1,2,3$.

(iv) $s = 2$, $t = 4$, *i.e. an embedding of the* GQ *with* 27 *points and* 45 *lines in* $AG(4,3)$. *Moreover, up to a collineation of the space* $AG(4,3)$, *there is just one embedding of the* GQ *of order* $(2,4)$ *in* $AG(4,3)$ *(so that the* GQ *is contained in no subspace* $AG(3,3)$*)*. This embedding may be described as follows. Let $PG(3,3)$ be the hyperplane at infinity of $AG(4,3)$, let ω_∞ be a plane of $PG(3,3)$, and let ℓ be a point of $PG(3,3)-\omega_\infty$. In ω_∞ choose points $m,n_x,n_y,n_z,n'_x,n'_y,n'_z,n''_x,n''_y,n''_z$, in such a way that m,n_x,n_y,n_z (resp., m, n'_x,n'_y,n'_z) (resp., m, n''_x,n''_y,n''_z) (resp., n_a,n'_b,n''_c with $\{a,b,c\} = \{x,y,z\}$) are collinear. Let L be an affine line through ℓ, and let x,y,z be the affine points of L. The plane defined by L and m is denoted by ω . The points of the GQ are the 27 affine points of the lines am, an_a, an'_a, an''_a, with $a = x,y,z$. The 45 lines of the GQ are the affine lines of ω with points at infinity ℓ and m, the affine lines of the (2-dimensional) hyperbolic quadric containing am, bn_b, cn_c (resp., am, bn'_b, cn'_c) (resp., am, bn''_b, cn''_c) (resp., an_a, bn'_b, cn''_c) with $\{a,b,c\} = \{x,y,z\}$.

Proof. Suppose that $s = 1$. Let $x_0,x_1,\ldots,x_t$, $y_0,y_1,\ldots,y_t$, $t \in \{2,\ldots,7\}$, be distinct points of AG(4,2) which are not contained in any hyperplane. Then the sets $P = \{x_i, y_j \parallel i,j \in \{0,\ldots,t\}\}$ and $B = \{\{x_i, y_j\} \parallel i,j \in \{0,\ldots,t\}\}$ define a GQ of order (1,t). From now on we suppose $s \geq 2$.

Let L,M be two nonconcurrent lines of S which are not parallel in AG(4,s+1), and suppose that AG(3,s+1) is the affine subspace containing these lines. By 2.3 the points and lines of S in AG(3,s+1) form a GQ $S' = (P',B',I')$ of order (s,t'). This GQ S' is embedded in AG(3,s+1) (and is not contained in any subplane AG(2,s+1)).

Suppose that S' is of type 7.3.1 (iii) or 7.3.1 (iv). Then $t' = s+2$. By 2.2.1 we have $st' \leq t$. Since $s \neq 1$, we also have $t \leq s^2$. Hence $s(s+2) \leq s^2$, an impossibility.

Next we suppose that S' is of type 7.3.1 (v). Then $s = t' = 2$. Since $st' \leq t \leq s^2$, we have $t = 4$. So S is the GQ with 27 points and 45 lines. For the points and lines of S' we use the notation introduced in 7.3.2. Let $N_x, N'_x, N''_x, M_x, L_0$ be the lines of B which contain x_0. The hyperplane AG(3,3) defined by ω and N_x is denoted by H, the hyperplane $\omega N'_x$ is denoted by H', and the hyperplane $\omega N''_x$ is denoted by H". It is clear that the subquadrangle $S'' = (P'',B'',I'')$ (resp., $S''' = (P''',B''',I''')$) induced in H' (resp., H") has order (2,2). Suppose that L_0, M_a, N'_a (resp., L_0, M_a, N''_a) are the lines of S'' (resp., S''') which are incident with a_0, a = y,z. Then each point of S is on one of the lines $L_0, M_a, N_a, N'_a, N''_a$, with a = x,y,z.

The point at infinity of the lines L_0, L_1, L_2 is denoted by ℓ, of the lines M_x, M_y, M_z by m, of the lines N_a by n_a, of the lines N'_a by n'_a and of the lines N''_a by n''_a (a = x,y,z). Then the points n_x, n_y, n_z, m are on a line N_∞, the points n'_x, n'_y, n'_z, m are on a line N'_∞, and the points n''_x, n''_y, n''_z, m are on a line N''_∞. Note that the lines $N_\infty, N'_\infty, N''_\infty$ are distinct.

Consider the lines N_a and N'_b, a, b $\in$ {x,y,z} and $a \neq b$. There are three lines $L_0, L_{abc}, L'_{abc} \in B$, {a,b,c} = {x,y,z}, concurrent with N_a and N'_b. Since all lines of S are regular (cf. 3.3.1), there are also lines N_a, N'_b, T''_c, {a,b,c} = {x,y,z}, concurrent with each of L_0, L_{abc}, L'_{abc}. Clearly we have $T''_c = N''_c$. Consequently the lines $N_a, N'_b, N''_c, L_0, L_{abc}, L'_{abc}$ form a GQ of order (s,1) which is embedded in the affine threespace defined by N_a and N'_b. So this GQ is type 7.3.1 (ii). It follows that n_a, n'_b, n''_c are on a line V_∞, that ℓ and the points at infinity ℓ_{abc} and ℓ'_{abc} of the lines L_{abc} and L'_{abc}, respectively, are on a line W_∞, and that V_∞ and W_∞ intersect. Now it is also clear that the points n_a, n'_a, n''_a, m, with a = x,y,z, are in a plane ω_∞. Since

S is not contained in a subspace AG(3,3), we have $\ell \notin \omega_\infty$.

If L_0, D_{ab}, E_{ab} (resp., L_0, D'_{ab}, E'_{ab}) (resp., L_0, D''_{ab}, E''_{ab}), $a \neq b$ and a, $b \in \{x,y,z\}$, are the lines of S which are concurrent with N_a, N_b (resp., N'_a, N'_b) (resp., N''_a, N''_b), then the lines $L_0, L_1, L_2, M_x, M_y, M_z, N_x, N_y, N_z, N'_x, N'_y, N'_z, N''_x, N''_y, N''_z, D_{ab}, E_{ab}, D'_{ab}, E'_{ab}, D''_{ab}, E''_{ab}, L_{abc}, L'_{abc}$ are the 45 lines of S.

Now we show that up to a collineation of AG(4,3) there is at most one GQ of this type. In ω_∞ choose a coordinate system as follows : m(1,0,0), $n_x(0,1,0)$, $n'_x(0,0,1)$, $n''_z(1,1,1)$. Then we have $n''_x(0,1,1)$, $n_y(1,1,0)$, $n_z(1,-1,0)$, $n'_y(1,0,1)$, $n'_z(1,0,-1)$, $n''_y(1,-1,-1)$. Hence in the hyperplane at infinity PG(3,3), the configuration formed by the points $m, n_x, n_y, n_z, n'_x, n'_y, n'_z, n''_x, n''_y, n''_z, \ell$ is unique up to a projectivity of PG(3,3). Now it easily follows that in AG(4,3) the configuration formed by the affine points of the lines $L_0, M_a, N_a, N'_a, N''_a$, with a = x,y,z, is unique up to a collineation of AG(4,3). Hence, up to a collineation of AG(4,3) there is at most one GQ S for which S' is of type 7.3.1 (v).

Finally, it is not difficult, but tedious, to check that the described GQ S does indeed exist. So case (iv) of 7.4.1 is completely handled.

Now suppose that every two noncoplanar lines of S define a subquadrangle of type 7.3.1 (ii). Let L and M be two concurrent lines of S. Choose a line N which is concurrent with L, but not coplanar with M (such a line N exists). The points and lines of S in the threespace MN form a subquadrangle of type 7.3.1 (ii). Hence the plane LM contains only the lines L,M of S.

Next let L be a line of S, let $p_0, p_1, \ldots, p_s$ be the points of L, and let $L, M_{i1}, \ldots, M_{it}$ be the t+1 lines of S through p_i. Clearly the t^2+s+1 hyperplanes $M_{0k}M_{1\ell}$, $LM_{i1}M_{i2}$ are distinct. The number of hyperplanes containing L equals $(s+1)^2+(s+1)+1$, implying $t^2 \leqslant (s+1)^2+1$. Hence $t \leqslant s+1$. Since each pair of distinct lines of S is regular, we have $t = 1$ or $t \geqslant s$ by 1.3.6. But $t \neq 1$, so $t \in \{s, s+1\}$. Since $s \neq 1$ and $(s+t) | st(s+1)(t+1)$ (cf. 1.2), it follows that s = t. Now by dualizing 5.2.1 we have $S \cong Q(4,s)$.

Let W be the threespace defined by three concurrent lines L_0, L_1, L_2 of S. The common point of these lines is denoted by p. By 2.3 all the lines of S in W contain p and any point of S in W is on one of these lines. The lines of S in W are denoted by $L_0, L_1, \ldots, L_{t'}$.

First suppose that $t' < t$, and let L_t be a line of S through p which is not in W. Clearly then $t > 2$. Let q I L_t, $q \neq p$. The t+1 lines of S through q are L_t and t lines in the threespace $\bar{W}$ through q and parallel to W. Analogously, the t+1 lines of S through p are L_t and t lines in W. So t' = t-1. Now consider the threespace $\overline{\overline{W}}$ defined by L_0, L_1, L_t. Notice that the plane $W \cap \overline{\overline{W}}$ contains only the lines L_0 and L_1 of S. Hence $L_{t'}$ is not in $\overline{\overline{W}}$, implying that $\overline{\overline{W}}$ contains exactly t lines of S through p. Since W and $\overline{\overline{W}}$ both contain t lines of S through p, their intersection contains t-1 lines of S through p. Consequently, t-1 = 2, implying s = t = 3. Let the points of L_t be denoted by p_0, p_1, p_2, p_3, and let $L_t, M_{i1}, M_{i2}, M_{i3}$ be the lines of S through p_i. The lines M_{i1}, M_{i2}, M_{i3} define a hyperplane which is parallel to W. The plane at infinity of W is denoted by ω_∞, the point at infinity of M_{ij} is denoted m_{ij}, and the point at infinity of L_t is denoted by ℓ ($\ell \notin \omega_\infty$). The points m_{i1}, m_{i2}, m_{i3} are not collinear, so they form a triangle V_i in ω_∞. If T is a line of ω_∞ which contains a vertex of V_i and V_j, $i \neq j$, then, since any two lines M_{ia} and M_{jb} define a subquadrangle of type 7.3.1 (ii), the line T also contains a vertex of V_k and V_ℓ, $\{i,j,k,\ell\} = \{0,1,2,3\}$. If these vertices on T are denoted by $m_{ia}, m_{jb}, m_{kc}, m_{\ell d}$, respectively, then clearly the cross-ratio $\{p_i, p_j; p_k, p_\ell\}$ equals the cross-ratio $\{m_{ia}, m_{jb}; m_{kc}, m_{\ell d}\}$. Further, a line which contains two vertices of V_i contains no vertex of V_j, $i \neq j$. The total number of lines of these two types equals 21, so that each line of ω_∞ has 2 or 4 points in common with the set V of all vertices of V_0, V_1, V_2, V_3. It follows that each line of ω_∞ has 1 or 3 points in common with $H = \omega_\infty - V$. Since $|H| = 9$, the set H is a hermitian curve [80] of ω_∞. Clearly the triangles V_0, V_1, V_2, V_3 are exactly the four triangles of ω_∞ whose vertices are exterior points of H and whose sides are secants (non-tangents) of H. Note that the 40 points of S are the affine points of the lines M_{ij} and that the 40 lines of S are the affine lines of the 9 subquadrangles defined by the pairs $\{M_{ia}, M_{jb}\}$, $i \neq j$. Moreover, the lines at infinity on the quadrics corresponding to these subquadrangles are the 9 tangents of H and the 9 lines which join ℓ to the points of H. From this detailed description of S it easily follows that up to a collineation (whose companion automorphism is the identity) of AG(4,4) there is at most one embedding of this type. Finally, it is not difficult to check that the GQ as described does exist. So case (iii) of 7.4.1 is handled.

Finally, suppose that for each point p of S, the lines of S through p are contained in a hyperplane. Our next goal is to show that s = 2. So assume s > 2. Let L and M be concurrent lines of S, and consider the s+2 threespaces which contain the plane LM. If W is such a threespace, then W contains only the lines L,M of S, or W contains each line of S through the common point p of L and M, or the points and lines of S in W form a subquadrangle of type 7.3.1 (ii). Clearly s of these hyperplanes through LM are of the third type, one is of the second type, and consequently one is of the first type. This hyperplane through LM which contains only the lines L,M of S is denoted by W'. Let N be a line of S through p which is not contained in W', and let q I N, $q \neq p$. The s+1 lines of S through q are N and s lines in the threespace W" through q and parallel to W'. Since $s > 2$, all the lines of S in W" contain q and any point of S in W" is on one of these lines. Analogously, the s+1 lines of S through p are N and s lines in W', a contradiction since $s > 2$. It follows that $s = t = 2$.

Let L be a line of S, let p_0,p_1,p_2 be the points of L, and let L, M_{i1}, M_{i2} be the lines of S through p_i. Through the plane $M_{01}M_{02}$ there is exactly one hyperplane W_0 which contains only the lines M_{01},M_{02} of S . It is clear that the lines $M_{11},M_{12},M_{21},M_{22}$ are parallel to W_0. The plane at infinity of W_0 is denoted by ω_∞, the point at infinity of M_{ij} by m_{ij}, and the point at infinity of L by ℓ ($\ell \notin \omega_\infty$). In the threespace $M_{ia}M_{jb}$, $i \neq j$, the points and lines of S form a subquadrangle of type 7.3.1 (ii), so we may assume that m_{01},m_{11},m_{21} are on a line N_1, that m_{02},m_{11},m_{22} are on a line N_2, that m_{02}, m_{12},m_{21} are on a line N_3, and that m_{01},m_{12},m_{22} are on a line N_4. The fourth point on the line N_i is denoted by n_i. We notice that the lines N_1,N_2,N_3,N_4 are contained in the plane ω_∞. Clearly the 15 points of S are the affine points of the lines M_{ij}, and the 15 lines of S are the affine lines of the 4 (2-dimensional) hyperbolic quadrics containing $p_0m_{0a},p_1m_{1b},p_2m_{2c}$, with m_{0a},m_{1b},m_{2c} collinear. The lines at infinity of these 4 subquadrangles are N_1,N_2,N_3,N_4 and the lines $\ell n_1,\ell n_2,\ell n_3,\ell n_4$. From this detailed description of S it easily follows that up to a collineation of AG(4,3) there is at most one GQ of this type. Finally, it is not difficult to check that the GQ as described does indeed exist. So case (ii) in the statement of 7.4.1 is handled, and this completes the embedding problem in AG(4,s+1). □

7.5. EMBEDDING IN AG(d,s+1), d ⩾ 5

7.5.1. *Suppose that the* GQ S = (P,B,I) *of order* (s,t) *is embedded in* AG(d,s+1), $d \geq 5$, *and that* P *is not contained in any hyperplane* AG(d-1,s+1). *Then one of the following cases must occur :*

(i) $s = 1$ *and* $t \in \{[d/2],\dots,2^{d-1}-1\}$, *with* $[d/2]$ *the greatest integer less than or equal to* d/2 (*trivial case*) *;*

(ii) $d = 5$, $s = 2$, $t = 4$, *i.e. an embedding of the* GQ *with* 27 *points and* 45 *lines in* AG(5,3). *Moreover, up to a collineation of the space* AG(5,3) *there is just one embedding of a* GQ *of order* (2,4) *in* AG(5,3) *so that it is contained in no subspace* AG(4,3). This embedding may be described as follows. Let PG(4,3) be the hyperplane at infinity of AG(5,3), let H_∞ be a hyperplane of PG(4,3) and let ℓ be a point of PG(4,3)-H_∞. In H_∞ choose points $m_x,m_y,m_z,$ $n_x,n_y,n_z,n'_x,n'_y,n'_z,n''_x,n''_y,n''_z$ in such a way that m_x,m_y,m_z are collinear, that m_x,m_y,m_z,n_x,n_y,n_z are in a plane ω_∞, that $m_x,m_y,m_z,n'_x,n'_y,n'_z$ are in a plane ω'_∞, that $m_x,m_y,m_z,n''_x,n''_y,n''_z$ are in a plane ω''_∞, that m_a,n_b,n_c (resp., $m_a,n'_b,$ n'_c) (resp., m_a,n''_b,n''_c) with $\{a,b,c\} = \{x,y,z\}$, are collinear, and that n_a,n'_b,n''_c, with $\{a,b,c\} = \{x,y,z\}$, are collinear. Let L be an affine line through ℓ, and let x,y,z be the affine points of L. The points of the GQ are the 27 affine points of the lines am_a,an_a,an'_a,an''_a, with $a = x,y,z$. The 45 lines of the GQ are the affine lines of the (2-dimensional) hyperbolic quadric containing xm_x,ym_y,zm_z (resp., am_a,bn_b,cn_c) (resp., am_a,bn'_b,cn'_c) (resp., am_a,bn''_b,cn''_c) (resp., an_a,bn'_b,cn''_c), with $\{a,b,c\} = \{x,y,z\}$.

Proof. Suppose that $s = 1$. Let $x_0,x_1,\ldots,x_t,y_0,y_1,\ldots,y_t$, with $t \in \{[d/2],\ldots,2^{d-1}-1\}$ and $[d/2]$ the greatest integer less than or equal to d/2, be distinct points of AG(d,2) which are not contained in a hyperplane. Then the sets $P = \{x_i,y_j \parallel i,j \in \{0,\ldots,t\}\}$ and $B = \{\{x_i,y_j\} \parallel i,j \in \{0,\ldots,t\}\}$ define a GQ of order (1,t). From now on we suppose $s \geqslant 2$.

Let L,M be two nonconcurrent lines of S which are not parallel in AG(d,s+1), and suppose that AG(3,s+1) is the affine threespace containing these lines. Suppose that p is a point of S which does not belong to AG(3,s+1), and call AG(4,s+1) the fourdimensional affine space defined by AG(3,s+1) and p. Assume that q is a point of S which does not belong to AG(4,s+1) and call AG(5,s+1) the affine space defined by AG(4,s+1) and q. By 2.3 the points and lines of S in AG(3,s+1) (resp., AG(4,s+1), AG(5,s+1)) form a GQ S' (resp., S'', S''') of order (s,t') (resp., (s,t''), (s,t''')). We have $t' < t'' < t''' \leqslant t \leqslant s^2$. From 2.2.2 (vi) it follows that $t' = 1$, $t'' = s$, $t''' = s^2$, implying that $t = t''' = s^2$ and $d = 5$. And from 7.4.1 it follows that $t'' = s = 2$ or $t'' = s = 3$.

Let us first assume that $s = 2$, $t = 4$, $d = 5$. By the preceding paragraph we know that there is a subquadrangle S' of order (2,2) of S which is embedded in a hyperplane H of AG(5,3) and which is not contained in a subspace AG(3,3). Let L_0 be a line of S', suppose that x_0,y_0,z_0 are the points of L_0, that N_a,M_a,L_0 are the lines of S' containing a_0, $a = x,y,z$, and that M_x,M_y,M_z belong to a threedimensional affine space T. Let N_x,N'_x,N''_x,M_x,L_0 be the lines of S which contain x_0. The hyperplane defined by T and N'_x is denoted by H', and the hyperplane defined by T and N''_x is denoted by H". The subquadrangle $S'' = (P'',B'',I'')$ (resp., $S''' = (P''',B''',I''')$) formed by the points and lines of S in H' (resp., H") has order (2,2). Suppose that N'_y, $N'_z \in B''$, $y_0 \; I \; N'_y$, $z_0 \; I \; N'_z$, $N'_y \notin \{M_y,L_0\}$, $N'_z \notin \{M_z,L_0\}$, and that N''_y, $N''_z \in B'''$, $y_0 \; I \; N''_y$, $z_0 \; I \; N''_z$, $N''_y \notin \{M_y,L_0\}$, $N''_z \notin \{M_z,L_0\}$. Any point of S is on one of the lines L_0,M_a,N_a,N'_a,N''_a, with $a = x,y,z$.

The point at infinity of the line L_0 is denoted by ℓ, that of the line M_a by m_a, that of the line N_a by n_a, that of the line N'_a by n'_a, and that of the line N''_a by n''_a, for $a = x,y,z$. Then m_x,m_y,m_z are on a line M_∞. Moreover, the points m_x,m_y,m_z,n_x,n_y,n_z are in a plane ω_∞, the points $m_x,m_y,m_z,n'_x,n'_y,n'_z$ are in a plane ω'_∞, and the points $m_x,m_y,m_z,n''_x,n''_y,n''_z$ are in a plane ω''_∞ (cf. 7.4.1 (ii)). Note that ω_∞, ω'_∞, ω''_∞ are distinct, and that ℓ is in none of these planes. Moreover, if $\{a,b,c\} = \{x,y,z\}$, then the points m_a,n_b,n_c (resp., m_a,n'_b,n'_c)(resp., m_a,n''_b,n''_c) are collinear. Further, there are three lines L_0,L_{abc},L'_{abc} of S, $\{a,b,c\} = \{x,y,z\}$, concurrent with N_a and N'_b, and since all lines of S are regular (cf. 3.3.1) there are also three lines N_a,N'_b,T''_c, $\{a,b,c\} = \{x,y,z\}$, concurrent with each of L_0,L_{abc},L'_{abc}. Clearly we have $T''_c = N''_c$, with $\{a,b,c\} = \{x,y,z\}$. It follows that n_a,n'_b,n''_c are on a line V_∞, that ℓ and the points at infinity ℓ_{abc} and ℓ'_{abc} of the lines L_{abc} and L'_{abc}, respectively, are on a line W_∞, and that V_∞ and W_∞ intersect. Now it is also clear that the points m_a,n_a,n'_a,n''_a are in a three-space H_∞. And since S is not contained in an AG(4,3), we have $\ell \notin H_\infty$.

If L_0,D_{ab},E_{ab} (resp., L_0,D'_{ab},E'_{ab}) (resp., L_0,D''_{ab},E''_{ab}), $a \neq b$ and $a, b \in \{x,y,z\}$, are the lines of S which are concurrent with N_a,N_b (resp., N'_a, N'_b) (resp. N''_a,N''_b), and if L_0,L'_0,L''_0 are the lines of S concurrent with M_x, M_y,M_z, then the lines $L_0,L'_0,L''_0,M_x,M_y,M_z,N_x,N_y,N_z,N'_x,N'_y,N'_z,N''_x,N''_y,N''_z,D_{ab},E_{ab}$, $D'_{ab},E'_{ab},D''_{ab},E''_{ab},L_{abc},L'_{abc}$ are the 45 lines of S.

Now we show that up to a collineation of AG(5,3) there is at most one GQ of this type. In H_∞ choose a coordinate system as follows : $m_x(1,0,0,0)$, $m_y(0,1,0,0)$, $n_x(0,0,1,0)$, $n'_x(0,0,0,1)$, $n''_z(1,1,1,1)$. Then necessarily we have $n_y(1,1,1,0)$, $n'_y(1,1,0,1)$, $n_z(0,1,1,0)$, $n'_z(0,1,0,1)$, $n''_y(0,1,1,1)$, $n''_x(1,-1,1,1)$, $m_z(1,1,0,0)$. Hence in the hyperplane at infinity

PG(4,3), the configuration formed by the points $\ell, m_a, n_a, n'_a, n''_a$, with $a = x,y,z$, is unique up to a projectivity of PG(4,3). Now it easily follows that in AG(5,3) the configuration formed by the affine points of the lines L_0, M_a, N_a, N_a, N_a, with $a = x,y,z$, is unique up to a collineation of AG(5,3). Hence up to a collineation of AG(5,3) there is at most one embedding of this type. Finally, it is not difficult but tedious, to check that the described GQ S does indeed exist. So case (ii) of 7.5.1 is completely handled.

Finally, we assume that $s = 3$, $t = 9$, $d = 5$. By the second paragraph of the proof we know that S has subquadrangles of order (3,3) of the type described in 7.4.1 (iii). So in S we may choose three concurrent lines L,M,N, with common point p, in such a way that L,M,N are the only lines of S in the threespace T defined by L,M,N. Let x I L, $x \neq p$, and x I V, $V \neq L$. The points and lines of S in the hyperplane defined by T and V form a subquadrangle S' of order (3,3). Since there are 9 choices for V, and since in the subquadrangle S' there are three such lines V, there are exactly 3 hyperplanes containing T in which the points and lines of S form a subquadrangle of order (3,3). Let H_1, H_2 be the other hyperplanes through T. The lines of S in H_i all contain p, and the number of lines of S in H_i equals $3+a_i$ with $a_1+a_2 = 4$. Let L_1 be a line of S through p and not in H_1, and let q I L_1, $q \neq p$. The 10 lines of S through q are L_1 and 9 lines $M_1,\ldots,M_9$ in the hyperplane H_3 through q and parallel to H_1. It is easy to see that any point of S in H_3 is on one of the 9 lines $M_1,\ldots,M_9$. Now it is clear that the 10 lines of S through p are L_1 and 9 lines in the hyperplane H_1. Consequently $3+a_1 = 9$, an impossibility. □

8 Elation generalized quadrangles and translation generalized quadrangles

8.1. WHORLS, ELATIONS, AND SYMMETRIES

Let $S = (P,B,I)$ be a GQ of order (s,t), $s \neq 1$, $t \neq 1$. A collineation θ of S is a *whorl about the point* p provided θ fixes each line incident with p. The following is an immediate consequence of 2.4.1 and 1.2.3.

8.1.1. *Let* θ *be a nonidentity whorl about* p. *Then one of the following must occur :*

(i) $y^\theta \neq y$ *for each* $y \in P - p^\perp$.

(ii) *There is a point* y, $y \not\sim p$, *for which* $y^\theta = y$. *Put* $T = \{p,y\}^\perp$, $U = \{p,y\}^{\perp\perp}$. *Then* $T \cup \{p,y\} \subset P_\theta \subset T \cup U$, *and* $L \in B_\theta$ *iff* L *joins a point of* T *with a point of* $U \cap P_\theta$.

(iii) *The substructure of elements fixed by* θ *forms a subquadrangle* S_θ *of order* (s',t), *where* $2 \leqslant s' \leqslant s/t \leqslant t$, *so* $t < s$.

Let θ be a whorl about p. If $\theta = \mathrm{id}$ or if θ fixes no point of $P - p^\perp$, then θ is an *elation about* p. If θ fixes each point of $p^\perp$, then θ is a *symmetry about* p. It follows from 8.1.1 that any symmetry about p is automatically an elation about p. The symmetries about p form a group. For each $x \sim p$, $x \neq p$, this group acts semiregularly on the set $\{L \in B \parallel x \mathrel{I} L, p \mathrel{\not{I}} L\}$, and therefore its order divides t. The point p is called a *center of symmetry* provided its group of symmetries has order t. It follows readily that any center of symmetry must be regular. Symmetries about lines are defined dually, and a line whose symmetry group has maximal order s is called an *axis of symmetry* and must be regular. There is an immediate corollary of 1.9.1.

8.1.2. *If* S *has a nonidentity symmetry* θ *about some line, then* $st(1+s) \equiv 0 \pmod{s+t}$.

The following simple result is occasionally useful.

8.1.3. *Let* σ, θ *be nonidentity symmetries about distinct lines* L,M, *respectively. Then*

(i) $\sigma\theta = \theta\sigma$ *iff* $L \sim M$.

(ii) $\sigma\theta$ *is not a symmetry about any line* (*or point*).

Proof. First suppose that L and M meet at a point x, and let $y \in P-x^{\perp}$. Let L' be the line through y meeting L and M' the line through y meeting M. It follows readily that both $y^{\sigma\theta}$ and $y^{\theta\sigma}$ must be the point at which $(M')^{\sigma}$ meets $(L')^{\theta}$. But if $\sigma\theta$ and $\theta\sigma$ have the same effect on points of $P-x^{\perp}$, clearly $\sigma\theta = \theta\sigma$. Now suppose that $L \not\sim M$. Clearly $L^{\theta} \not\sim L$, so that $L^{\theta\sigma} \neq L^{\theta}$, but $L^{\sigma\theta} = L^{\theta}$. This proves (i).

For the proof of (ii) note that if L I x I M, then $x^{\sigma\theta} = x$, $y^{\sigma\theta} \sim y \neq y^{\sigma\theta}$ iff $y \in x^{\perp}-\{x\}$, and $y^{\sigma\theta} \not\sim y$ iff $y \notin x^{\perp}$. And if $L \not\sim M$, then $y^{\sigma\theta} \not\sim y$ for all y not incident with any line of $\{L,M\}^{\perp}$. It follows readily that $\sigma\theta$ is not a symmetry about any line (or point). □

8.2. ELATION GENERALIZED QUADRANGLES

In general it seems to be an open question as to whether or not the set of elations about a point must be a group. One of our goals is to show that this is the case as generally as possible, and to study those GQ for which it holds. If there is a group G of elations about p acting regularly on $P-p^{\perp}$, we say S is an *elation generalized quadrangle* (EGQ) with *elation group* G and *base point* p. Briefly, we say that $(S^{(p)},G)$ or $S^{(p)}$ is an EGQ. Most known examples of GQ are EGQ, the notable exceptions being those of order (s-1,s+1) and their duals. In this chapter we will be concerned primarily with the following special kind of EGQ : if $(S^{(p)},G)$ is an EGQ for which G contains a full group of s symmetries about each line through p, then S is a *translation generalized quadrangle* (TGQ) with *base point* p and *translation group* G. Briefly, we say $(S^{(p)},G)$ or $S^{(p)}$ is a TGQ.

A TGQ of order (s,t) must have $s \leqslant t$ since it has some regular line. At the opposite end of the spectrum is the following kind of EGQ which will be studied in more detail in Chapter 10 : if $(S^{(p)},G)$ is an EGQ for which G contains a full group C of t symmetries about p, we say $S^{(p)}$ is a *skew-translation generalized quadrangle* (STGQ) with *base point* p and *skew-translation group* G. Briefly, we say $(S^{(p)},G)$ is a STGQ. Since a STGQ $(S^{(p)},G)$ has a regular point p, $t \leqslant s$.

Until further notice let $(S^{(p)},G)$ be an EGQ of order (s,t), and let y be a fixed point of $P-p^{\perp}$. Let $L_0,\ldots,L_t$ be the lines incident with p, and define z_i and M_i by L_i I z_i I M_i I y, $0 \leqslant i \leqslant t$. Put $S_i = \{\theta \in G \parallel M_i^{\theta} = M_i\}$,

$S_i^* = \{\theta \in G \parallel z_i^\theta = z_i\}$, and $J = \{S_i \parallel 0 \leq i \leq t\}$. Then $|G| = s^2t$; J is a collection of 1+t subgroups of G, each of order s; for each i, $0 \leq i \leq t$, S_i^* is a subgroup of order st containing S_i as a subgroup. Moreover, the following two conditions are satisfied :

K1. $S_iS_j \cap S_k = 1$, for distinct i,j,k.

K2. $S_i^* \cap S_j = 1$, for distinct i,j.

Conversely, suppose that K1 and K2 are satisfied, along with the restrictions on the orders of the groups G, S_i, S_i^* given above. Then it was first noted by W.M. Kantor [88] that the incidence structure $\mathcal{S}(G,J)$ described below is an EGQ with base point (∞).

Points of $\mathcal{S}(G,J)$ are of three kinds :

(i) elements of G ,

(ii) right cosets S_i^*g, $g \in G$, $i \in \{0,\ldots,t\}$,

(iii) a symbol (∞).

Lines of $\mathcal{S}(G,J)$ are of two kinds :

(a) right cosets S_ig, $g \in G$, $i \in \{0,\ldots,t\}$,

(b) symbols $[S_i]$, $i \in \{0,\ldots,t\}$.

A point g of type (i) is incident with each line S_ig, $0 \leq i \leq t$. A point S_i^*g of type (ii) is incident with $[S_i]$ and with each line S_ih contained in S_i^*g. The point (∞) is incident with each line $[S_i]$ of type (b). There are no further incidences.

It is a worthwhile exercise to check that indeed $\mathcal{S}(G,J)$ is a GQ of order (s,t). Moreover, if we start with an EGQ $(\mathcal{S}^{(p)},G)$ to obtain the family J as above, then we have the following.

8.2.1. $(\mathcal{S}^{(p)},G) \cong \mathcal{S}(G,J)$.

Proof. Of course y^g corresponds to g, z_i^g corresponds to S_i^*g, p corresponds to (∞), M_i^g corresponds to S_ig, and L_i corresponds to $[S_i]$. □

Now start with a group G and families $\{S_i\}$ and $\{S_i^*\}$ as described above satisfying K1 and K2, so that $\mathcal{S}(G,J)$ is a GQ. It follows rather easily (cf. 10.1) that $S_i^* = S_i \cup \{g \in G \parallel S_ig \cap S_j = \phi \text{ for } 0 \leq j \leq t\}$, from which part (iii) of the following theorem follows immediately.

8.2.2. (i) G *acts by right multiplication as a (maximal) group of elations about* (∞).

(ii) S_i *is the subgroup of* G *fixing the line* S_i *of* $\mathcal{S}(G,J)$.

(iii) *Any automorphism of* G *leaving* J *invariant induces a collineation of* $S(G,J)$ *fixing* (∞).

(iv) S_i *is a group of symmetries about* $[S_i]$ *iff* $S_i \lhd G$ (*so that* $S(G,J)$ *is a* TGQ *if* $S_i \lhd G$ *for each* i) *only if* $[S_i]$ *is a regular line iff* $S_iS_j = S_jS_i$ *for all* $S_j \in J$.

(v) $C = \cap\{S_i^* \parallel 0 \leqslant i \leqslant t\}$ *is a group of symmetries about* (∞) *iff* $C \lhd G$. *Moreover, if* $C \lhd G$ *and* $|C| = t$, *then* $S(G,J)$ *is an* STGQ *with base point* (∞) *and skew-translation group* G.

Proof. The details are all straightforward, so we give a proof only of part (iv), assuming that the first three parts have been proved. Then $h \in G$ determines a symmetry about $[S_i]$ iff the collineation it determines by right multiplication fixes each line of the form S_ig iff $S_igh = S_ig$ for all $g \in G$ iff $ghg^{-1} \in S_i$ for all $g \in G$. Hence h is a symmetry about $[S_i]$ iff all conjugates of h lie in S_i. It follows that S_i is a group of symmetries about $[S_i]$ iff $S_i \lhd G$, in which case $[S_i]$ is a regular line. Now let g be an arbitrary point not collinear with (∞). The set S_iS_jg consists of those points not collinear with (∞) which lie on lines of $\{[S_i],S_jg\}^\perp$, $i \neq j$. Similarly, the set S_jS_ig consists of those points not collinear with (∞) which lie on lines of $\{[S_j], S_ig\}^\perp$. Hence $([S_i],S_jg)$ is regular iff $S_iS_jg = S_jS_ig$ iff $S_iS_j = S_jS_i$. So $[S_i]$ is regular iff $S_iS_j = S_jS_i$ for all $j = 0,1,\ldots,t$. □

There is an immediate corollary.

8.2.3. *If* $(S^{(p)},G)$ *is an* EGQ *with* G *abelian, then it is a* TGQ.

8.2.4. *Let* $S = (P,B,I)$ *be a* GQ *of order* (s,t) *with* $s \leqslant t$, *and let* p *be a point for which* $\{p,x\}^{\perp\perp} = \{p,x\}$ *for all* $x \in P-p^\perp$. *And let* G *be a group of whorls about* p.

(i) *If* $y \sim p$, $y \neq p$, *and* θ *is a nonidentity whorl about both* p *and* y, *then all points fixed by* θ *lie on* py *and all lines fixed by* θ *meet* py.

(ii) *If* θ *is a nonidentity whorl about* p, *then* θ *fixes at most one point of* $P-p^\perp$.

(iii) *If* G *is generated by elations about* p, *then* G *is a group of elations, i.e. the set of elations about* p *is a group*.

(iv) *If* G *is transitive on* $P-p^\perp$ *and* $|G| > s^2t$, *then* G *is a Frobenius group on* $P-p^\perp$, *so that the set of all elations about* p *is a normal subgroup of*

G *of order* s^2t *acting regularly on* $P-p^{\perp}$, *i.e.* $S^{(p)}$ *is an* EGQ *with some normal subgroup of* G *as elation group.*

(v) *If* G *is transitive on* $P-p^{\perp}$ *and* G *is generated by elations about* p, *then* $(S^{(p)},G)$ *is an* EGQ.

Proof. Both (i) and (ii) are easy consequences of 8.1.1. Suppose there is some point $x \in P-p^{\perp}$ for which $|G| \neq |G_x| \neq 1$. Then by (ii) G is a Frobenius group on x^G (cf. [87]). So the Frobenius kernel of G acts regularly on x^G. If G is generated by elations about p (so trivially $|G| \neq |G_x|$ if $|G| > 1$), then G itself must act regularly on x^G. Since this holds for each $x \in P-p^{\perp}$, each element of G is an elation about p. Parts (iii), (iv) and (v) of the theorem are now easy consequences. □

8.2.5. *If* $S^{(p)}$ *is an* EGQ *of order* (s,t) *with elation group* G, $s \leq t$ *and* $|\{x,p\}^{\perp\perp}| = 2$ *for all* $x \in P-p^{\perp}$, *then* G *is the set of all elations about* p.

Proof. Let θ be an elation about p, and put $G_1 = \langle G,\theta\rangle$. Then $G = G_1$ by 8.2.4, implying $\theta \in G$. □

TGQ were first introduced by J.A. Thas [189] only for the case s = t, and the definition was equivalent to but different from that given here. An EGQ $(S^{(p)},G)$ of order (s,s) was defined in [189] to be a TGQ provided p is coregular, in which case it was shown that G is abelian, so the two definitions are indeed equivalent. Moreover, if p is a coregular point of S, the set E of elations about p was shown to be a group. Some of the technical details were isolated and sharpened slightly by S.E. Payne in [128], from which we take the following.

8.2.6. *Let* (p,L) *be an incident point-line pair of the* GQ S *of order* s. *Let* E *be the set of elations about* p, *and let* $\theta \in E$. *Then the following hold :*

(i) *The collineation* $\bar{\theta}$ *induced by* θ *on the projective plane* π_L *(as in the dual of* 1.3.1) *is an elation of* π_L *with axis* p, *if* L *is regular.*

(ii) E *is a group if* L *is regular.*

(iii) *If* p *is regular and* E *is a group, then the collineation* $\bar{\theta}$ *induced by* θ *on the projective plane* π_p *(as in* 1.3.1) *is an elation with center* p.

Proof. Suppose L is regular. Then θ clearly induces a central collineation $\bar{\theta}$ on π_L with axis p. The problem is to show that the center of $\bar{\theta}$ must be incident with p in π_L. Suppose otherwise, i.e. there is a line M of S that as a point of π_L is the center of $\bar{\theta}$, and M meets L at a point y, $y \neq p$. Then $M^\theta = M^{\bar{\theta}} = M$, so θ permutes the points of M different from y. As θ fixes no point w of $P-p^\perp$, any power of θ with a fixed point w on M, $w \neq y$, must fix at least two points of M different from y, and hence by 8.1.1 must be the identity. Hence θ splits points of M different from y into cycles of length n, where n is the order of θ. So $n \mid s$. The same argument applied to the s-1 points of L different from p and y shows that $n \mid (s-1)$. Hence n = 1. Consequently, if $\bar{\theta} \neq id$, then the center of $\bar{\theta}$ must be on p in π_L, proving (i).

For the proof of (ii) it suffices to show that E is closed. Let θ_1, $\theta_2 \in E$, and suppose that $\theta_1\theta_2$ fixes a point y, $y \not\sim p$. Let y I M I z I L, with L regular. Then θ_1 and θ_2 induce elations $\bar{\theta}_1$ and $\bar{\theta}_2$, respectively, on π_L, with axis p. Hence $\theta_1\theta_2$ induces an elation $\overline{\theta_1\theta_2} = \bar{\theta}_1\bar{\theta}_2$ with axis p. But clearly $\overline{\theta_1\theta_2} = \bar{\theta}_1\bar{\theta}_2$ fixes M, so must be the identity on π_L. Hence $\theta_1\theta_2$ fixes y, $y \not\sim p$, and also fixes every line meeting L. By 8.1.1 $\theta_1\theta_2 = id$, completing the proof of (ii).

Finally, suppose that p is regular and that E is a group. Clearly θ induces a central collineation $\bar{\theta}$ of π_p with center p. Moreover, E must act semiregularly on the s^3 points of $P-p^\perp$, so the order of θ is prime to s-1. But if $\bar{\theta}$ is an homology of π_p, its order must divide s-1, implying $\bar{\theta} = id$. Hence if $\bar{\theta} \neq id$, then $\bar{\theta}$ is an elation of π_p. □

For the last theorem of this section we adopt the following notation. $(S^{(p)},G)$ is an EGQ identified with $S(G,J)$ as in 8.2.1, and 1 denotes the identity of G. Further, E is the set of all elations about $p = (\infty)$, W the group of all whorls about (∞), $H = W_1$ = the group of whorls about (∞) fixing 1, and A the group of automorphisms of G for which $S_i^\alpha = S_i$ for all $i = 0,1,\ldots,t$. Finally, the elation group of $S(G,J)$ which corresponds to G will also be denoted by G.

8.2.7. (i) $N_W(G) \cap H = N_H(G) = A \in H$.

(ii) $E = G$ *iff* E *is a group, in which case* $A = H$.

Proof. Here we are identifying an element $g \in G$ with the elation θ_g defined by $h^{\theta_g} = hg$, $(S_ih)^{\theta_g} = S_ihg$, etc. As mentioned above, S_i^* is the union of S_i together with those cosets of S_i which are disjoint from all S_j. Hence if $\alpha \in A$, then $S_i^\alpha = S_i$ implies $(S_i^*)^\alpha = S_i^*$, so that α defines a whorl about (∞) with fixed point 1, with $(S_jg)^\alpha = S_ig^\alpha$ and $(S_i^*g)^\alpha = S_i^*g^\alpha$. Hence $A \subset H$. Now suppose $\alpha \in H$ and $\alpha^{-1}G\alpha = G$. We must show $\alpha \in A$. Clearly α defines a permutation of the elements of G, and since $S_i^\alpha = S_i$ for all $i = 0,\ldots,t$ we need only show that α preserves the operation of G. By hypothesis, if $g \in G$, then $\alpha^{-1}\theta_g\alpha \in G$. But $1^{\alpha^{-1}\theta_g\alpha} = g^\alpha$, so $\alpha^{-1}\theta_g\alpha = \theta_{g^\alpha}$ (or, by identification of g and θ_g, $\alpha^{-1}g\alpha = g^\alpha$). Hence $(gh)^\alpha = (1^{\theta_g\theta_h})^\alpha = 1^{\alpha.\alpha^{-1}\theta_g\alpha.\alpha^{-1}\theta_h\alpha} = 1^{\theta_{g^\alpha}\cdot\theta_{h^\alpha}} = g^\alpha h^\alpha$. This shows that $N_H(G) \subset A$. Now suppose $\alpha \in A$. We claim $\alpha^{-1}G\alpha = G$. For $g, h \in G$, $h^{\alpha^{-1}\theta_g\alpha} = (h^{\alpha^{-1}}g)^\alpha = h^{\alpha^{-1}\alpha}g^\alpha = h^{\theta_{g^\alpha}}$, implying $\alpha^{-1}\theta_g\alpha = \theta_{g^\alpha} \in G$. This essentially completes the proof of (i).

For the proof of (ii), clearly E is a group iff $E = G$. So suppose $E = G$ and let $\alpha \in H$. Then $\alpha^{-1}G\alpha \subset E = G$, implying $\alpha \in N_H(G) = A$. □

8.3. RECOGNIZING TGQ

8.3.1. *Let* $S = (P,B,I)$ *be a* GQ *of order* (s,t). *Suppose each line through some point* p *is an axis of symmetry, and let* G *be the group generated by the symmetries about the lines through* p. *Then* G *is abelian and* $(S^{(p)},G)$ *is a* TGQ.

Proof. For $s = t = 2$, $S \cong W(2)$, so since $s \leq t$ we may assume $t > 2$. Let $L_0,\ldots,L_t$ be the lines through p, with S_i the group of symmetries about L_i, $0 \leq i \leq t$, so that $|S_i| = s \leq t$. For $i \neq j$, each element of S_i commutes with each element of S_j (cf. 8.1.3). For each i, $0 \leq i \leq t$, put $G_i = \langle S_j \parallel 0 \leq j \leq t, j \neq i\rangle$. So $[S_i,G_i] = 1$ and $G = S_iG_i$. One goal is to show that $G = G_i$, from which it follows that S_i is abelian and G is abelian.

The first step is to show that G_i is transitive on $P-p^\perp$, and with no loss in generality we consider $i = 0$. Let $x_1,\ldots,x_s$ be the points on L_0 different from p. If a point y of $P-p^\perp$ is collinear with x_j, there is a symmetry about L_1 moving y to a point collinear with x_1. Hence we need only to show that G_0 is transitive on $x_1^\perp \cap (P-p^\perp)$. Let M_1,M_2 be two distinct lines through x_1, $L_0 \neq M_i$, and let y_i I M_i, $y_i \neq x_1$, $i = 1,2$. It suffices to show that y_1 and y_2 are in the same G_0-orbit. First suppose some point $u \in \{y_1,y_2\}^\perp$, $u \neq x_1$, is collinear with x_j, $2 \leq j \leq s$. Let L_{j_i} be the line through p meeting the line y_iu, $i = 1,2$ (note $j_i \neq 0$). As y_i and u are in the same S_{j_i}-orbit, $i = 1,2$, it follows that y_1 and y_2 are in the same G_0-orbit. On the other hand, if each point in $\{y_1,y_2\}^\perp$ is in $p^\perp$, let y_3 be a point of

$P-p^{\perp}$ for which (y_1,y_2,y_3) is a triad with center x_1 and $y_3 \notin \{y_1,y_2\}^{\perp\perp}$. (Such a point exists since $t > 2$ and $s > 1$.) Hence by the previous case y_3 and y_i are in the same G_0-orbit, $i = 1,2$. It follows that G_0 (and hence also G) is transitive on $P-p^{\perp}$.

The next step is to show that $G = G_i$, where again we may take $i = 0$. As $|P-p^{\perp}| = s^2t$, if $y \in P-p^{\perp}$, $|G| = s^2tk$, where $k = |G_y|$, and $|G_0| = s^2tm$, where $m = |(G_0)_y|$. Clearly $m|k$, say $mr = k$. Then $s^2tk = |G| = |S_0G_0| = |S_0|.|G_0|/|S_0 \cap G_0| = s^3tm/|S_0 \cap G_0|$, implying $r|S_0 \cap G_0| = s$. Hence $r|s$ and $r|k$. Let q be a prime dividing r. Then there must be a collineation $\theta \in G_y$ having order q. Let M be the line through y meeting L_0 at x_i. Clearly θ fixes L_0 and M. The orbits of θ on M consist of cycles of length q and fixed points including y and x_i. As $q|s$, there are at least $q+1$ points of M fixed by θ. Moreover, each point of $\{y,p\}^{\perp}$ is fixed by θ. Considering the possible substructures of fixed elements allowed by 8.1.1 if $\theta \neq \mathrm{id}$, we have a contradiction. Hence $r = 1$, implying $G = G_0$.

At this point we know that G is an abelian group transitive on $P-p^{\perp}$, and hence by elementary permutation group theory must be regular on $P-p^{\perp}$. By 8.2.3 the proof is complete. □

8.3.2. *The translation group of a* TGQ *is uniquely defined and is abelian.*
Proof. Let $(S^{(p)},G)$ be a TGQ. If G' is the group generated by the symmetries about lines through p, then by 8.3.1 we have $s^2t = |G'|$. As also $s^2t = |G|$ and $G' \leqslant G$, clearly $G = G'$. □

If $(S^{(p)},G)$ is a TGQ, the elements of G are called the *translations about* p.

8.3.3. (*J.A. Thas* [189]). *If* $(S^{(p)},G)$ *is an* EGQ *with* $s = t$ *and* p *coregular, then* $(S^{(p)},G)$ *is a* TGQ. *Moreover,* $G = E$.
Proof. By 8.2.6 (i) the elations in G fixing a line M not through p are symmetries about the line through p meeting M. Hence each line through p is an axis of symmetry and all these symmetries are in G, implying $(S^{(p)},G)$ is a TGQ. By 8.2.6 (ii) and 8.2.7 (ii) we have $G = E$, which finishes the proof. □

8.4. FIXED SUBSTRUCTURES OF TRANSLATIONS

Let $(S^{(p)},G)$ be a TGQ, so that G is abelian and $s \leqslant t$. As above let $L_0,\ldots,L_t$

be the lines through p and S_i the group of symmetries about L_i, $0 \leq i \leq t$. With $J = \{S_0,\ldots,S_t\}$, recall the coset geometry notation of 8.2. Then $\theta \in G$ fixes a point S_i^*g of $[S_i]$ iff $\theta \in g^{-1}S_i^*g = S_i^*$ iff θ fixes all points of $[S_i]$. In other words, $\theta \in G$ fixes one point ($\neq p$) of L_i iff θ fixes all points of L_i, and S_i^* is the *point stabilizer* of L_i.

8.4.1. *The substructure* $S_\theta = (P_\theta, B_\theta, I_\theta)$ *of the fixed elements of the non-identity translation* θ *must be given by one of the following :*

(i) P_θ *is the set of all points on* r *lines through* p *and* B_θ *is the set of all lines through* p, $1 \leq r \leq 1+t$.

(ii) $P_\theta = \{p\}$ *and* B_θ *is the set of lines through* p.

(iii) P_θ *is the set of all points on one line* L_i *through* p *and* B_θ *is the set of all lines concurrent with* L_i *, i.e.* θ *is a symmetry about* L_i.

Proof. By the remark preceding 8.4.1 and by 8.1.1 we have possibilities (i), (ii) or P_θ is the set of all points on one line L_i through p, and B_θ consists of at least t+2 lines concurrent with L_i. In the last case let $L^\theta = L$, $p \not\mathrm{I}\, L$, and assume $x^\theta = y$ with x I L, $x \not\sim p$. Since the translation group acts regularly on $P-p^\perp$, θ must be the unique symmetry about L_i with $x^\theta = y$. □

There is an easy corollary.

8.4.2. *Let* $x \in P-p^\perp$. *For each* $z \in P-p^\perp$ *there is a unique* $\theta \in G$ *with* $x^\theta = z$. *Moreover,* (p,x,z) *is a triad iff* θ *is not a symmetry about some line through* p, *in which case the number of centers of* (p,x,z) *is the number* r *of lines of fixed points of* θ.

8.4.3. (i) $|S_i^* \cap S_j^*| = t$, *if* $0 \leq i < j \leq t$.

(ii) $|S_i^* \cap S_j^* \cap S_k^*| \geq t/s$, *if* $0 \leq i < j < k \leq t$.

Proof. With the notation of 8.2, $S_i^* \cap S_j^*$ acts regularly on $\{z_i,z_j\}^\perp-\{p\}$, proving (i). And for i,j,k distinct, we have $|(S_i^* \cap S_j^*)S_k^*| = |S_i^* \cap S_j^*|.|S_k^*|/|S_i^* \cap S_j^* \cap S_k^*| \leq |G|$, implying (ii). □

Part (ii) of the preceding result has the following corollary.

8.4.4. *If* $S^{(p)}$ *is a* TGQ, *any triad of points with at least two centers and having* p *as center must have at least* 1+t/s *centers.*

Proof. A triad having p as center and having g as center must be of the form (S_i^*g, S_j^*g, S_k^*g). But then $g^{-1}(S_i^* \cap S_j^* \cap S_k^*)g$ is a subgroup fixing the triad and whose orbit containing g provides at least t/s centers ($\neq$p) of the triad. □

8.5. THE KERNEL OF A TGQ

Let $(S^{(p)},G)$ be a TGQ with S_i, S_i^*, J, etc., as above. The *kernel* K of $S^{(p)}$ (or of $(S^{(p)},G)$ or of J) is the set of all endomorphisms α of G for which $S_i^\alpha \subset S_i$, $0 \leq i \leq t$. With the usual addition and multiplication of endomorphisms, K is a ring.

As the only GQ with s = 2 and t > 1 are W(2) and Q(5,2), we may assume in this section that 2 < s.

8.5.1. K *is a field, so that* $S_i^\alpha = S_i$, $(S_i^*)^\alpha = S_i^*$ *for all* $i = 0,1,\ldots,t$ *and all* $\alpha \in K^\circ = K - \{0\}$.

Proof. If each $\alpha \in K^\circ$ is an automorphism of G, then clearly K is a field. So suppose some $\alpha \in K^\circ$ is not an automorphism. Then $\langle S_0,\ldots,S_t\rangle = G \supsetneq G^\alpha = \langle S_0^\alpha,\ldots,S_t^\alpha\rangle$, implying $S_i^\alpha \neq S_i$ for some i. Let $g^\alpha = 1$, $g \in S_i - \{1\}$. If i,j,k are mutually distinct and $g' \in S_j$ with $\{g'\} \neq S_j \cap S_k^* g^{-1}$, then $gg' = hh'$ with $h \in S_k$, $h' \in S_\ell$, for a uniquely defined $\ell, \ell \neq k,j$. (This holds because $S_k^*, S_kS_0 - S_k, S_kS_1 - S_k, \ldots, S_kS_t - S_k$ (omitting the term $S_kS_k - S_k$) is a partition of the set G.) Hence $h^\alpha h'^\alpha = g'^\alpha$, implying that $h^\alpha = h'^\alpha = g'^\alpha = 1$ (by K1). Since g' was any one of s-1 elements of S_j, $|\ker(\alpha) \cap S_j| \geq s-1 > s/2$, implying $S_j \subset \ker(\alpha)$. This implies $S_j \subset \ker(\alpha)$ for each j, $j \neq i$, so that $G = G_i \subset \ker(\alpha)$, recalling G_i from the proof of 8.3.1. This says $\alpha = 0$, a contradiction. Hence we have shown that K is a field and $S_i^\alpha = S_i$ for $i = 0,\ldots,t$ and $\alpha \in K^\circ$. Since S_i^* is the set-theoretic union of S_i together with all those cosets of S_i disjoint from $\cup\{S_i \parallel 0 \leq i \leq t\}$ (cf. the remark preceding 8.2.2) we also have $(S_i^*)^\alpha = S_i^*$. □

For each subfield F of K there is a vector space (G,F) whose vectors are the elements of G, and whose scalars are the elements of F. Vector addition is the group operation in G, and scalar multiplication is defined by $g\alpha = g^\alpha$, $g \in G, \alpha \in F$. It is easy to verify that (G,F) is indeed a vector space. There is an interesting corollary.

8.5.2. G *is elementary abelian, and* s *and* t *must be powers of the same prime. If* s < t, *then there is a prime power* q *and an odd integer* a *for which* $s = q^a$ *and* $t = q^{a+1}$.

Proof. Let |F| = q, so q is a prime power. Since G is the additive group of a vector space, it must be elementary abelian. Moreover, S_i and S_i^* may be viewed as subspaces of (G,F). Hence $|S_i| = s = q^n$ and $|S_i^*| = st = q^{n+m}$. By 8.1.2 $q^{n+m}(1+q^n) \equiv 0 \pmod{q^n+q^m}$, implying $1+q^n \equiv 0 \pmod{1+q^{m-n}}$, if $s < t$, i.e. $m \neq n$. Since $n < m \leqslant 2n$ we may write $m = n+v$, with $0 < v \leqslant n$, so $(1+q^v) \mid (1+q^n)$. Put $n = av+r$, $0 \leqslant r < v$. Then $1+q^n = 1+(q^v)^a q^r \equiv 1+(-1)^a q^r \equiv 0 \pmod{1+q^v}$. This is possible only if $r = 0$ and a is odd, in which case $s = q^n = (q^v)^a$ and $t = q^m = q^{n+v} = (q^v)^{a+1}$. □

The kernel of a TGQ is useful in describing the given GQ in terms of an appropriate projective space. Before pursuing this idea, however, we obtain some additional combinatorial information.

8.6. THE STRUCTURE OF TGQ

Let $(S^{(p)},G)$ be a TGQ of order (s,t). If s = t, then p is regular when s is even, antiregular when s is odd (cf. 1.5.2), so that a triad containing p has 1 or 1+s centers when s is even and 0 or 2 centers when s is odd. For the remainder of this section we suppose $s = q^a$, $t = q^{a+1}$, where q is a prime power and a is odd. And we continue to use the notation S_i, S_i^*, J, etc., of the preceding sections.

8.6.1. *Let* x *be a fixed point of* $P-p^\perp$, *and let* N_i *be the number of triads* (p,x,y) *having exactly* i *centers,* $0 \leqslant i \leqslant 1+t$. *Then the following hold :*

(i) $N_0 = t(s-1)(s^2-t)/(s+t)$,

(ii) $N_{1+q} = (t^2-1)s^2/(s+t)$,

(iii) $N_i = 0$ *for* $i \notin \{0,1+q\}$,

(iv) $t = s^2$ *if* q *is even.*

Proof. Suppose (p,x,y) is a triad with r centers, and let θ be the unique translation for which $x^\theta = y$. Then using 8.4.2 and 1.9.1 with $f = 1+rs$ and $g = (t+1-r)s$, we have $r \equiv 0 \pmod{1+q}$. Hence $N_i \neq 0$ implies $i \equiv 0 \pmod{1+q}$. In particular $N_i = 0$ for $0 < i < 1+q$, so that by 1.7.1 , (iii) must hold, as well as (i) and (ii). Finally, from 1.5.1 (iii), if t is even then $N_0 = 0$, so (iv) follows from (i). □

Of course from 1.7.1 (i) we also have the following.

8.6.2. *Each triad of points in* $p^\perp$ *has exactly* 1+q *centers.*

Interpreting these results for G, S_i, S_i^*, etc., we have

8.6.3. (i) *If* i,j,k *are distinct, then* $|S_i^* \cap S_j^* \cap S_k^*| = q$ (*cf.* 8.4.3 *and* 8.4.4).

(ii) *If* $\theta \in G$ *belongs to no* S_i, *then it belongs to* S_i^* *for exactly* $1+q$ *values of* i *or for no value of* i, *with the latter actually occuring precisely when* $a > 1$, *i.e. when* $t < s^2$.

8.6.4. *If* $(S^{(p)},G)$ *is a* TGQ *of order* (s,t), $s \leq t$, *then* G *is the complete set of all elations about* p.

Proof. For s = t see 8.3.3. For s < t it follows from 8.6.1 that $|\{p,x\}^{\perp\perp}| = 2$ for each $x \in P-p^{\perp}$, so the proof is complete by 8.2.5. □

8.6.5. *The multiplicative group* K° *of the kernel is isomorphic to the group of all whorls about* p *fixing a given* $y, y \not\sim p$.

Proof. This is an immediate corollary of 8.6.4 and 8.2.7. □

8.7. <u>T(n,m,q)</u>

In PG(2n+m-1,q) consider a set O(n,m,q) of q^m+1 (n-1)-dimensional subspaces $PG^{(0)}(n-1,q),\dots,PG^{(q^m)}(n-1,q)$, every three of which generate a PG(3n-1,q), and such that each element $PG^{(i)}(n-1,q)$ of O(n,m,q) is contained in a $PG^{(i)}(n+m-1,q)$ having no point in common with any $PG^{(j)}(n-1,q)$ for $j \neq i$. It is easy to check that $PG^{(i)}(n+m-1,q)$ is uniquely determined, $i = 0,\dots,q^m$. The space $PG^{(i)}(n+m-1,q)$ is called the *tangent space* of O(n,m,q) at $PG^{(i)}(n-1,q)$. Embed PG(2n+m-1,q) in a PG(2n+m,q), and construct a point-line geometry T(n,m,q) as follows.

Points are of three types :

(i) The points of PG(2n+m,q)-PG(2n+m-1,q).

(ii) The (n+m)-dimensional subspaces of PG(2n+m,q) which intersect PG(2n+m-1,q) in one of the $PG^{(i)}(n+m-1,q)$.

(iii) The symbol (∞).

Lines are of two types :

(a) The n-dimensional subspaces of PG(2n+m,q) which intersect PG(2n+m-1,q) in a $PG^{(i)}(n-1,q)$.

(b) The elements of O(n,m,q).

Incidence in T(n,m,q) is defined as follows : A point of type (i) is incident only with lines of type (a); here the incidence is that of PG(2n+m,q). A point of type (ii) is incident with all lines of type (a) contained in it and with the unique element of O(n,m,q) contained in it. The point (∞) is incident with no line of type (a) and with all lines of type (b).

8.7.1. T(n,m,q) *is a* TGQ *of order* (q^n,q^m) *with base point* (∞) *and for which* GF(q) *is a subfield of the kernel. Moreover, the translations of* T(n,m,q) *induce the translations of the affine space* AG(2n+m,q) = PG(2n+m,q)-PG(2n+m-1,q). *Conversely, every* TGQ *for which* GF(q) *is a subfield of the kernel is isomorphic to a* T(n,m,q). *It follows that the theory of* TGQ *is equivalent to the theory of the sets* O(n,m,q).

Proof. It is routine to show that T(n,m,q) is a GQ of order (q^n,q^m). A translation of AG(2n+m,q) defines in a natural way an elation about (∞) of T(n,m,q). It follows that T(n,m,q) is an EGQ with abelian elation group G, where G is isomorphic to the translation group of AG(2n+m,q), and hence T(n,m,q) is a TGQ. (With the q^n translations of AG(2n+m,q) having center in $PG^{(i)}(n-1,q)$ there correspond q^n symmetries of T(n,m,q) about the line $PG^{(i)}(n-1,q)$ of type (b).) It also follows that GF(q) is a subfield of the kernel of T(n,m,q) : with the group of all homologies of PG(2n+m,q) having center $y \notin PG(2n+m-1,q)$ and axis PG(2n+m-1,q) there corresponds in a natural way the multiplicative group of a subfield of the kernel (cf. 8.6.5).

Conversely, consider a TGQ $(S^{(p)},G)$ for which GF(q) = F is a subfield of the kernel. If $s = q^n$ and $t = q^m$, then $[(G,F):F] = 2n+m$. Hence with $S^{(p)}$ there corresponds an affine space AG(2n+m,q). The cosets S_ig of a fixed S_i are the elements of a parallel class of n-dimensional subspaces of AG(2n+m,q), and the cosets S_i^*g of a fixed S_i^* are the elements of a parallel class of (n+m)-dimensional subspaces of AG(2n+m,q). The interpretation in PG(2n+m,q) together with K1 and K2 prove the last part of the theorem. □

8.7.2. *The following hold for any* O(n,m,q) *:*

(i) n = m *or* n(a+1) = ma *with* a *odd* .

(ii) *If* q *is even* , *then* n = m *or* m = 2n.

(iii) *If* n ≠ m (*resp*., 2n = m), *then each point of* PG(2n+m-1,q) *which is not contained in an element of* O(n,m,q) *belongs to* 0 *or* $1+q^{m-n}$ (*resp*., *to exactly* $1+q^n$) *tangent spaces of* O(n,m,q).

(iv) *If* n ≠ m, *the* q^m+1 *tangent spaces of* O(n,m,q) *form an* $O^*(n,m,q)$ *in the dual space of* PG(2n+m-1,q). *So in addition to* T(n,m,q) *there arises a* TGQ $T^*(n,m,q)$.

(v) *If* n ≠ m (*resp*., 2n = m), *then each hyperplane of* PG(2n+m-1,q) *which does not contain a tangent space of* O(n,m,q) *contains* 0 *or* $1+q^{m-n}$ *(resp., contains exactly* $1+q^n$*) elements of* O(n,m,q).

Proof. Since T(n,m,q) is a TGQ of order (q^n,q^m), by 8.5.2 n = m or ma = n(a+1) with a odd, which proves (i).

Let q be even. Then $t = s^2$ by 8.6.1 (iv), i.e. m = 2n.

Next, let $n \neq m$ and let x be a point of PG(2n+m-1,q) which is not contained in an element of O(n,m,q). Consider distinct points y,z of type (i) of T(n,m,q), chosen so that x,y,z are collinear in PG(2n+m,q). Then $|\{(\infty),y,z\}^{\perp}|$ is the number of tangent spaces of O(n,m,q) that contain x. By 8.6.1 (iii) $|\{(\infty),y,z\}^{\perp}| \in \{0,1+q^{m-n}\}$. If 2n = m, i.e. $t = s^2$, then clearly $|\{(\infty),y,z\}^{\perp}| = s+1 = q^n+1$, so that (iii) is completely proved.

Now consider the tangent spaces $PG^{(i)}(n+m-1,q) = \pi_i$, $PG^{(j)}(n+m-1,q) = \pi_j$, $PG^{(k)}(n+m-1,q) = \pi_k$ of O(n,m,q), with i,j,k distinct. If x is a point of type (i) of T(n,m,q), then the spaces $x\pi_i$, $x\pi_j$, $x\pi_k$ are points of type (ii) of T(n,m,q). Clearly $|\{x\pi_i,x\pi_j,x\pi_k\}^{\perp}| = q^r+1$, with r-1 the dimension of $\pi_i \cap \pi_j \cap \pi_k$. By 8.6.1 and 1.7.1 (i) $q^r = q^{m-n}$, i.e. r = m-n. Now (iv) easily follows. (The tangent spaces of $O^*(n,m,q)$ are the elements of O(n,m,q).)

Finally, by applying (iii) to $O^*(n,m,q)$, (v) is obtained. □

8.7.3. *Let* $(S^{(p)},G)$ *be a* TGQ *of order* (s,t).

(i) *If* s *is prime, then* $S \cong Q(4,s)$ *or* $S \cong Q(5,s)$.

(ii) *If all lines are regular, then* $S \cong Q(4,s)$ *or* $t = s^2$.

Proof. If s is a prime, then either n = 1 = a, m = 2, or n = m = 1. Moreover, T(1,m,s) is a $T_{m+1}(O)$ of J. Tits, m = 1,2 (cf. 3.1.2). If s is prime, then the oval or ovoid, respectively, is a conic or elliptic quadric, so that $S \cong Q(4,s)$ or $S \cong Q(5,s)$ (cf. 3.2.2 and 3.2.4).

Now assume all lines are regular. If s = t, then $S \cong Q(4,s)$ by 5.2.1. So suppose s < t. By 1.5.1 (iv) s+1 divides $(t^2-1)t^2$, but with $s = q^n$, $t = q^m$, s+1 must divide t^2-1. Since ma = n(a+1), a odd, we have $t^2-1 = q^{2m}-1 = q^{2n+2n/a}-1 = (q^n)^2q^{2n/a}-1 \equiv q^{2n/a}-1 \pmod{q^n+1}$, implying n < 2n/a, or a < 2, i.e. a = 1 and $t = s^2$. □

The only known TGQ are the $T_2(O)$ and $T_3(O)$ of J. Tits, and it is useful to have characterizations of these among all TGQ.

8.7.4. *Let* $(S^{(p)},G)$ *be a* TGQ *arising from the set* O(n,2n,q). *Then* $S^{(p)} \cong T_3(O)$ *if and only if any one of the following holds :*

(i) *For a fixed point* $y, y \not\sim p$, *the group of all whorls about* p *fixing* y *has order* s-1.

(ii) *For each point* z *not contained in an element of* O(n,2n,q), *the* q^n+1 *tangent spaces containing* z *have exactly* $(q^n-1)/(q-1)$ *points in common.*

(iii) *Each* PG(3n-1,q) *containing at least three elements of* O(n,2n,q) *contains exactly* q^n+1 *elements of* O(n,2n,q).

Proof. In view of 8.6.5, the condition in (i) is just that the kernel has order s, which means $S^{(p)}$ is a $T(1,2,q^n)$, i.e. a $T_3(O)$. We now show that the condition in (ii) is equivalent to the 3-regularity of the point p. Consider a triad (p,x,y). Then all points of $\{p,x,y\}^{\perp}$, which clearly are s+1 points of the second type, are obtained as follows. Let z be the intersection of the line xy of PG(4n,q) and the hyperplane PG(4n-1,q). If $PG^{(i_1)}(3n-1,q),\ldots,PG^{(i_{s+1})}(3n-1,q)$ are the tangent spaces of O(n,2n,q) through z, then $\{p,x,y\}^{\perp}$ consists of the 3n-dimensional spaces $zPG^{(i_j)}(3n-1,q)$, $j = 1,\ldots,s+1$. Notice that every point of the line xy which is not in PG(4n-1,q) is in $\{p,x,y\}^{\perp\perp}$, so that $|\{p,x,y\}^{\perp\perp}| \geq 1+q$. Finally, $|\{p,x,y\}^{\perp\perp}| = q^n+1$ iff the q^n+1 spaces $PG^{(i_j)}(3n-1,q)$ have an (n-1)-dimensional space in common, which proves (ii).

Condition (iii) for O(n,2n,q) is merely condition (ii) for $O^*(n,2n,q)$. Hence (iii) is satisfied iff $T^*(n,2n,q) \cong T_3(O^*)$ for some ovoid O^* of $PG(3,q^n)$. If $T^*(n,2n,q) \cong T_3(O^*)$ and O^* is not an elliptic quadric, then the point (∞) of $T_3(O^*)$ is the only coregular point of $T_3(O^*)$ (cf. 3.3.3 (iii)), and consequently the points (∞) of $T^*(n,2n,q)$ and $T_3(O^*)$ correspond to each other under any isomorphism between these GQ. On the other hand, if $T^*(n,2n,q) \cong T_3(O^*)$ and O^* is an elliptic quadric, then there is always an isomorphism between these GQ mapping the point (∞) of $T^*(n,2n,q)$ onto the point (∞) of $T_3(O^*)$. Suppose that O(n,2n,q) satisfies (iii). Since $T_3(O^*)$ has q^n-1 whorls about (∞) fixing any given point $y \not\sim (\infty)$, also $T^*(n,2n,q)$ has q^n-1 whorls about (∞) fixing any given point $z \not\sim (\infty)$. As $T^*(n,2n,q)$ is the interpretation of $T_3(O^*)$ in the 4n-dimensional space over the subfield GF(q) of the kernel $GF(q^n)$, it is clear that $O^*(n,2n,q)$ satisfies (iii). Hence O(n,2n,q) satisfies (ii), and then by the preceding paragraph $T(n,2n,q) \cong T_3(O)$. Conversely, assume that $T(n,2n,q) \cong T_3(O)$ for some ovoid of $PG(3,q^n)$. Again by the preceding argument O(n,2n,q) satisfies (iii). □

8.7.5. *Consider a* $T(n,2n,q)$ *with all lines regular. Then* $T(n,2n,q) \cong Q(5,q^n)$ *if the following conjecture is true :*

<u>Conjecture</u> : *In* $PG(4n-1,q)$ *let* $PG^{(i)}(n-1,q)$, $i = 0,1,\ldots,q^n$, *be* q^n+1 $(n-1)$-*dimensional subspaces, any three of which generate a* $PG(3n-1,q)$. *Suppose that each* $PG^{(i)}(n-1,q)$ *is contained in a* $PG^{(i)}(n,q)$, *in such a way that* $PG^{(i)}(n,q) \cap PG^{(j)}(n-1,q) = \phi$, *that* $PG^{(i)}(n,q) \cap PG^{(j)}(n,q)$ *is a point, and that the* $(2n-1)$-*dimensional space spanned by* $PG^{(i)}(n-1,q)$ *and* $PG^{(j)}(n-1,q)$ *contains a point of* $PG^{(k)}(n,q)$ *whenever* i,j,k *are distinct. Then the* q^n+1 *spaces* $PG^{(i)}(n-1,q)$ *are contained in a* $PG(3n-1,q)$.

Proof. Consider the TGQ $T(n,2n,q)$ arising from the set $O(n,2n,q) = \{PG^{(0)}(n-1,q),\ldots,PG^{(q^{2n})}(n-1,q)\}$, and assume that all lines of $T(n,2n,q)$ are regular. Let L_0,L_1 be two nonconcurrent lines of type (a), with $L_0 \sim PG^{(i_0)}(n-1,q)$, $L_1 \sim PG^{(i_1)}(n-1,q)$, and $i_1 \neq i_0$. Further, let $\{L_0,L_1\}^{\perp} = \{M_0,M_1,\ldots,M_{q^n}\}$ and $\{L_0,L_1\}^{\perp\perp} = \{L_0,L_1,\ldots,L_{q^n}\}$, with $L_j \sim PG^{(i_j)}(n-1,q) \sim M_j$, $j = 0,1,\ldots,q^n$. If $PG^{(i_j)}(n+1,q)$ is the space spanned by M_j and L_j, then let $PG^{(i_j)}(n+1,q) \cap PG(4n-1,q) = PG^{(i_j)}(n,q)$, with $PG(4n-1,q)$ the projective space containing the elements of $O(n,2n,q)$. Clearly $PG^{(i_j)}(n-1,q) \subset PG^{(i_j)}(n,q) \subset PG^{(i_j)}(3n-1,q)$, with $PG^{(i_j)}(3n-1,q)$ the tangent space of $O(n,2n,q)$ at $PG^{(i_j)}(n-1,q)$. Since $PG^{(i_j)}(n+1,q)$ and $PG^{(i_k)}(n+1,q)$, $j \neq k$, have a line in common, clearly $PG^{(i_j)}(n,q) \cap PG^{(i_k)}(n,q)$ is a point. Further, the $2n$-dimensional space containing M_j and $PG^{(i_k)}(n-1,q)$ (and hence also L_k), $j \neq k$, has a line in common with $PG^{(i_r)}(n+1,q)$, j,k,r distinct. Hence the $(2n-1)$-dimensional space spanned by $PG^{(i_j)}(n-1,q)$ and $PG^{(i_k)}(n-1,q)$ contains a point of $PG^{(i_r)}(n,q)$.

If the conjecture is true, then the q^n+1 spaces $PG^{(i_k)}(n-1,q)$, $k = 0,1,\ldots,q^n$, are contained in a $PG(3n-1,q)$. By 8.7.2 (v) $PG(3n-1,q)$ contains exactly q^n+1 elements of $O(n,2n,q)$. Now it follows from 8.7.4 (iii) that $T(n,2n,q) \cong Q(5,q^n)$. □

8.7.6. *If* $s = q^n = p^2$ *with* p *a prime, then any* $T(n,2n,q)$ *with all lines regular must be isomorphic to* $Q(5,s)$.

Proof. Clearly $n = 1$ or $n = 2$. If $n = 1$, the resulting $T(1,2,p^2)$ is clearly a Tits quadrangle T(O). Since all lines are regular it must be isomorphic to $Q(5,s)$ (cf. 3.3.3 (iii)). Now suppose $n = 2$, so $q = p$, and let $O(2,4,p) = \{PG^{(0)}(1,p),\ldots,PG^{(p^4)}(1,p)\}$. Use the notation of the proof of 8.7.5. Then, from $PG^{(i_0)}(2,p)$ project the lines $PG^{(i_j)}(1,p)$, $j = 1,\ldots,p^2$, onto a $PG(4,p) \subset PG(7,p)$ skew to $PG^{(i_0)}(2,p)$. There arise p^2 lines $PG^{(t_j)}(1,p)$, $j = 1,\ldots,p^2$, having in pairs exactly one point in common. So these lines either have a point in common or are contained in a plane. If $PG(3,q)$ contains $PG^{(i_0)}(2,p)$ but is not contained in $PG^{(i_0)}(5,p)$, then we know that $PG(3,p)$ has a point in common with p elements of $O(2,4,p)-\{PG^{(i_0)}(1,p)\}$ (every plane of $PG(3,p)$ through $PG^{(i_0)}(1,p)$, but different from $PG^{(i_0)}(2,p)$, contains exactly one point of some element of $O(2,4,p)$). Hence each point of $PG(4,p)$ is contained in at most p of the lines $PG^{(t_j)}(1,p)$, $j = 1,\ldots,p^2$. It follows that the p^2 lines $PG^{(t_j)}(1,p)$, $j = 1,\ldots,p^2$, are contained in a plane . Hence the p^2+1 lines $PG^{(i_j)}(1,p)$, $j = 0,\ldots,p^2$, are contained in a $PG(5,p)$. By 8.7.2 (v) $PG(5,p)$ contains exactly p^2+1 lines of $O(2,4,p)$. Now it follows from 8.7.4 (iii) that $T(2,4,p) \cong Q(5,p^2)$. □

For the remainder of this section we assume $n = m$, i.e. $s = t$.

Consider a line $L = PG^{(i)}(n-1,q)$ of type (b) of $T(n,n,q)$. Then L is regular. So with L corresponds a projective plane π_L of order q^n (cf. 1.3.1). By projection from $PG^{(i)}(n-1,q)$ onto a $PG(2n,q)$ skew to it in $PG(3n,q)$, it is seen that π_L is isomorphic to the plane π described as follows : points of π are the points of $PG(2n,q)-PG(3n-1,q)$, with $PG(3n-1,q)$ the $(3n-1)$-dimensional space containing $O(n,n,q)$, and the $(n-1)$-dimensional spaces $PG^{(i)}(2n-1,q) \cap PG(2n-1,q) = \Delta_0$, $\langle PG^{(i)}(n-1,q), PG^{(j)}(n-1,q)\rangle \cap PG(2n-1,q) = \Delta_j$, for all $j \neq i$, with $PG(2n-1,q) = PG(3n-1,q) \cap PG(2n,q)$; lines of π are $PG(2n-1,q)$ and the n-dimensional spaces in $PG(2n,q)$ which contain a Δ_k, $k = 0,\ldots,q^n$, and are not contained in $PG(2n-1,q)$; incidence is containment. Hence up to an isomorphism π_L is the projective completion of the affine translation plane defined by the $(n-1)$-spread [49] $\{\Delta_0,\ldots,\Delta_{q^n}\} = V_i$ of $PG(2n-1,q)$.

Let q be even. Then by 1.5.2 the coregular point (∞) is regular. It follows that all tangent spaces of $O(n,n,q)$ have a space $PG(n-1,q)$ in common (cf. also [179]). This space is called the *nucleus* or *kernel* of $O(n,n,q)$. By projection from $PG(n-1,q)$ onto a $PG(2n,q)$ skew to it (in $PG(3n,q)$), it is seen that the projective plane $\pi_{(\infty)}$ arising from the regular point (∞) is isomorphic to the plane π described as follows : points of π are $PG(2n-1,q) = PG(2n,q) \cap PG(3n-1,q)$ (with $PG(3n-1,q)$ the space containing $O(n,n,q)$) and the n-dimensional spaces in $PG(2n,q)$ which contain a $\Gamma_k = \langle PG(n-1,q), PG^{(k)}(n-1,q)\rangle \cap PG(2n-1,q)$, $k = 0,\dots,q^n$, and are not contained in $PG(2n-1,q)$; lines of π are the points of $PG(2n,q)-PG(2n-1,q)$ and the spaces $\Gamma_0,\dots,\Gamma_{q^n}$; and incidence is containment. Hence up to an isomorphism $\pi_{(\infty)}$ is the dual of the projective completion of the affine translation plane defined by the (n-1)-spread $\{\Gamma_0,\dots,\Gamma_{q^n}\} = V$ of $PG(2n-1,q)$.

Now let q be odd. Then by 1.5.2 the coregular point (∞) is antiregular. It follows that any point of $PG(3n-1,q)$ which is not contained in $O(n,n,q)$ is in exactly 0 or 2 tangent spaces of $O(n,n,q)$ (cf. also [142]). Let $PG(2n,q)$ be a 2n-dimensional subspace of $PG(3n,q)$ which contains the tangent space $PG^{(i)}(2n-1,q)$ of $O(n,n,q)$ and is not contained in $PG(3n-1,q)$. Then the affine plane $\pi((\infty),PG(2n,q))$ (cf. 1.3.2) is easily seen to be isomorphic to the following structure π : points of π are the n-dimensional spaces of $PG(2n,q)$ intersecting $PG^{(i)}(2n-1,q)$ in an element $PG^{[j]}(n-1,q) = PG^{(i)}(2n-1,q) \cap PG^{(j)}(2n-1,q)$, $j \neq i$; lines of π are the spaces $PG^{[j]}(n-1,q)$, $j \neq i$, and the points in $PG(2n,q)-PG^{(i)}(2n-1,q)$; incidence is containment. The projective completion of π is the dual of the projective translation plane arising from the (n-1)-spread $V_i^* = \{PG^{(i)}(n-1,q)\} \cup \{PG^{[j]}(n-1,q), j \neq i\}$ of $PG^{(i)}(2n-1,q)$.

If $T(n,n,q)$ is isomorphic to a $T_2(O)$ of J. Tits, then all corresponding projective or affine planes are desarguesian, and hence all corresponding (n-1)-spreads are regular [49].

8.7.7. (*L.R.A. Casse, J.A. Thas and P.R. Wild* [37]). *Consider an* $O(n,n,q)$ *with* q *odd. Then at least one of the* (n-1)*-spreads* $V_0,\dots,V_{q^n}$ *is regular iff at least one of the* (n-1)*-spreads* $V_0^*,\dots,V_{q^n}^*$ *is regular. In such a case all* (n-1)*-spreads* $V_0,\dots,V_{q^n}$, $V_0^*,\dots,V_{q^n}^*$ *are regular and* $T(n,n,q)$ *is isomorphic to* $Q(4,q^n)$.

Proof. Let V_i^*, $i \in \{0,\dots,q^n\}$, be regular. Then by 5.2.7 $T(n,n,q) \cong Q(4,q^n)$. Consequently all (n-1)-spreads $V_0,\dots,V_{q^n}$, $V_0^*,\dots,V_{q^n}^*$ are regular. Next, let V_i, $i \in \{0,\dots,q^n\}$, be regular. Since q is odd, the set

$\{PG^{(0)}(2n-1,q),\ldots,PG^{(q^n)}(2n-1,q)\}$ of all tangent spaces of $O(n,n,q)$ is a set $\hat{O}(n,n,q)$ relative to the dual space $\widehat{PG}(3n-1,q)$ of $PG(3n-1,q)$. The elements of $O(n,n,q)$ are the tangent spaces of $\hat{O}(n,n,q)$. Clearly the (n-1)-spreads V_j and $\hat{V}^*_j$ (resp., $\hat{V}_j$ and V^*_j), $j = 0,\ldots,q^n$, may be identified. Since $\hat{V}^*_i$ is regular, by the first part of the proof all (n-1)-spreads $\hat{V}_0,\ldots,\hat{V}_{q^n}$, $\hat{V}^*_0,\ldots,\hat{V}^*_{q^n}$ are regular. Hence $V^*_0,\ldots,V^*_{q^n}$, $V_0,\ldots,V_{q^n}$ are regular, and the theorem is completely proved. □

9 Moufang conditions

Most of the results and/or details of proofs in this chapter either came from or were directly inspired by the following works of J.A. Thas and/or S.E. Payne : [144, 199, 214, 215].

9.1. DEFINITIONS AND AN EASY THEOREM

Let S = (P,B,I) be a GQ of order (s,t). For a fixed point p define the following condition.

$(M)_p$: *For any two lines* A *and* B *of* S *incident with* p, *the group of collineations of* S *fixing* A *and* B *pointwise and* p *linewise is transitive on the lines* (≠ A) *incident with a given point* x *on* A (x ≠ p).

S is said to satisfy condition (M) provided it satisfies $(M)_p$ for all points p ∈ P. For a fixed line L ∈ B let $(\hat{M})_L$ be the condition that is the dual of $(M)_p$, and let $(\hat{M})$ be the dual of (M). If s ≠ 1 ≠ t and S satisfies both (M) and $(\hat{M})$ it is said to be a *Moufang* GQ. A celebrated result of J. Tits [220] is that all Moufang GQ are classical or dual classical. His proof uses deep results from algebra and group theory, and it is one of our goals to approach this theorem using only rather elementary geometry and algebra. At the same time we are able to study Moufang conditions locally and obtain fairly strong results, and there are some intermediate Moufang conditions that have proved useful. We say S satisfies $(\hat{M})_p$ provided it satisfies $(\hat{M})_L$ for all lines incident with p. The dual condition is denoted $(M)_L$. A somewhat weaker condition is the following :

$(\bar{M})_p$: *For each line* L *through* p *and each point* x *on* L, x ≠ p, *the group* S_x *of collineations of* S *fixing* L *pointwise and* p *and* x *linewise is transitive on the points* (*not* p *or* x) *of each line* (≠ L) *through* p *or* x .

A main use of this condition is the following.

9.1.1. *If* S *satisfies* $(\bar{M})_p$ *for some point* p, *then* p *has property* (H).

Proof. We must show that if (x,y,z) is a triad of points in $p^\perp$ with x ∈ cl(y,z), then y ∈ cl(x,z). So suppose x ~ w ∈ $\{y,z\}^{\perp\perp}$. By $(\bar{M})_p$ there is a collineation θ which is a whorl about p, a whorl about pz, and which maps w to x. It follows that $y^\theta \in \{x,z\}^{\perp\perp}$, so that $y \in cl(x,z)$. □

An immediate corollary of this result and 5.6.2 is the following.

9.1.2. *If S satisfies condition* $(\hat{M})$, *then one of the following must occur :*

(i) *All points of S are regular* (*so* $s = 1$ *or* $s \geq t$).

(ii) $|\{x,y\}^{\perp\perp}| = 2$ *for all points* x,y, *with* $x \not\sim y$.

(iii) $S \cong H(4,s)$.

9.2. THE MOUFANG CONDITIONS AND PROPERTY (H)

Let $S = (P,B,I)$ be a GQ of order (s,t), $s \neq 1$ and $t \neq 1$.

9.2.1. *Let* θ *be a nonidentity collineation of S for which* θ *is a whorl about each of* A,p,B, *where* A *and* B *are distinct lines through the point* p. *Then the following hold :*

(i) θ *is an elation about* p.

(ii) *A line* L *is fixed by* θ *iff* L I p.

(iii) *If* x I A, y I B, $x \not\sim y$, *and* $z \in \{x,y\}^{\perp\perp}$, *then* $z^\theta = z$.

(iv) *If* $z^\theta = z$ *for some z not on A or B, then there are points x,y on A,B respectively, for which x,y,z are three centers of some triad containing p.*

(v) *If* p *is regular, then* θ *is a symmetry about p.*

(vi) *If* p *is antiregular, a point* z *is fixed by* θ *iff* z *is on* A *or* B .

Proof. This is an easy exercise starting with 8.1.1 (and its dual). □

There is an immediate corollary.

9.2.2. (i) *A point* p *of S is a center of symmetry iff* p *is a regular point for which S satisfies* $(M)_p$.

(ii) $S^{(p)}$ *is a* TGQ *iff* p *is a coregular point for which S satisfies* $(\hat{M})_p$.

9.2.3. *Suppose S satisfies* $(M)_p$ *for some point* p . *Let* A *and* B *be distinct lines through* p *with* x I A, y I B, $x \not\sim y$. *Let* θ *be a nonidentity collineation of S which is a whorl about each of* A,p,B, *and with* P_θ *as its set of fixed points. Then the following hold :*

(i) *If* $z \in P_\theta - \{p\}$, so $z \in p^\perp$, *then each point on* pz *is in* P_θ.

(ii) $cl(x,y) \cap p^\perp \subset P_\theta$.

(iii) *For any* x',y' *with* x' I A, y' I B, $x' \not\sim y'$, $cl(x,y) \cap p^\perp =$

$cl(x',y') \cap p^{\perp}$.

Proof. Let z be a point of P_θ not incident with A or B. Let L be any line through z different from pz. By $(M)_p$ there is a collineation θ' which is a whorl about each of B, p, and pz, and for which $(L^{\theta})^{\theta'} = L$. Then $\theta\theta'$ is a whorl about both B and p, and $L^{\theta\theta'} = L$, $z^{\theta\theta'} = z$. Clearly each point of L is fixed by $\theta\theta'$, and by 8.1.1 we have $\theta\theta' = id$. Hence θ fixes each point of pz, proving (i). From 9.2.1 (iii) it follows that $cl(x,y) \cap p^{\perp} \subset P_\theta$, proving (ii).

Now suppose x,x' are points of A, y,y' are points of B, with $x \not\sim y$, $x' \not\sim y'$. We claim $cl(x',y') \cap p^{\perp} \subset cl(x,y) \cap p^{\perp}$. So let $z' \in cl(x',y') \cap p^{\perp}$. Clearly we may assume that z' is not on A or B. Let v_1, v_2 be any two points of $\{x,y\}^{\perp} - \{p\}$, and let z be the point on pz' collinear with v_1. By $(M)_p$ there is a collineation θ which is a whorl about A, p and B, and which maps yv_1 to yv_2. It follows that $v_1^{\theta} = v_2$. Since $z' \in cl(x',y') \cap p^{\perp}$, by the preceding paragraph θ fixes each point of pz'. Hence $(zv_1)^{\theta} = zv_2$, implying that $v_2 \sim z$. It follows that $z \in \{x,y\}^{\perp\perp}$, so that $z' \in cl(x,y) \cap p^{\perp}$. This shows that $cl(x',y') \cap p^{\perp} \subset cl(x,y) \cap p^{\perp}$, and (iii) follows. □

As an immediate corollary of part (iii) of 9.2.3 we have the following.

9.2.4. *If S satisfies* $(M)_p$ *for some point* p, *then* p *has property* (H).

Hence if *S* satisfies condition (M), then each point has property (H). This result and its dual along with 9.1.2 and its dual yield the following approximation to the result of J. Tits.

9.2.5. *Let S be Moufang with* $1 < s \leq t$. *Then one of the following holds :*

(i) *Either S or its dual is isomorphic to* W(s) *and* $(s,t) = (q,q)$ *for some prime power* q.

(ii) $S \cong H(4,s)$ *and* $(s,t) = (q^2,q^3)$ *for some prime power* q.

(iii) $S^{(p)}$ *is a* TGQ *for each point* p, *and* $(s,t) = (q,q^2)$ *for some prime power* q.

(iv) $|\{x,y\}^{\perp\perp}| = |\{L,M\}^{\perp\perp}| = 2$ *for all* $x,y \in P$, $x \not\sim y$, *and all* $L,M \in B$, $L \not\sim M$. (*In Section 9.5 we shall show that case* (iv) *cannot arise*).

Proof. In listing the cases allowed by 9.1.2 and 9.2.4 and their duals, the cases that arise are (i), (ii), (iv) and the following : All lines are regular, $s < t$, and $|\{x,y\}^{\perp\perp}| = 2$ for all points x,y with $x \not\sim y$. But in this case 9.2.2 implies $S^{(p)}$ is a TGQ for each point p, and $t = s^2$ with s a prime power by 8.7.3 and 8.5.2.

9.3. MOUFANG CONDITIONS AND TGQ

Let $S = (P,B,I)$ be a GQ of order (s,t), $s \neq 1$ and $t \neq 1$.

9.3.1. *If* $S^{(p)}$ *is a* TGQ *then* S *satisfies* $(M)_p$.

Proof. Let L_0 I p I L_1, $L_0 \neq L_1$, $p \neq x$ I L_0, A I x I B, $A \neq B \neq L_0 \neq A$. On L_1 choose a point y, $y \neq p$, and define points z and u by A I $z \sim y$, B I $u \sim y$. If θ is the (unique) translation for which $z^\theta = u$, then $x^\theta = x$, $y^\theta = y$, $A^\theta = B$, and 8.4.1 implies that θ fixes L_0 and L_1 pointwise. It follows that S satisfies $(M)_p$. □

At this point we know that if $S^{(p)}$ is a TGQ, then p is a coregular point for which S satisfies both $(M)_p$ and $(\hat{M})_p$. Conversely, we seek minimal Moufang type conditions on S at p that will force $S^{(p)}$ to be a TGQ. Let G_p be a minimal group of whorls about p containing all the elations about p of the type guaranteed by $(M)_p$. Without some further hypothesis on S it is not possible to show even that G_p is transitive on $P-p^\perp$. For example, if p is regular then $(M)_p$ implies that p is a center of symmetry so that G_p is just the group of symmetries about p. And there are examples (e.g. W(s) with s odd) for which $S^{(p)}$ is not a TGQ and p is a center of symmetry. Moreover, notice that a GQ with a regular point p and $s \neq t$ has $s > t$, and hence is not a TGQ. As the regularity of p does not seem to be helpful, we try something that gets away from regularity.

<u>For the remainder of this section (with the exception of 9.3.6) we assume that p is a point of S for which S satisfies $(M)_p$ with G_p defined as above, and that p belongs to no unicentric triad.</u>

Then $\{p,x\}^{\perp\perp} = \{p,x\}$ for all $x \in P-p^\perp$.

9.3.2. *The point* p *is coregular, so that* $s \leqslant t$.

Proof. Let $L,M \in B$, $L \not\sim M$, p I L. Let N_1,N_2,N_3 be distinct lines in $\{L,M\}^\perp$ with p I N_1. If there were a line concurrent with N_1 and N_2 but not concurrent with N_3, there would be points y_i on N_i, $i = 1,2,3$, with $y_1 \sim y_2$, $y_1 \sim y_3$, $y_2 \not\sim y_3$. Then (p,y_2,y_3) would be a triad with center y_1, and hence by hypothesis would have an additional center u. Since S satisfies $(M)_p$

there must be a $\theta \in G_p$ fixing p linewise, py_1 and pu pointwise, and mapping y_1y_2 to y_1y_3. Define $v_i \in P$ by N_i I v_i I M, i = 2,3. It follows that $y_2^\theta = y_3$, $N_2^\theta = N_3$, $M^\theta = M$, and $v_2^\theta = v_3$. Define $w \in P$ by $v_2 \sim w$ I pu. Then $(wv_2)^\theta = wv_3$, giving a triangle with vertices w,v_2,v_3. This impossibility shows that the pair (L,M) must be regular, and p must be coregular. □

9.3.3. *$(S^{(p)},G_p)$ is an* EGQ *and* G_p *is the set of all elations about* p.
Proof. By 8.2.4 and 8.2.5 we need only show that G_p is transitive on $P-p^\perp$. First, let (p,x,y) be a centric triad, hence with at least two centers u and v. By $(M)_p$ there is a $\theta \in G_p$ for which θ is a whorl about each of pu, p, pv, and $(ux)^{\theta} = uy$. Clearly $x^\theta = y$. Second, let $x,y \in P-p^\perp$ with $x \sim y$. Put M = xy, and let p I L, $L \not\sim M$. Define u_i by $p \sim u_1$ I M, $y \sim u_2$ I M, and u_3 is any element of $\{u_1,u_2\}^\perp-\{p,y\}$. As (p,x,u_3) and (p,y,u_3) are centric triads, x and u_3, respectively y and u_3, are in the same G_p-orbit. Hence x and y are in the same G_p-orbit. Finally, suppose that (p,x,y) is an acentric triad. Let $u \in \{x,y\}^\perp$, so $u \notin p^\perp$. Then x,u,y are all in the same G_p-orbit by the preceding case. Hence G_p is transitive on $P-p^\perp$. □

9.3.4. *If* $\theta \in G_p$ *fixes a line* M *not incident with* p, *then* θ *is a whorl about the point on* M *collinear with* p .
Proof. Let θ be any nonidentity elation about p. First suppose there is some point $x \in P-p^\perp$ for which (p,x,x^θ) is a centric triad, and hence has at least two centers u and v. It follows that θ is the unique element of G_p mapping x to x^θ, i.e. θ is a whorl about each of pu, p, pv. By 9.2.1 θ fixes no line not incident with p. Second, suppose no triad of the form (p,x,x^θ) is centric. And suppose $M^\theta = M$ for some line M not through p. Let z be the point on M collinear with p. If some line N through z is moved by θ, let y be any point on N, $y \neq z$. Then (p,y,y^θ) would be a centric triad. Hence θ must be a whorl about z. □

9.3.5. *$(S^{(p)},G_p)$ is a* TGQ. *If* t *is even, then* $t = s^2$.
Proof. By 9.3.3. $(S^{(p)},G_p)$ is an EGQ, so we may shift to the coset geometry description $S(G_p,J)$, $J = \{S_0,\dots,S_t\}$, etc., of Section 8.2. Since p is coregular, by 8.2.2 we know that $S_iS_j = S_jS_i$ for $0 \leq i, j \leq t$, implying S_iS_j is a subgroup of order s^2 if $i \neq j$. Moreover, the condition in 9.3.4, when interpreted for S_i and S_i^*, says that $S_i \lhd S_i^*$, $0 \leq i \leq t$. Put $T_{ij} = S_i^* \cap S_j^*$, $i \neq j$. Then $G_p = S_iS_j^* = S_iS_jT_{ij}$, if $i \neq j$. Since $T_{ij} \subset N_{G_p}(S_i) \cap N_{G_p}(S_j)$,

clearly $S_iS_j \lhd \langle S_iS_j, T_{ij}\rangle = G_p$, if $i \neq j$. Hence each conjugate of S_i is contained in S_iS_j, if $i \neq j$. But if i,j,k are distinct, $S_iS_j \cap S_iS_k = S_i$, by K1. It follows that $S_i \lhd G_p$, $0 \leq i \leq t$, and by 8.2.2 S_i is a (full) group of symmetries about the line $[S_i]$. From 8.3.1 it follows that $(S^{(p)},G_p)$ is a TGQ. By 8.6.1, if t is even either $t = s$ or $t = s^2$. Clearly $t \neq s$, because then p would be regular and hence belong to some unicentric triad. □

9.3.6. *Let* $S = (P,B,I)$ *be a* GQ *of order* (s,t), *and suppose that* S *satisfies* $(M)_p$ *for some point* p, *with* G_p *a minimal group of whorls about* p *containing all elations of the type guaranteed to exist by* $(M)_p$.

(i) *If* p *is coregular and* t *is odd, then* $(S^{(p)},G_p)$ *is a* TGQ.

(ii) *If each triad containing* p *has at least two centers, then* $(S^{(p)},G_p)$ *is a* TGQ *and* $t = s^2$.

(iii) *If* $t = s^2$, *then* $(S^{(p)},G_p)$ *is a* TGQ.

Proof. In each case the hypotheses guarantee that p is in no unicentric triad, so that the results of this section apply. To complete the proof of (ii), use part (i) of 8.6.1 if $s < t$. And if $s = t$, p must belong either to an acentric or a unicentric triad according as t is odd or even (i.e. according as p is antiregular or regular (cf. 1.5.2, (iv) and (v))). □

9.4. AN APPLICATION OF THE HIGMAN-SIMS TECHNIQUE

For any GQ $S = (P,B,I)$ of order (s,t), $s \neq 1 \neq t$, let O be a set of points with $|O| = q \geq 2$. A line of S will be called a *tangent*, *secant* or *exterior* line according as it is incident with exactly 1, at least 2, or no point of O. Let Δ be a set of tangent lines, and put $\Delta_0 = B-\Delta$. Suppose $\{\Delta_1,\dots,\Delta_f\}$ is a partition of Δ satisfying the following :

A1. $f \geq 2$.

A2. For $1 \leq i \leq f$, each point of O is incident with θ lines of Δ_i, θ a nonzero constant.

A3. If x and y are noncollinear points of O, then each line of Δ_k through x meets a line of Δ_k through y, $1 \leq k \leq f$.

Put $\delta_i = |\Delta_i|$, $0 \leq i \leq f$. Then the following is clear :

$$q\theta = \delta_i,\ 1 \leq i \leq f,\ \text{so}\ \delta_0 = (1+t)(1+st)-qf\theta. \tag{1}$$

Put $\delta_{ij} = |\{(L,M) \in \Delta_i \times \Delta_j \parallel L \not\sim M\}|$, $0 \leq i, j \leq f$. For $0 \neq i \neq j \neq 0$, each line of Δ_i meets θ lines of Δ_j, so that each line of Δ_i misses $(q-1)\theta$ lines of Δ_j. Hence

$$\delta_{ij}/\delta_i = (q-1)\theta,\ 1 \leqslant i,\ j \leqslant f,\ i \neq j. \tag{2}$$

Let $O = \{x_1,\ldots,x_q\}$, and suppose x_i is collinear with b_i points ($\neq x_i$) of O (necessarily on secants through x_i), $1 \leqslant i \leqslant q$. Let L be a fixed line of Δ_j meeting O at x_i ($1 \leqslant j \leqslant f$, $1 \leqslant i \leqslant q$). There are $q-1-b_i$ points of O lying on at least one line ($\neq$ L) of Δ_j that meets L at a point not in O. So L meets $\theta+q-1-b_i$ lines of Δ_j and misses $q\theta-(\theta+q-1-b_i) = (q-1)(\theta-1)+b_i$ lines of Δ_j. It follows that $\delta_{jj} = \sum_{i=1}^{q} \theta((q-1)(\theta-1)+b_i) = q\theta(q-1)(\theta-1)+\theta\sum_{i=1}^{q} b_i$. Put $\bar{b} = \sum_{i=1}^{q} b_i/q$. Then

$$\delta_{jj}/\delta_j = (q-1)(\theta-1)+\bar{b},\ 1 \leqslant j \leqslant f. \tag{3}$$

Since $\sum_{j=0}^{f} \delta_{ij}/\delta_i = st^2$, we may calculate

$$\delta_{i0}/\delta_i = st^2-\bar{b}-(q-1)(\theta f-1),\ 1 \leqslant i \leqslant f. \tag{4}$$

Since δ_{i0}/δ_i is independent of i for $1 \leqslant i \leqslant f$, so is $\delta_{0i}/\delta_0 = (\delta_{i0}/\delta_i)(\delta_i/\delta_0)$. Write $e = \delta_{00}/\delta_0$, $a = \delta_{i0}/\delta_i$, $b = \delta_{0i}/\delta_0$, $1 \leqslant i \leqslant f$; $c = \delta_{ij}/\delta_i$, $1 \leqslant i, j \leqslant f$, $i \neq j$; $d = \delta_{jj}/\delta_j$, $1 \leqslant j \leqslant f$. Put $B^\Delta = (\delta_{ij}/\delta_i)_{0 \leqslant i,\ j \leqslant f}$. It follows that

$$B^\Delta = \begin{pmatrix} e & b & \ldots & b \\ a & & & \\ \vdots & & cJ+(d-c)I & \\ a & & & \end{pmatrix}, \tag{5}$$

where J is the $f \times f$ matrix of 1's, and I is the $f \times f$ identity matrix.

For each j, $2 \leqslant j \leqslant f$, define $\bar{v}_j = (v_0,v_1,\ldots,v_f)^T$ by $v_1 = 1$, $v_j = -1$, $v_k = 0$ otherwise. Then $\bar{v}_j$ is an eigenvector of B^Δ associated with the eigenvalue $d-c = \bar{b}-q+1$. By the theorem of Sims as applied in 1.4 (but dualized so as to use lines instead of points, and interchanging s and t), we have $-t \leq \bar{b}-q+1$, or

$$q \leq 1+t+\bar{b}. \tag{6}$$

Moreover, if equality in (6) *holds,* $\bar{v}_j$ *may be extended to an eigenvector of the matrix* B (*dually defined in* 1.4) *associated with the eigenvalue* $-t$, *by repeating* v_i δ_i *times,* $0 \leq i \leq f$.

Suppose in fact that equality does hold in (6). Then writing out the inner product of a row of B indexed by a line of Δ_1 meeting O at x_i and the extension of $\bar{v}_j$, $1 \leq i \leq q$, $2 \leq j \leq f$, we find that $b_i = \bar{b}$.

$$\textit{If } q = 1+t+\bar{b}, \textit{ then } b_i = \bar{b} = q-1-t,\ 1 \leq i \leq q. \tag{7}$$

The following theorem gives a general version of the setting in which the basic inequality (6) is to be applied.

9.4.1. *Let* O *and* Ω *be disjoint sets of points of S for which there is a group* G *of collineations of S satisfying the following :*

(i) $|O| \geq 2$; $\Omega^G = \Omega$; *G is not transitive on* Ω.

(ii) $|G_y|$ *is independent of* y *for* $y \in \Omega$.

(iii) *Each element of* Ω *is collinear with a constant number* r ($r > 0$) *of points of* O.

(iv) *If* $x \sim y$, $x \in O$, $y \in \Omega$, *and* z *is any point of the line* xy *different from* x, *then* $z \in y^G$.

(v) *If* $x, y \in O$, $x \neq y$, *there is a sequence* $x = x_0, x_1, \ldots, x_n = y$ *of points of* O *for which* $x_{i-1} \not\sim x_i$, $1 \leq i \leq n$.

Then $|O| \leq 1+t+\bar{b}$, *where* $\bar{b}$ *is the average number of points* ($\neq x$) *of* O *collinear with a given point* x *of* O.

Proof. Let $O = \{x_1, \ldots, x_q\}$, and let b_i be the number of points ($\neq x_i$) of O collinear with x_i, $1 \leq i \leq q$. If $y \in \Omega$ and L is a line through y meeting O in a point x_i, then L has s points of Ω and is tangent to O by (i) and (iv). Let Δ be the set of all tangents to O containing points of Ω. By hypothesis G splits Ω into orbits $\Omega_1, \ldots, \Omega_f$, $f \geq 2$. Put Δ_i equal to the set of tangents to O containing points of Ω_i, $1 \leq i \leq f$. By (iv) Δ_i consists of tangents each of whose points outside O is in Ω_i. Then $\{\Delta_1, \ldots, \Delta_f\}$ is a partition of Δ, and we claim $(0,\Delta_1,\ldots,\Delta_f)$ satisfies the conditions A1,A2,A3. Clearly A1 holds by (i) and A3 holds by (iv).

Let $x \in 0$ and suppose Δ_i has θ_i lines $L_1^{(i)},\ldots,L_{\theta_i}^{(i)}$, incident with x. Next let $x' \in 0$ with $x \not\sim x'$, and suppose Δ_i has θ_i' lines through x'. The θ_i lines through x' meeting $L_1^{(i)},\ldots,L_{\theta_i}^{(i)}$ must lie in Δ_i by A3, so $\theta_i \leqslant \theta_i'$. Similarly, $\theta_i' \leqslant \theta_i$, so by (v) θ_i is independent of x in 0. Then for any $y \in \Omega$, $|G| = |\Omega_i||G_y| = q\theta_i sr^{-1}|G_y|$ (making use of (iii) and (iv)), implying that $\theta_i = r|G|/qs|G_y|$ is independent of i. Hence A2 is satisfied. Then (6) finishes the proof. □

Remark : If $|0| = 1+t+\bar{b}$, then (7) has an obvious consequence in the context of 9.4.1.

We now specialize the setting of 9.4.1.

9.4.2. *Let* $L_0,L_1,\ldots,L_r$ *be* $r+1$ *lines* $(r \geqslant 1)$ *incident with a point* p *of S . Let* 0 *be the set of points different from* p *on the lines* $L_0,\ldots,L_r$, *and put* $\Omega = P-p^\perp$. *Suppose* G *is a group of elations about* p *with the property that* G *is transitive on the set of points of* Ω *incident with a line tangent to* 0. *If* $r > t/s$, *then* G *must be transitive* (*and hence regular*) *on* Ω.
Proof. Suppose G has f orbits on Ω with $f \geqslant 2$. Since $r \geqslant 1$, the hypotheses of 9.4.1 are all satisfied with $\bar{b} = b_i = s-1$. Hence $q = |0| = s(r+1) \leqslant 1+t+(s-1)$, i.e. $r \leqslant t/s$.

There are two corollaries.

9.4.3. *Let S be a* GQ *of order* (s,t), $t \geqslant s$, *with a point* p *for which* $|\{p,x\}^{\perp\perp}| = 2$ *for all* $x \in P-p^\perp$. *If S satisfies* $(\bar{M})_p$ *and* G_p *is the group generated by all the elations guaranteed to exist by* $(\bar{M})_p$, *then* $(S^{(p)},G_p)$ *is an* EGQ *with* G_p *consisting of all elations about* p. *If* G *is the complete group of whorls about* p, *either* $G = G_p$ *or* G *is a Frobenius group on* $P-p^\perp$.
Proof. Use 8.2.4, 8.2.5 and 9.4.2. □

9.4.4. *If* $(S^{(p)},G)$ *is a* TGQ *and* $r > t/s$, *then* G *is generated by the symmetries about any* 1+r *lines through* p.
Proof. Immediate. □

9.5. THE CASE (iv) OF 9.2.5

The fact that case (iv) of 9.2.5 cannot arise is an immediate corollary of the theorem of this section. Hence to complete a proof of the theorem of J. Tits it would be sufficient to show that if $S^{(p)}$ is a TGQ of order (s,s^2), $s \neq 1$, for each point p of S, then $S \cong Q(5,s)$.

9.5.1. *There is no GQ* $S = (P,B,I)$ *of order* (s,t), $1 < s \leq t$, *with a point* p *for which the following hold :*

(i) S *satisfies* $(\bar{M})_p$.

(ii) $|\{p,x\}^{\perp\perp}| = 2$ *whenever* $x \in P-p^{\perp}$.

(iii) S *satisfies* $(M)_p$.

(iv) $|\{L,M\}^{\perp\perp}| = 2$ *whenever* p I L *and* $M \in B-L^{\perp}$.

Proof. From hypotheses (i) and (ii) and 9.4.3 we know $(S^{(p)},G_p)$ is an EGQ, where G_p is the set of all elations about p. Hence we recall the group coset geometry description $S \cong (G_p,J)$, with $J = \{S_0,\ldots,S_t\}$, $S_0^*,\ldots,S_t^*$, etc. (cf. 8.2).

Suppose some $\theta \in G_p$ fixes a line M not through p, and define the point y by p ~ y I M. If z I M, $z \neq y$, then θ must be the unique element of G_p mapping z to z^θ. Hence θ must be the collineation guaranteed by $(\bar{M})_p$ to map z to z^θ and is therefore a whorl about y (and also about p and py). In terms of J, this means that $S_i \lhd S_i^*$ for each $i = 0,\ldots,t$. Now suppose $p \neq y = y^\theta$ for some $\theta \in G_p$. If θ fixes some line M through y, p $\not\mathrm{I}$ M, then θ is a whorl about py as in the preceding case. If $M^\theta \neq M$ for some M through y, use $(M)_p$ to obtain a φ in G_p which is a whorl about py and maps M^θ to M. Hence $\theta\varphi \in G_p$ is a whorl about py and about y, forcing θ to be a whorl about py. The fact that any $\theta \in G_p$ fixing a point y, $y \neq p$, must be a whorl about py, may be interpreted for J to say that $S_i^* \lhd G_p$ for all i.

We claim that $N_{G_p}(S_i) = S_i^*$. For suppose $g \in N_{G_p}(S_i)-S_i^*$. Any coset of S_i not in S_i^* must meet some member of J, since $\{S_i^*, S_iS_0-S_i, S_iS_1-S_i,\ldots,S_iS_t-S_i\}$ (omitting $S_iS_i-S_i$) is a partition of the set G_p. Hence there is a j ($\neq$ i) for which there is a $\sigma_j \in S_j \cap S_ig$, say $\sigma_j = \sigma_ig$ for some $\sigma_i \in S_i$. Then $\sigma_j = \sigma_ig \in N_{G_p}(S_i)$. For any $\sigma \in S_i$, $(S_i\sigma_j)\sigma = S_i\sigma_j$, since $\sigma \in S_i = \sigma_j^{-1}S_i\sigma_j$. But as σ fixes the line $S_i\sigma_j$ through $S_i^*\sigma_j$, it must be a whorl about $S_i^*\sigma_j$. Hence each element of S_i fixes each line meeting any one of p, S_i^*, $S_i^*\sigma_j$, and it follows that the lines S_i, $S_i\sigma_j$, and $[S_j]$ are all concurrent with $[S_i]$ and with the s images of S_j under the action of S_i. This says that

$|\{[S_j], S_i\}^{\perp\perp}| > 2$, contradicting hypothesis (iv) of the theorem. This shows that $N_{G_p}(S_i) = S_i^*$.

For convenience, specialize $i = 0$, $j = 1$. As $S_0^* \lhd G_p$, $S_1^* \lhd G_p$, clearly $S_0^* \cap S_1^* \lhd S_0^*$. And $S_0^* = S_0(S_0^* \cap S_1^*)$ with $S_0 \lhd S_0^*$. Hence S_0^* is the direct product of S_0 and $S_0^* \cap S_1^*$, implying that each element of $S_0^* \cap S_1^*$ commutes with each element of S_0 (and also with each element of S_1). Put $H = \langle S_0,S_1\rangle \cap (S_0^* \cap S_1^*)$. Clearly each element of H commutes with each element of $\langle S_0,S_1\rangle$ and with each element of $S_0^* \cap S_1^*$, hence also with each element of $G_p = S_0S_1^* = S_0S_1(S_0^* \cap S_1^*) \subset \langle S_0,S_1\rangle(S_0^* \cap S_1^*)$. So $H \subset Z(G_p) \subset \bigcap_i N_{G_p}(S_i) = \bigcap_i S_i^* = \{e\}$, where this last equality holds since any nonidentity element of $\bigcap_i S_i^*$ would be a symmetry about p (cf. 8.2.2) and force a contradiction of hypothesis (ii). Then $H = \{e\}$ and $G_p = \langle S_0, S_1\rangle(S_0^* \cap S_1^*)$ imply $|\langle S_0,S_1\rangle| = s^2$, i.e. $\langle S_0,S_1\rangle = S_0S_1$. Similarly, S_iS_j is a group whenever $i \neq j$, so $S_iS_j = S_jS_i$, implying each line through p (or (∞)) is regular (by 8.2.2), contradicting hypothesis (iv). This completes the proof. □

9.6. THE EXTREMAL CASE $q = 1+s^2+\bar{b}$

Recall the setting and notation of Section 9.4 and let $(O,\Delta_1,\ldots,\Delta_f)$ be a system satisfying A1, A2, and A3. Moreover, suppose that $|O| = q = 1+s^2+\bar{b}$ $(t > 1,\ s > 1)$. As $t \leqslant s^2$ by D.G. Higman's inequality, it follows from (6) that $t = s^2$ and $b_i = \bar{b}$ for $1 \leqslant i \leqslant q$. By 1.10.1 applied to the set O, $\bar{b}+1 \leqslant s+q/(1+s)$. So $\bar{b}+1 = q-s^2 \leqslant s+q/(1+s)$, which is equivalent to $q \leqslant (1+s)^2$. And of course $q = 1+\bar{b}+s^2$ implies $q \geqslant 1+s^2$. Hence

$$1+s^2 \leqslant q \leqslant (1+s)^2. \tag{8}$$

Let $L \in \Delta_j$, $1 \leqslant j \leqslant f$. Let P_L be the set of points in O, together with the points on lines of Δ_j meeting L and the points off O lying on at least two secant lines. The number of such points is $v' = s\theta+1+\bar{b}+s(q-1-\bar{b})+\delta$, where δ is the number of points off O but lying on at least two secants. Hence

$$|P_L| = v' = s\theta+q-s^2+s^3+\delta. \tag{9}$$

9.6.1. *Suppose that* P_L *is the pointset of a subquadrangle* S'. *Then* S' *has order* (s,s) *and one of the following three cases must occur :*

(i) $q = 1+s^2$, $\theta = 1+s$, $\delta = 0$, $\bar{b} = 0$, *and* O *is an ovoid of* S' (*i.e. each line of* S' *is incident with a unique point of* O).

(ii) $q = s(1+s)$, $\theta = s$, $\delta = 1$, $\bar{b} = s-1$, *and* O *is the set of all points different from a given point* x *but incident with one of a set of* $1+s$ *lines all concurrent at* x.

(iii) $q = (1+s)^2$, $\theta = s-1$, $\delta = 0$, $\bar{b} = 2s$, *and* O *is the set of points on a grid.*

Moreover, each of the above cases does arise.

Proof. Since each point on a line of Δ_j meeting L is in P_L by definition, S' has order (s,t') for some t'. Since $f \geq 2$, S' must be a proper subquadrangle, so $t' < t$, implying $t' \leq s$ by 2.2.2. We claim each line of Δ_j is a line of S'. Let L meet O at x_i and suppose $M \in \Delta_j$. If x_i is on M, then M is a line of S'. So suppose M is incident with $x_r \in O$, $x_r \neq x_i$. If $x_i \not\sim x_r$, let y be the point on M collinear with x_i. By A3 $x_i y \in \Delta_j$, and as both y and x_r belong to S' so does M. So suppose $x_i \sim x_r$. Each point off O on M is collinear with -on average- $1+(q-1-\bar{b})/s = 1+s$ points of O. Hence some point z of M, $z \notin O$ (i.e. $z \neq x_r$) is collinear with at least s points of O different from x_r, say $u_1,\ldots,u_s$. If $u_1,\ldots,u_s$ are all collinear with x_i , then $u_1x_i,\ldots,u_sx_i$, x_rx_i would be $s+1$ lines of S' through x_i, giving a total of at least $1+\theta+s$ lines of S' through x_i, an impossibility since $1+\theta+s > 1+s \geq 1+t'$. Hence we may suppose that $u_s \not\sim x_i$. Then u_sz belongs to S' by a previous argument, implying $zx_r = M$ belongs to S'. Thus each line of Δ_j belongs to S'.

Let $M \in \Delta_j$ and recall that the points off O on M are collinear with -on average- $1+s$ points of O. But no such point is collinear with more than $1+s$ points of O since $t' \leq s$. Hence each point off O on M is collinear with exactly $1+s$ points of O, and $t' = s$. Hence $|P_L| = v' = 1+s+s^2+s^3$, and from (9) we have

$$\theta = 2s+1-(q+\delta-1)/s \, . \tag{10}$$

From (8) we have $q \geq 1+s^2$, so that (10) implies

$$\theta \leq 1+s. \tag{11}$$

Since each point of O is on θ lines of Δ_j, and each point off O on some line of Δ_j is collinear with 1+s points of O, it follows that $v' = q\theta s/(1+s)+q+\delta = s\theta+q-s^2+s^3+\delta$. Solving for θ we find

$$\theta = s(s-1)(s+1)/(q-s-1). \tag{12}$$

As $q \leqslant (1+s)^2$ from (8), $q-s-1 \leqslant (1+s)s$, and $\theta \geqslant s-1$. This proves

$$s-1 \leqslant \theta \leqslant s+1. \tag{13}$$

Setting $\theta = s+1,s,s-1$, respectively, in (12) and solving for q, yields $q = 1+s^2,s(1+s),(1+s)^2$, respectively. Then (9) may be used to solve for δ in each case, since $v' = 1+s+s^2+s^3$, and (7) may be used to determine $\bar{b}$ as stated in the theorem.

In case (i), $\bar{b} = 0$ and $q = 1+s^2$ force O to be an ovoid of S'.

In case (ii), $\bar{b} = s-1$ and $\delta = 1$. Since $\theta = s$ and S' has order (s,s), the s-1 points $x_j \in O$ different from but collinear with a fixed point x_i of O must all lie on the only line M_k of S' through x_i and not tangent to O. So there arise 1+s lines $M_0,\ldots,M_s$, each incident with s points of O, no two having a point of O in common, and no point of M_i in O being collinear with a point of M_j in O, $i \neq j$. Hence each point of M_i in O must be collinear with that point of M_j not in O. It follows that $M_0,\ldots,M_s$ all meet at a point x not in O, which is evidently the unique point lying on two (or more) intersecting secants.

In case (iii), $\bar{b} = 2s$. Since $\theta = s-1$ and S' has order (s,s), the 2s points $x_j \in O$ for which x_j is collinear with but distinct from a given point x_i in O must lie on two lines through x_i. From $q = (1+s)^2$ it follows readily that O is the pointset of a grid.

To complete the proof of 9.6.1 we give several examples to show that each of the above cases does arise.

Examples 1.

Let S be the GQ Q(5,s) of order (s,s^2) obtained from an elliptic quadric Q in PG(5,s). Let P_3 be a fixed PG(3,s) contained in PG(5,s).

(i) If $Q \cap P_3$ is an elliptic quadric O, let $P_4^1,\ldots,P_4^f$ be f ($\geqslant 2$) PG(4,s)'s containing O and not containing an intersection of Q and the polar line of P_3 with respect to Q (i.e. $P_4^i \cap Q$ is not a cone).
Then the linesets $\Delta_1,\ldots,\Delta_f$ of $P_4^1 \cap Q,\ldots,P_4^f \cap Q$, respectively, yield an

example with $|O| = 1+s^2$.

(ii) If $P_3 \cap Q$ is a cone O' with vertex x_0, then f (≥ 2) PG(4,s)'s containing O' and intersecting Q in a nonsingular quadric yield an example with $O = O'-\{x_0\}$, $|O| = s(1+s)$.

(iii) If $P_3 \cap Q$ is an hyperbolic quadric O, then f (≥ 2) PG(4,s)'s containing O will yield an example with $|O| = (1+s)^2$.

Examples 2.

Consider the GQ $T_3(\Omega)$ with Ω an ovoid of $PG(3,q) = P_3$ and P_3 an hyperplane of $PG(4,q) = P_4$.

(i) $q = 1+s^2$. Let L be a line of P_3 containing no point of Ω. Let π be a plane of P_4 meeting P_3 in L. Let $\pi_1,\ldots,\pi_f$ ($2 \leq f \leq s-1$) be distinct planes of P_3 containing L and meeting Ω in an oval. Put $P_3^i = \langle\pi,\pi_i\rangle$. Let $O = (\pi-P_3) \cup \{(\infty)\}$. Then Δ_i is to be the set of lines of P_3^i meeting P_3 in a point of $\pi_i \cap \Omega$ together with the points of $\pi_i \cap \Omega$ considered as lines of type (b) of $T_3(\Omega)$. Here $\theta = s+1$ and O is an ovoid in the subquadrangle whose lineset is Δ_i.

(ii) $q = s(1+s)$. (a) Let π be a plane of P_3 meeting Ω in an oval Ω'. Let O be the set consisting of the $s(1+s)$ points of type (ii) or $T_3(\Omega)$ that are incident with the $1+s$ elements of Ω' considered as lines of type (b) of $T_3(\Omega)$. Let $P_3^1,\ldots,P_3^f$ be distinct PG(3,s)'s meeting P_3 in π, $2 \leq f \leq s$. Then Δ_i is the set of lines of P_3^i meeting P_3 in a point of Ω'.

(b) Let L be a line of P_3 which is tangent to Ω at the point x. Let π be a plane of P_4 meeting P_3 in L. Let $\pi_1,\ldots,\pi_f$ be distinct planes of P_3 containing L and meeting Ω in an oval, $2 \leq f \leq s$, and put $P_3^i = \langle\pi,\pi_i\rangle$. There is one point P_3^* of type (ii) containing the plane π. Set $O = (\pi-P_3) \cup$ {points of type (ii) distinct from P_3^* and incident with the point x considered as a line of $T_3(\Omega)$} $\cup \{(\infty)\}$. Then Δ_i is the set of lines of P_3^i not contained in π and meeting P_3 in a point of $\pi_i \cap \Omega$, together with the points of $(\pi_i \cap \Omega)-\{x\}$ considered as lines of $T_3(\Omega)$.

(iii) $q = (1+s)^2$. Let L be a line of P_3 containing two points of Ω. Let $\pi_1,\ldots,\pi_f$ be distinct planes of P_3 containing L, $2 \leq f \leq s+1$, so necessarily π_i meets Ω in an oval Ω_i. Let x_0 be a fixed point of P_4-P_3, and put $P_3^i = \langle x_0,\pi_i\rangle$. So $P_3^i \cap P_3^j = \langle x_0,L\rangle$ if $i \neq j$. Put $O = (\langle x_0,L\rangle-L) \cup \{(\infty)\} \cup \{P_3' \mid P_3'$ is a hyperplane of P_4 meeting P_3 in a plane tangent to Ω at one of the two points of $L \cap \Omega\}$. Then Δ_i is the set of lines of P_3^i meeting P_3 in a point of Ω_i but not contained in $\pi = \langle x_0,L\rangle$, together with the points of Ω_i-L considered as lines of $T_3(\Omega)$.

Notice that in all these examples the line L may be chosen arbitrarily in $\Delta_1 \cup \ldots \cup \Delta_f$. This completes the proof of 9.6.1. □

Remark : Suppose that f = 4 in Example 2 (i). Put $\Delta_1' = \Delta_1 \cup \Delta_2$, $\Delta_2' = \Delta_3 \cup \Delta_4$, and let O be the same as in that example. Then we have $q-1-\bar{b} = s^2 = t$ and each set Δ_i' of tangents is a union of linesets of subquadrangles of order (s,s) containing O, and $\theta' = 2(1+s)$. For $f = mk \leq s-1$ it is easy to see how to generalize this example so as to obtain $\theta' = k(1+s)$.

Moreover, there is a kind of converse of the preceding theorem which is obtained as an application of the theory (1) - (6) : *In a situation sufficiently similar to one of the cases* (i), (ii), (iii) *considered above, a* GQ *S' of order* (s,s) *must arise in the manner hypothesized in* 9.6.1. We make this precise as follows.

Let O and $\Delta_1,\ldots,\Delta_f$ be given with A1, A2, A3 satisfied, assuming as always that $|O| \geq 2$ and $s > 1$, $t > 1$.

(i)' Suppose O consists of pairwise noncollinear points, so $\bar{b} = 0$. Then $|O| = q \leq 1+t$ by (6). Suppose $|O| = 1+s^2$, implying $t = s^2$. For each $L \in \Delta_j$ and each z on L, $z \notin O$, suppose that z is on at most (or at least) s+1 lines of Δ_j, so that in fact z is on exactly 1+s lines of Δ_j . The number of points on lines of Δ_j is $v' = (s^2+1)\theta s/(s+1)+s^2+1$, so that s+1 divides θ. Fix a line $L \in \Delta_j$ and consider all lines of Δ_j concurrent with L. Counting points on these lines we have $s^3+\theta s+1$, which equals v' iff $\theta = s+1$. If $\theta = s+1$, then $v' = (1+s)(1+s^2)$ and each of the v' points is incident with 1+s lines of Δ_j. It follows that there is a subquadrangle S' of order (s,s) whose lines are just those of Δ_j. If $\theta = k(s+1)$ with $k > 1$, it is tempting to conjecture that Δ_j must be the union of linesets of k subquadrangles having O as an ovoid, as is the case in the first paragraph of this remark.

(ii)' Suppose O consists of those points different from a point x incident with r lines $L_1,\ldots,L_r$ concurrent at x. From (6) it follows that $r \leq 1+t/s$. Now suppose $r = 1+t/s = 1+s$. Fix a line $L \in \Delta_j$. For each point z on L, $z \notin O$, z is collinear with exactly 1+s points of O on 1+s lines of Δ_j. The number of points on lines of Δ_j together with x is $v' = 1+s(s+1)+ s(s+1)\theta s/(s+1) = 1+s+s^2+\theta s^2$. And the number of points on lines of Δ_j concurrent with L, together with the points on the line L_i meeting L, is $1+s+\theta s+s^3$, which must be at most v'. Then $1+s+\theta s+s^3 \leq 1+s+s^2+\theta s^2$ implies

$s \leqslant \theta$. If $\theta = s$, there arises a subquadrangle S' of order (s,s).

(iii)' Let $L_1,\ldots,L_m$ (resp., $M_1,\ldots,M_n$), $2 \leqslant m, n$, be pairwise nonconcurrent lines with $L_i \sim M_j$, $1 \leqslant i \leqslant m$, $1 \leqslant j \leqslant n$. Suppose O consists of the $q = mn$ points at which an L_i meets an M_j. Then (6) implies $(m-1)(n-1) \leqslant t$. Suppose that $m = n = s+1$, implying $t = s^2$ and $q = (1+s)^2$. Fix a line $L \in \Delta_j$. For each point z on L, $z \notin O$, z is collinear with $1+s$ points of O. The number of points on lines of Δ_j is $v' = (1+s)^2\theta s/(1+s)+(1+s)^2 = (1+s)(1+s+\theta s)$. And the number of points on lines of Δ_j concurrent with L, together with the points of O, is $1+2s+\theta s+s^3$. As this number cannot exceed v', it follows that $s-1 \leqslant \theta$. If $\theta = s-1$, there arises a subquadrangle S' of order (s,s).

9.7. A THEOREM OF M. RONAN

M.A. Ronan [150] gives a characterization of $Q(4,q)$ and $Q(5,q)$ which utilizes the work of J. Tits [220, 223] on Moufang GQ. M.A. Ronan's treatment includes infinite GQ and relies on topological methods. We offer here an "elementary" treatment which, although it still depends on the theorem of J. Tits, is combinatorial rather than topological, and which corrects a slight oversight in the case $t = 2$.

Let S be a GQ of order (s,t), $s > 1$ and $t > 1$. A *quadrilateral* of S is just a subquadrangle of order $(1,1)$. A quadrilateral Σ is said to be *opposite* a line L if the lines of Σ are not concurrent with L. If Σ is opposite L, the four lines incident with the points of Σ and concurrent with L are called the *lines of perspectivity of* Σ *from* L. Two quadrilaterals Σ and Σ' are *in perspective* from L if either $\Sigma = \Sigma'$ is opposite L, or $\Sigma \neq \Sigma'$ and Σ, Σ' are both opposite L and the lines of perspectivity of Σ from L are the same as the lines of perspectivity of Σ' from L.

9.7.1. *Let* L *be a given line of the* GQ $S = (P,B,I)$ *of order* (s,t), $s > 1$ *and* $t > 2$. *Then* L *is an axis of symmetry iff the following condition holds : Given any quadrilateral* Σ *opposite the line* L *and any point* x', $x' \not\mathrm{I}\, L$, *incident with a line of perspectivity of* Σ *from* L, *there is a quadrilateral* Σ' *containing* x' *and in perspective with* Σ *from* L.

Proof. Let L be an axis of symmetry. Suppose that Σ is a quadrilateral opposite L and that x', $x' \not\mathrm{I}\, L$, is incident with a line of perspectivity of Σ from L. Let $x' \mathrm{I}\, X \sim L$ and $x \mathrm{I}\, X$ with x in Σ. By hypothesis there is a symmetry θ of S with axis L and mapping x onto x'. Clearly θ maps Σ onto a quadrilateral Σ' containing x and in perspective with Σ from L.

Conversely, suppose that given any quadrilateral Σ opposite L and any point x', $x' \not I L$, incident with a line of perspectivity of Σ from L, there is always a quadrilateral Σ' containing x' and in perspective with Σ from L. We shall prove that L is an axis of symmetry of S.

First of all we show that L is regular. Let $L_1 \not\sim L$, let M_0, M_1, M_2 be distinct lines of $\{L, L_1\}^{\perp}$, and let $L_2 \in \{M_0, M_1\}^{\perp} - \{L, L_1\}$. We must show that $L_2 \sim M_2$. So suppose $L_2 \not\sim M_2$. If $L_2 \text{ I } y \text{ I } M_1$, then let V be defined by $y \text{ I } V$ and $V \sim M_2$. Further, let $V \text{ I } z \text{ I } M_2$ and $L_2 \text{ I } u \text{ I } M_0$. Since $t > 2$, there is a quadrilateral Σ containing u, y, z, L_2, V and which is opposite L. Clearly there is no quadrilateral Σ' containing $M_0 \cap L_1$ and which is in perspective with Σ from L, a contradiction. Hence $L_2 \sim M_2$ and L must be regular.

We introduce the following notation : If x (resp., y,z,u,...) is not incident with L, then the line which is incident with x (resp., y,z,u,...) and concurrent with L is denoted by X (resp., Y,Z,U,...). Let $z \sim z'$, $z \neq z'$, $z \not I L \not I z'$ and $Z = zz' \sim L$. Then we define as follows a permutation $\theta(z,z')$ of $P \cup B$. First, put $x^{\theta(z,z')} = x$ for all $x \text{ I } L$ and $z^{\theta(z,z')} = z'$. Now let $y \sim z$, $y \not I Z$. Then $y^{\theta(z,z')} = y'$ is defined by $y' \sim z'$ and $y' \text{ I } Y$. Next, let $d \not\sim z$ and $d \not I L$. If $u \in \{z,d\}^{\perp}$, with $u \not I Z$ and $u \not I D$, then $d' = d^{\theta(z,z')}$ is defined by $d' \text{ I } D$ and $d' \sim u'$ where $u' = u^{\theta(z,z')}$. We show that d' is independent of the choice of u. For let $w \in \{z,d\}^{\perp}$, with $w \neq u$ and $Z \not I w \not I D$. Then the quadrilateral Σ containing z,u,d,w is opposite the line L. Hence there is a quadrilateral Σ' containing z' and in perspective with Σ from L. It follows immediately that w defines the same point d'. (Note : Since $t \geq 2$, d' is uniquely defined.)

Let $d \not I L$, $d \not I Z$ and $d' = d^{\theta(z,z')}$. Then clearly $z' = z^{\theta(d,d')}$. Now we show that for any point u, with $u \not I Z$, $u \not I D$, $u \not I L$, we have $u^{\theta(z,z')} = u^{\theta(d,d')}$.

First let $z \sim d$. If $u \in zd$, then by the regularity of L it is clear that $u^{\theta(z,z')} = u^{\theta(d,d')}$. Now suppose that $u \in z^{\perp} \cup d^{\perp}$, $u \notin zd$, e.g. assume $u \in d^{\perp}$. Then $d \in \{z,u\}^{\perp}$, implying that $u^{\theta(z,z')}$ is incident with U and is collinear with d'. Hence $u^{\theta(z,z')} = u^{\theta(d,d')}$. Finally, let $u \notin z^{\perp} \cup d^{\perp}$. Suppose that w is the point which is incident with zd and collinear with u. Since L is regular, the line W is concurrent with the line z'd'. If $U \neq W$, then $u^{\theta(z,z')}$ as well as $u^{\theta(d,d')}$ is the point which is incident with U and collinear with $w' = W \cap z'd'$. So assume $U = W$. Let $D \sim R \sim L$, $R \neq L$

and $R \neq D$, let r be incident with R and collinear with z, and let $h \in \{r,d\}^{\perp}$ with $rh \notin \{R, rz\}$ and $rh \not\sim U$. (This is possible since $t > 2$.) By preceding cases we have : $u^{\theta(z,z')} = u^{\theta(r,r')}$, with $r' = r^{\theta(z,z')}$; $u^{\theta(r,r')} = u^{\theta(h,h')}$, with $h' = h^{\theta(r,r')} = h^{\theta(z,z')}$; $u^{\theta(h,h')} = u^{\theta(d,d'')}$, with $d'' = d^{\theta(h,h')} = d^{\theta(r,r')} = d^{\theta(z,z')} = d'$. Hence $u^{\theta(z,z')} = u^{\theta(d,d')}$.

Now suppose that $z \not\sim d$. If $u \in z^{\perp} \cup d^{\perp}$, e.g. $u \in d^{\perp}$, then $z^{\theta(d,d')} = z^{\theta(u,u')}$ with $u' = u^{\theta(d,d')}$. Hence $z' = z^{\theta(d,d')} = z^{\theta(u,u')}$, and $u' = u^{\theta(z,z')}$, proving that $u^{\theta(z,z')} = u^{\theta(d,d')}$. So assume now that $u \notin z^{\perp} \cup d^{\perp}$. Let $w \in \{z,d\}^{\perp}$, $w \not\mathrm{I}\ Z$ and $w \not\mathrm{I}\ D$, and let $w' = w^{\theta(z,z')} = w^{\theta(d,d')}$. Since $t \geqslant 3$ we may assume that $w \not\mathrm{I}\ U$. Then $u^{\theta(z,z')} = u^{\theta(w,w')} = u^{\theta(d,d')}$. Hence again $u^{\theta(z,z')} = u^{\theta(d,d')}$.

At this point the action of $\theta(z,z')$ is defined on all points except those of Z different from z and not on L. So let $c\ \mathrm{I}\ Z$ and $c \not\mathrm{I}\ L$. If $d \not\mathrm{I}\ Z$ and $d \not\mathrm{I}\ L$, then define $c' = c^{\theta(z,z')}$ by $c' = c^{\theta(d,d')}$, with $d' = d^{\theta(z,z')}$. We show that c' is independent of the choice of d . Let $u \not\mathrm{I}\ Z$, $u \not\mathrm{I}\ L$, $u \neq d$, and $u' = u^{\theta(z,z')}$. If $U \neq D$, then $u' = u^{\theta(z,z')} = u^{\theta(d,d')}$, and $c^{\theta(d,d')} = c^{\theta(u,u')}$. If $U = D$, then choose a point w with $w \not\mathrm{I}\ Z$, $w \not\mathrm{I}\ L$, $W \neq D$. We have $c^{\theta(d,d')} = c^{\theta(w,w')}$, with $w' = w^{\theta(z,z')}$, and $c^{\theta(u,u')} = c^{\theta(w,w')}$. Hence $c^{\theta(d,d')} = c^{\theta(u,u')}$.

It is now clear that $\theta(z,z')$ defines a permutation of the pointset P of S. We next define the action of $\theta(z,z')$ on the lineset B of S.

For all $M \sim L$ we define $M^{\theta(z,z')} = M$. Now let $N \not\sim L$ and $N \not\sim Z$. The point which is incident with N and collinear with z is denoted by d . Further, let $u\ \mathrm{I}\ N$ with $u \neq d$. If $d' = d^{\theta(z,z')}$ and $u' = u^{\theta(z,z')}$, then since $d \in \{z,u\}^{\perp}$, we have $d' \sim u'$. We define $N^{\theta(z,z')} = N'$ to be the line $d'u'$, and we show that N' is independent of the choice of u. To this end, let $w\ \mathrm{I}\ N$, $d \neq w \neq u$, and $w' = w^{\theta(z,z')}$. By the regularity of L there holds $W \sim d'u'$. Since $d \in \{z,w\}^{\perp}$, we have $w' = W \cap d'u'$. Hence it is now clear that N' is independent of the choice of u. Finally, let $N \not\sim L$ and $N \sim Z$. If $c = Z \cap N$ and $d\ \mathrm{I}\ N$, $d \neq c$, then $c^{\theta(z,z')} = c^{\theta(d,d')} = c'$, with $d' = d^{\theta(z,z')}$. Hence $c' \sim d'$. Define $N^{\theta(z,z')} = N'$ to be the line $c'd'$. We show that N' is independent of the choice of d. Let $u\ \mathrm{I}\ N$, $c \neq u \neq d$, and $u' = u^{\theta(z,z')}$. By the regularity of L we have $U \sim c'd'$. Clearly $u' = u^{\theta(z,z')} = u^{\theta(d,d')} = U \cap c'd'$. Consequently N' is independent of the choice of d.

In this way $\theta(z,z')$ defines a permutation of the lineset B of S. It is also clear that for all $h \in P$ and $R \in B$, $h \, I \, R$ is equivalent to $h^{\theta(z,z')} \, I \, R^{\theta(z,z')}$. Hence $\theta(z,z')$ is an automorphism of S. Since $M^{\theta(z,z')} = M$ for all $M \sim L$, $\theta(z,z')$ is a symmetry with axis L and mapping z onto z'. It follows that L is an axis of symmetry. □

9.7.2. (*M.A. Ronan* [150]). *The* GQ $S = (P,B,I)$ *of order* (s,t), $s > 1$ *and* $t > 2$, *is isomorphic to* $Q(4,q)$ *or* $Q(5,q)$ *iff given a quadrilateral* Σ *opposit a line* L *and a point* x', $x' \not{I} L$, *incident with a line of perspectivity of* Σ *from* L, *there is a quadrilateral* Σ' *containing* x' *and in perspective with* Σ *from* L.

Proof. Let $S \cong Q(4,q)$ or $S \cong Q(5,q)$, so S has order (q,q) or (q,q^2), respectively. Each line is an axis of symmetry (recall that S is a TGQ with base point any point of S (cf. 8.7)), and the conclusion follows from 9.7.1.

Conversely, suppose the quadrilateral condition holds, with $t > 2$, $s > 1$. Then by 9.7.1 each line of S is an axis of symmetry. By 8.3.1 $S^{(p)}$ is a TGQ for each point p. Now by 9.2.2 and 9.3.1 S satisfies $(\hat{M})_L$ and $(M)_p$ for each line L and each point p, i.e. S is a Moufang GQ. By the theorem of J. Tits [220] S is classical or dual classical. Since all lines of S are regular $S \cong Q(4,q)$ or $S \cong Q(5,q)$. □

Remark : The case t = 2. If $t = 2$ and $s > 1$, then $S \cong Q(4,2)$ or $S \cong H(3,4)$ (cf. 5.2.3 and 5.3.2). Let L be a line of S and assume the quadrilateral Σ is opposite the line L. The points of Σ are denoted by x,y,z,u, with $x \sim y \sim z \sim u \sim x$. Since $t = 2$, it is easy to show that $X \cap L = Z \cap L$ and $Y \cap L = U \cap L$. Also, it is easy to verify that given a line L there is always at least one quadrilateral Σ opposite L. Now let $x' \, I \, X$, $x' \not{I} L$ and $x \neq x'$. Since S is Moufang, there is an automorphism θ of S fixing L pointwise, $X \cap L$ and $Y \cap L$ linewise, and mapping x onto x'. Then θ maps Σ onto a quadrilateral Σ' containing x' and in perspective with Σ from L. Hence for $t = 2$ and $s > 1$, i.e. for $Q(4,2)$ and $H(3,4)$, M.A. Ronan's quadrilateral condition of the preceding theorem is satisfied.

9.8. OTHER CLASSIFICATIONS USING COLLINEATIONS

In this section we state three results that are in the spirit of this chapter but for whose proofs we direct the reader elsewhere.

Let $S = (P,B,I)$ be a finite GQ of order (s,t), $1 < s$, $1 < t$. The first result, which has appeared so far only in [55], answers affirmatively a conjecture of E.E. Shult.

9.8.1. (*C.E. Ealy, Jr.* [55]). *Let the group of symmetries about each point of* S *have even order. Then* s *is a power of* 2 *and one of the following must hold :* (i) $S \cong W(s)$, (ii) $S \cong H(3,s)$, (iii) $S \cong H(4,s)$.

9.8.2. (*M. Walker* [229]). *Let* G *be a group of collineations of* S *leaving no point or line of* S *fixed. Suppose that* S *has a point* p *and a line* L *for which the group of symmetries about* p *(respectively, about* L*) has order at least* 3 *and is a subgroup of* G. *Then* S *contains a* G*-invariant subquadrangle* S' *isomorphic to* $W(2^n)$ *(for some integer* $n \geq 2$*) such that the restriction of* G *to this subquadrangle contains* $PSp(4,2^n)$.

For the third result we need a couple definitions. Let $x,y \in P$, $x \not\sim y$. A *generalized homology* with centers x,y is a collineation θ of S which is a whorl about x and a whorl about y. The group of all generalized homologies with centers x,y is denoted $H(x,y)$. S is said to be (x,y)-*transitive* if for each $z \in \{x,y\}^{\perp}$ the group $H(x,y)$ is transitive on $\{x,z\}^{\perp\perp}-\{x,z\}$ and on $\{y,z\}^{\perp\perp}-\{y,z\}$.

9.8.3. (*J.A. Thas* [210]). *Let* S *be* (x,y)-*transitive for all* $x,y \in P$ *with* $x \not\sim y$. *Then one of the following must hold :* (i) $S \cong W(s)$, (ii) $S \cong Q(4,s)$, (iii) $S \cong Q(5,s)$, (iv) $S \cong H(3,s)$, (v) $S \cong H(4,s)$.

10 Generalized quadrangles as group coset geometries

10.1. 4-GONAL FAMILIES

Let G be a group of order s^2t, $1 < s$, $1 < t$. Let $J = \{S_0,\ldots,S_t\}$ be a family of t+1 subgroups of G, each of order s. We say J is a *weak 4-gonal family for* G provided J satisfies condition K1 of Section 8.2.

K1. $S_iS_j \cap S_k = 1$ for distinct i,j,k.

Given a weak 4-gonal family J, we seek conditions on J that will guarante the existence of an associated family $J^* = \{S_0^*,\ldots,S_t^*\}$ of subgroups, each of order st, with $S_i \subset S_i^*$, and for which condition K2 is satisfied.

K2. $S_i^* \cap S_j = 1$ for distinct i,j.

Clearly the family J^* is desired so that W.M. Kantor's construction of the GQ $S(G,J)$ is possible.

So suppose J is a weak 4-gonal family for G. Put $\Omega = \bigcup_{i=0}^{t} S_i$. In the t members of $J-\{S_i\}$ there are t(s-1) nonidentity elements, no two of which may belong to the same coset of S_i by condition K1. Hence there are $st-t(s-1)-1 = t-1$ cosets of S_i disjoint from Ω. Let S_i^* be the union of these t-1 cosets together with S_i, i.e. $S_i^* = \cup\{S_ig \parallel g \in G \text{ and } S_ig \cap \Omega \subset S_i\}$. It is clear that if there is a subgroup A_i^* of order st containing S_i and for which $A_i^* \cap S_j = 1$ whenever $j \neq i$, then necessarily $A_i^* = S_i^*$. Put $J^* = \{S_i^* \parallel 0 \leq i \leq t\}$.

If a construction similar to that given by W.M. Kantor actually yields a GQ, it follows that S_i^* must be a group for each i. Hence we make the following definition : the weak 4-gonal family J for G is called a *4-gonal family for* G provided S_i^* is a subgroup for each i. In any case the set S_i^* is called the *tangent space of* Ω *at* S_i.

10.1.1. (*S.E. Payne* [135] *and J.A. Thas* [189]). *Let* $J = \{S_0, \ldots, S_t\}$ *be a weak* 4*-gonal family for the group* G, $|G| = s^2t$, $|S_i| = s$, $1 < s$, $1 < t$, $0 \leqslant i \leqslant t$.

(i) *If there is a subgroup* C *of order* t *for which* $C \lhd G$ *and* $S_iC \cap S_j = 1$ *for* $i \neq j$, *then* $S_i^* = S_iC$; *hence* S_i^* *is a subgroup for each* i *and* J *is a* 4*-gonal family. If* $S = S(G,J)$ *is the corresponding* GQ *of order* (s,t), *then* $S^{(\infty)}$ *is a* STGQ .

(ii) *If* $s = t$ *and each member of* J *is normal in* G, *then* J *is a* 4*-gonal family. If* $S = S(G,J)$ *is the corresponding* GQ *of order* (s,s), *then* $S^{(\infty)}$ *is a* TGQ.

Proof. First suppose there is a subgroup C satisfying the hypothesis of part (i). As S_iC contains t cosets of S_i whose union meets Ω in S_i, clearly $S_i^* = S_iC$, so that S_i^* is a group. As C acts as a full group of symmetries about (∞), $S^{(\infty)}$ must be an STGQ (implying $s \geqslant t$).

Now suppose that each member of J is normal in G and that $s = t$. Suppose that $\phi : G \to G/S_0$ is the natural homomorphism, and put $\bar{S}_0 = \phi(S_0^*) = \{\bar{g}_1, \ldots, \bar{g}_s\}$, with $\bar{g}_1 = S_0$, and $\bar{S}_i = \phi(S_i)$, $1 \leqslant i \leqslant s$. Clearly $\bar{S}_i \cong S_i$, $1 \leqslant i \leqslant s$. As $\{S_0^*, S_0S_1 - S_0, \ldots, S_0S_s - S_0\}$ is a partition of G, $\{\bar{S}_0, \bar{S}_1, \ldots, \bar{S}_s\}$ is a partition of G/S_0. We will show that $\bar{S}_0$ is closed under multiplication and hence is a group, forcing $S_0^* = \phi^{-1}(\bar{S}_0)$ to be a group. Similarly, each S_i^* is a group, forcing J to be a 4-gonal family.

So suppose $\bar{g}_i, \bar{g}_j$ are arbitrary nonidentity elements of $\bar{S}_0$ for which $\bar{g}_i\bar{g}_j \notin \bar{S}_0$. Hence $\bar{g}_i\bar{g}_j \in \bar{S}_k$ for some $k > 0$. For $m \neq n$, $1 \leqslant m, n \leqslant s$, $\bar{S}_m . \bar{S}_n = G/S_0$. In particular, for $m \neq 0, k$, $\bar{S}_m . \bar{S}_k = G/S_0$. Hence for each $m \in \{1, \ldots, s\} - \{k\}$, $\bar{g}_i = u_m v_m$, with $u_m \in \bar{S}_m - \{\bar{g}_1\}$, $v_m \in \bar{S}_k - \{\bar{g}_1\}$. Suppose $v_m = v_{m'}$ with $m \neq m'$. Then $u_m = \bar{g}_i v_m^{-1} = \bar{g}_i v_{m'}^{-1} = u_{m'} \in (\bar{S}_m \cap \bar{S}_{m'}) - \{\bar{g}_1\}$, an impossibility. Hence each of the nonidentity elements of $\bar{S}_k$ serves as a unique v_m. In particular, $\bar{g}_i\bar{g}_j = v_m$ for some $m \neq 0, k$. So $\bar{g}_i = u_m v_m = u_m(\bar{g}_i\bar{g}_j)$, implying $\bar{g}_j = \bar{g}_i^{-1} u_m^{-1} \bar{g}_i \in \bar{S}_m$ ($S_m \lhd G$ implies $\bar{S}_m \lhd G/S_0$), i.e. $\bar{g}_j \in \bar{S}_0 \cap \bar{S}_m - \{\bar{g}_1\}$, an impossibility. Hence $\bar{S}_0$ must be closed, completing the proof that J is a 4-gonal family for G. Then because $S_i \lhd G$, S_i is a full group of symmetries about the line $[S_i]$ of $S = S(G,J)$, and $S^{(\infty)}$ is a TGQ by Section 8.2. (cf. 8.3 also). □

It is frustrating that for $s < t$ we have no satisfactory criterion for deciding just when a weak 4-gonal family is in fact a 4-gonal family.

10.2. 4-GONAL PARTITIONS

Let J be a family of s+2 subgroups of the group G, each of order s, $|G| = s^3$ with $AB \cap C = 1$ for distinct $A,B,C \in J$. Then J is called a *4-gonal partition of* G.

10.2.1. (*S.E. Payne* [128]). *Let* J *be a 4-gonal partition of the group* G *with order* $s^3 > 1$.

(i) *A* GQ $S = S(G,J)$ *of order* (s-1,s+1) *is constructed as follows : the points of* S *are the elements of* G*; the lines of* S *are the right cosets of members of* J*; incidence is containment.*

(ii) *If* $J = \{C,S_0,\ldots,S_s\}$ *with* $C \lhd G$, *then* $J = \{S_0,\ldots,S_s\}$ *is a 4-gonal family for* G. *Moreover,* S(G,J) *is a* STGQ *of order* s *with base point* (∞), *and* $S(G,J)$ *is the* GQ $P(S(G,J),(\infty))$ (cf. 3.1.4).

(iii) *If two members of* J *are normal in* G, *say* $C \lhd G$, $S_0 \lhd G$, *then* G *is elementary abelian and* s *is a power of* 2.

Proof. $S(G,J)$ is readily seen to be a tactical configuration with s points on each line, s+2 lines through each point, and for which any two distinct points are incident with at most one common line. The condition $AB \cap C = 1$ for distinct $A,B,C \in J$ says there are no triangles. Hence a given point x is on s+2 lines and collinear with unique points on each of (s+2)(s-1)(s+1) other lines, accounting for all lines of $S(G,J)$. It follows that $S(G,J)$ is a GQ of order (s-1,s+1), completing the proof of (i). Part (ii) is an immediate corollary of 11.1.1 (i) and 3.1.4.

For part (iii), suppose $J = \{C,S_0,\ldots,S_s\}$ with $C \lhd G$, $S_0 \lhd G$. So $J = \{S_0,\ldots,S_s\}$ is a 4-gonal family with $S_i^* = S_iC$, $0 \leqslant i \leqslant s$. Since $S_0 \lhd G$, $[S_0]$ is an axis of symmetry with symmetry group S_0. Moreover, if θ_g is the collineation of $S(G,J)$ derived from right multiplication by g, $g \in G$, then by 8.2.6 (i) θ_g induces an elation $\bar{\theta}_g$ (with axis (∞)) of the plane π_0 derived from the regularity of $[S_0]$. The map $\theta_g \mapsto \bar{\theta}_g$ into the group of elations of π_0 with axis (∞) has kernel $\{\theta_g \| g \in S_0\}$ and image of order s^2. Hence the plane π_0 is a translation plane with elementary abelian translation group. By 8.2.6 (ii) and (iii) the collineations θ_g are mapped to elations $\bar{\bar{\theta}}_g$ of the plane π_∞ derived from the regularity of the point (∞) of $S(G,J)$. The map $\theta_g \mapsto \bar{\bar{\theta}}_g$ into the group of elations of π_∞ with center (∞) has kernel $\{\theta_g \| g \in C\}$ and image of size s^2. Hence the plane π_∞ is a dual translation plane, so the corresponding (dual) translation group is elementary abelian. Let g_1, g_2 be distinct elements of G, and put $g = g_1 g_2 g_1^{-1} g_2^{-1}$. By the previous remarks, θ_g must fix all points of $(\infty)^{\perp}$ and all lines of $[S_0]^{\perp}$, i.e. $g \in C \cap S_0$. Hence g must be the identity, implying that G is abelian. Hence each $[S_j]$ is an axis of symmetry and $S(G,J)$ is a TGQ with (∞) a regular base point, forcing s to be a power of 2 (cf. 1.5.2 (iv) and 8.5.2).

10.3. EXPLICIT DESCRIPTION OF 4-GONAL FAMILIES FOR TGQ

10.3.1. $T_2(O)$

Let $s = t = q = p^e$, p a prime. Let $F = GF(q)$, $G = \{(a,b,c) \parallel a,b,c \in F\}$ with the usual vector (pointwise) addition. Put $A(\infty) = \{(0,0,c) \parallel c \in F\}$. Let $\alpha : F \to F$ be a function, and for $t \in F$ put $A(t) = \{(\lambda, \lambda t, \lambda t^{\alpha}) \parallel \lambda \in F\}$. Put $J = \{A(\infty)\} \cup \{A(t) \parallel t \in F\}$. As $A(\infty)$ is just the set of all scalar multiples of (0,0,1) and A(t) is the set of all scalar multiples of $(1,t,t^{\alpha})$, it follows readily that J is a weak 4-gonal family (and hence a 4-gonal family by 10.1.1) provided the matrix $\begin{pmatrix} 1 & t & t^{\alpha} \\ 1 & u & u^{\alpha} \\ 1 & v & v^{\alpha} \end{pmatrix}$ is nonsingular for distinct $t,u,v \in F$. The determinant Δ of this matrix is $\Delta = (u-t)(v^{\alpha}-t^{\alpha}) - (v-t)(u^{\alpha}-t^{\alpha})$, which is nonzero iff $\frac{t^{\alpha}-v^{\alpha}}{t-v} \neq \frac{t^{\alpha}-u^{\alpha}}{t-u}$ for distinct $t,u,v \in F$. In this case J is an oval O in PG(2,q), and $S(G,J)$ is isomorphic to $T_2(O)$. It is easy to see that each $T_2(O)$ can be obtained in such a way . By B. Segre's result [158] we may assume $\alpha : x \mapsto x^2$ if q is odd. When q is even it is necessary that α be a permutation. Then $C = \{(0,b,0) \parallel b \in F\}$ is the subgroup (the nucleus of the oval J) for which $\{C\} \cup J$ is a 4-gonal partition of G (i.e. a (q+2)-arc of PG(2,q)). In this case several examples in addition to $\alpha : x \mapsto x^2$ are known, and much more will be said on the subject in Chapter 12.

10.3.2. $T_3(O)$

(i) $s^2 = t = q^2$, q a power of an odd prime.

Let c be a nonsquare in $F = GF(q)$. Put $G = \{(x_0,x_1,x_2,x_3) \parallel x_i \in F\}$, with the usual pointwise addition. Put $A(\infty) = \{(0,\lambda,0,0) \parallel \lambda \in F\}$. For $a, b \in F$, put $A(a,b) = \{(\lambda, -\lambda(a^2-b^2c), \lambda a, \lambda b) \parallel \lambda \in F\}$. Then $J = \{A(\infty)\} \cup \{A(a,b) \parallel a,b \in F\}$ is a 4-gonal family for G. Clearly J is an ovoid O of PG(3,q), in fact an elliptic quadric, and $\mathcal{S}(G,J) \cong T_3(O) \cong Q(5,q)$.

(ii) $s^2 = t = q^2$, q a power of 2.

Let δ be an element of F for which $x^2+x+\delta$ is irreducible over F. Put $G = \{(x_0,x_1,x_2,x_3) \parallel x_i \in F\}$, with the usual addition. Put $A(\infty) = \{(0,\lambda,0,0) \parallel \lambda \in F\}$. For $a,b \in F$ put $A(a,b) = \{(\lambda,\lambda(a^2+ab+\delta b^2),\lambda a,\lambda b) \parallel \lambda \in F\}$. Then $J = \{A(\infty)\} \cup \{A(a,b) \parallel a,b \in F\}$ is a 4-gonal family of G. Clearly J is an ovoid O of $PG(3,q)$, in fact an elliptic quadric, and $\mathcal{S}(G,J) \cong T_3(O) \cong Q(5,q)$.

(iii) $s^2 = t = q^2$, $q = 2^{2e+1}$, and $e \geqslant 1$.

For $F = GF(q)$ let σ be the automorphism of F defined by $\sigma : x \mapsto x^{2^{e+1}}$. Put $A(\infty) = \{(0,\lambda,0,0) \parallel \lambda \in F\}$. For $a,b \in F$, put $A(a,b) = \{(\lambda,\lambda(ab+a^{2\sigma}+b^{2\sigma+2}),\lambda a,\lambda b) \parallel \lambda \in F\}$. Put $J = \{A(\infty)\} \cup \{A(a,b) \parallel a,b \in F\}$. As in the preceding examples $G = F^4$ with pointwise addition. Then J is a 4-gonal family for G. In fact, J is a Tits ovoid O in $PG(3,2^{2e+1})$, the only known type of ovoid in $PG(3,q)$ not a quadric, and $\mathcal{S}(G,J) \cong T_3(O)$.

10.4. A MODEL FOR STGQ

Let $F = GF(q)$, $q = p^e$, p prime. For m and n positive integers, let $f : F^m \times F^m \to F^n$ be a fixed biadditive map. Put $G = \{(\alpha,c,\beta) \parallel \alpha,\beta \in F^m, c \in F^n\}$. Define a binary operation on G by

$$(\alpha,c,\beta).(\alpha',c',\beta') = (\alpha+\alpha',c+c'+f(\beta,\alpha'),\beta+\beta'). \tag{1}$$

This makes G into a group that is abelian if f is trivial and whose center is $C = \{(0,c,0) \in G \parallel c \in F^n\}$ is f is nonsingular. Suppose that for each $t \in F^n$ there is an additive map $\delta_t : F^m \to F^m$ and a map $g_t : F^m \to F^n$. Put $A(\infty) = \{(0,0,\beta) \in G \parallel \beta \in F^m\}$. For $t \in F^n$ put $A(t) = \{(\alpha, g_t(\alpha), \alpha^{\delta_t}) \in G \parallel \alpha \in F^m\}$. We want $A(t)$ to be closed under the product in G, so that it will be a subgroup of order q^m. Writing out the product of two elements of $A(t)$ we find that $A(t)$ is a subgroup of G iff

$$g_t(\alpha+\beta)-g_t(\alpha)-g_t(\beta) = f(\alpha^{\delta_t},\beta) \text{ for all } \alpha,\beta \in F^m,\ t \in F^n. \tag{2}$$

Put $\beta = 0$ in (2) to obtain

$$g_t(0) = 0 \text{ for all } t \in F^n. \tag{3}$$

From now on we suppose that condition (2) holds, and put $J = \{A(\infty)\} \cup \{A(t) \parallel t \in F^n\}$. With $A^* = AC$ for $A \in J$, we seek conditions on J and $J^* = \{A^* \parallel A \in J\}$ that will force K1 and K2 to hold, i.e. that will force J to be a 4-gonal family. Clearly A^* is a group of order q^{n+m} containing A as a subgroup. We note that

$$\begin{aligned} A^*(\infty) &= \{(0,c,\beta) \in G \parallel c \in F^n,\ \beta \in F^m\}, \\ A^*(t) &= \{(\alpha,c,\alpha^{\delta_t}) \in G \parallel \alpha \in F^m,\ c \in F^n\},\ t \in F^n. \end{aligned} \tag{4}$$

It is easy to check that $A^*(\infty) \cap A(t) = 1 = A^*(t) \cap A(\infty)$ for all $t \in F^n$.

An element of $A^*(t) \cap A(u)$ has the form $(\alpha,c,\alpha^{\delta_t}) = (\alpha,g_u(\alpha),\alpha^{\delta_u})$. For $t \neq u$ this must force $\alpha = 0$, so $A^*(t) \cap A(u) = 1$ iff

$$\delta(t,u) : \alpha \mapsto \alpha^{\delta_t} - \alpha^{\delta_u} \text{ is nonsingular if } t \neq u. \tag{5}$$

From now on we assume that (5) holds. Then J will be a 4-gonal family for G iff $AB \cap D = 1$ for distinct $A,B,D \in J$. Before investigating this condition we need a little more information about g_t.

Put $\beta = -\alpha$ in (2) to obtain $-g_t(\alpha)-g_t(-\alpha) = f(\alpha^{\delta_t},-\alpha) = -f(\alpha^{\delta_t},\alpha) = -(g_t(2\alpha)-2g_t(\alpha))$, implying

$$g_t(2\alpha) = 3g_t(\alpha)+g_t(-\alpha). \tag{6}$$

Using (2) and (6) we obtain $g_t((n+1)\alpha) = (n+1)g_t(\alpha)+g_t(n\alpha)+ng_t(-\alpha)$, from which an induction argument may be used to show that

$$g_t(n\alpha) = \binom{n+1}{2}g_t(\alpha)+\binom{n}{2}g_t(-\alpha). \tag{7}$$

<u>Note</u> : If $g_t(-\alpha) = -g_t(\alpha)$, then $g_t(n\alpha) = ng_t(\alpha)$.
If $g_t(-\alpha) = g_t(\alpha)$, then $g_t(n\alpha) = n^2g_t(\alpha)$.

Let $g \in A(\infty).A(t) \cap A(u)$, $t \neq u$, so g has the form $g = (0,0,\beta).(\alpha,g_t(\alpha),\alpha^{\delta_t}) = (\alpha,g_t(\alpha)+f(\beta,\alpha),\beta+\alpha^{\delta_t}) = (\alpha,g_u(\alpha),\alpha^{\delta_u})$. So $g_t(\alpha)-g_u(\alpha) = -f(\beta,\alpha)$, with $\beta = \alpha^{\delta(u,t)}$, should imply $\alpha = 0$. That is : $g_u(\alpha)-g_t(\alpha) = f(\alpha^{\delta_u},\alpha)-f(\alpha^{\delta_t},\alpha) = (g_u(2\alpha)-2g_u(\alpha))-(g_t(2\alpha)-2g_t(\alpha)) = (g_u(\alpha)+g_u(-\alpha))-(g_t(\alpha)+g_t(-\alpha))$ should imply $\alpha = 0$. Hence $A(\infty).A(t) \cap A(u) = 1$ (for $t \neq u$) iff

$$g_t(\alpha) = g_u(\alpha),\ t \neq u,\ \text{implies } \alpha = 0\ . \tag{8}$$

It is routine to check that, for $t \neq u$, also $A(t).A(\infty) \cap A(u) = 1$ iff $A(t).A(u) \cap A(\infty) = 1$ iff (8) holds. Hence we assume (8) holds and proceed to the hard case : What does $A(t).A(u) \cap A(v) = 1$ mean when t,u,v are distinct ? An element of this intersection would be of the form $(\alpha+\beta, g_t(\alpha)+g_u(\beta)+f(\alpha^{\delta t},\beta), \alpha^{\delta t}+\beta^{\delta u}) = (\alpha+\beta,\ g_v(\alpha+\beta), (\alpha+\beta)^{\delta v})$. Hence the intersection is trivial provided

$$\left.\begin{array}{l} g_t(\alpha)+g_u(\beta)+f(\alpha^{\delta t},\beta) = g_v(\alpha+\beta) \\ \alpha^{\delta t}+\beta^{\delta u} = (\alpha+\beta)^{\delta v} \\ t,u,v \text{ distinct} \end{array}\right\} \Rightarrow \alpha = \beta = 0. \tag{9}$$

Solving for β in (9) we find $\beta = \alpha^{\delta(t,v)\delta^{-1}(v,u)}$. Put $\gamma = \alpha^{\delta(t,v)} = \beta^{\delta(v,u)}$. The first equality of (9) becomes

$$\begin{aligned} 0 &= g_t(\alpha)+g_u(\beta)+f(\alpha^{\delta t},\beta)-g_v(\alpha)-g_v(\beta)-f(\alpha^{\delta v},\beta) \\ &= g_t(\alpha)-g_v(\alpha)+g_u(\beta)-g_v(\beta)+f(\alpha^{\delta(t,v)},\beta) \\ &= g_t(\alpha)-g_v(\alpha)+g_u(\beta)-g_v(\beta)+f(\beta^{\delta(v,u)},\beta) \\ &= g_t(\alpha)-g_v(\alpha)+g_u(\beta)-g_v(\beta)+(g_v(\beta)+g_v(-\beta))-(g_u(\beta)+g_u(-\beta)) \\ &= g_t(\alpha)-g_v(\alpha)+g_v(-\beta)-g_u(-\beta). \end{aligned}$$

Hence (9) is equivalent to :

$$\left.\begin{array}{l} g_t(\gamma^{\delta^{-1}(t,v)})-g_v(\gamma^{\delta^{-1}(t,v)})+g_v(-\gamma^{\delta^{-1}(v,u)})-g_u(-\gamma^{\delta^{-1}(v,u)}) = 0 \\ \text{implies } \gamma = 0 \text{ if } t,u,v \text{ are distinct.} \end{array}\right\} \tag{10}$$

We summarize these results as follows.

10.4.1. (*S.E. Payne* [135]). J *is a* 4*-gonal family for* G *provided the following hold :*

(i) $g_t(\alpha+\beta)-g_t(\alpha)-g_t(\beta) = f(\alpha^{\delta t},\beta) = f(\beta^{\delta t},\alpha)$ *for all* $\alpha,\beta \in F^m$, $t \in F^n$.

(ii) $\delta(t,u) : \alpha \mapsto \alpha^{\delta t}-\alpha^{\delta u}$ *is nonsingular for* $t \neq u$.

(iii) $g_t(\alpha) = g_u(\alpha)$, $t \neq u$, *implies* $\alpha = 0$.

(iv) (10) *holds.*

If J is a 4-gonal family, the resulting GQ $S = S(G,J)$ has order $(s,t) = (q^m,q^n)$. As C is a group of t symmetries about (∞) (cf.8.2.2),it follows that

$(S^{(\infty)},G)$ is a STGQ and $m \geq n$. By 8.1.2 $q^{m+n}(1+q^n) \equiv 0$ (modulo q^m+q^n). Then exactly the same argument as the one used in the proof of 8.5.2 shows that either $s = t$ or there is an *odd* integer a and a prime power q^v for which $s = q^m = (q^v)^{a+1}$, $t = q^n = (q^v)^a$. *Hence* $s = t$ *or* $s^a = t^{a+1}$ *with* a *odd*. It may be that there is a theory of the kernel of a STGQ analogous to that for TGQ which will lead to $s = t$ or $s^a = t^{a+1}$ with a odd for all STGQ, but we have been unable to show this. In any case the known examples of STGQ have $s = t$ or $s = t^2$. Hence we complete this section with the known examples of STGQ having $s = t$ and devote the next section to the case $s = t^2$.

10.4.2. *Examples of* STGQ *of order* (s,s)

First note that if $(S^{(p)},G)$ is any TGQ of order s with s even, then p must be regular and G induces a group of elations of the plane π_p with center p. The kernel of this representation of G must have order s and hence be a full group of symmetries about p. Therefore $(S^{(p)},G)$ is also a STGQ.

The known GQ of odd order s also provide examples as follows.

In the notation of this section let $m = n = 1$, q odd or even. Put $f(a,b) = -2ab$, $a^{\delta t} = -at$, and $g_t(a) = a^2 t$ for all $a,b,t \in F$. It is easy to check that the first three conditions of 10.4.1 are satisfied. We have $g_t(a) = g_t(-a)$ and $\delta^{-1}(t,v) : a \mapsto -a/(t-v)$. Hence (10) becomes : $(-a/(t-v))^2(t-v) = (-a/(v-u))^2(u-v)$ implies $a = 0$ if t,u,v are distinct. As this clearly holds, we have a STGQ which, in fact, turns out to be the dual of the example of 10.3.1 where α is defined by $\alpha : x \mapsto x^2$ (i.e., turns out to be isomorphic to W(q)). That these two examples are duals of each other may be seen as follows.

Let S be the STGQ described in the preceding paragraph. Since the point (∞) is regular, it is clear that $S \cong W(q)$ if all points of S not in $(\infty)^\perp$ are regular. Since S is an EGQ with base point (∞) it is sufficient to show that the point (0,0,0) is regular. By 1.3.6 (ii) the point (0,0,0) is regular iff each triad containing (0,0,0) is centric. Before proving this we note that $(a,c,b) \sim (a',c',b')$ iff $c-c'-a'b'+ab+a'b-ab' = 0$.

Consider a triad $((0,0,0),(a,c,b),(a_1,c_1,b_1))$. This triad has a center of the form (a',c',b') iff $-c'-a'b' = 0$, $c-c'-a'b'+ab+a'b-ab' = 0$, and $c_1-c'-a'b'+a_1b_1+a'b_1-a_1b' = 0$. So the triad has a unique center of the form (a',c',b') if $ba_1 \neq ab_1$. Now assume $ba_1 = ab_1$. If $a = a_1 = 0$, then

$A^*(\infty)$ is a center of the triad. If $a = 0 = b$ (resp., $a_1 = 0 = b_1$), then each $A^*(t)$, $t \in F \cup \{\infty\}$, is a center of the triad $((\infty),(0,0,0),(a,c,b))$ (resp., $((\infty),(0,0,0),(a_1,c_1,b_1))$), hence $((0,0,0),(a,c,b))$ (resp. $((0,0,0),(a_1,c_1,b_1))$) is regular and $((0,0,0),(a,c,b),(a_1,c_1,b_1))$ is centric. If $a \neq 0 \neq a_1$, then $A^*(-b/a)$ is a center of the triad.

Next consider a triad $((0,0,0),(a,c,b),A^*(\infty).(a_1,c_1,b_1))$. This triad has a center (a',c',b') iff $a' = a_1$, $-c'-a'b' = 0$ and $c-c'-a'b'+ab+a'b-ab' = 0$. If $a \neq 0$, there is a unique solution in a',c',b'. If $a = 0$, then $A^*(\infty)$ is a center of the triad.

Now consider a triad $((0,0,0)(a,c,b),A^*(t).(a_1,c_1,b_1))$, $t \in F$. This triad has a center (a',c',b') iff $-c'-a'b' = 0$, $c-c'-a'b'+ab+a'b-ab' = 0$ and $(a',c',b') \in A^*(t)(a_1,c_1,b_1)$. If $b+at \neq 0$ there is a unique center of this type. If $b+at = 0$, then the triad has center $A^*(t)$.

Since (∞) is regular, any triad $((0,0,0),x,y)$ with $x,y \in (\infty)^{\perp}$ is centric. Hence each triad containing $(0,0,0)$ is centric, i.e. $(0,0,0)$ is regular, which proves that $S \cong W(q)$.

10.5. A MODEL FOR CERTAIN STGQ WITH $(s,t) = (q^2,q)$

Throughout this section $[f]$ will denote a certain 2×2 matrix over $F = GF(q)$ subject to appropriate restrictions to be developed below. Put $f(\alpha,\beta) = \alpha[f]\beta^T$, for $\alpha,\beta \in F^2$. For each $t \in F$ let K_t be a 2×2 matrix over F and put $\alpha^{\delta t} = \alpha K_t$, for $\alpha \in F^2$. Then $f(\alpha^{\delta t},\beta) = \alpha K_t[f]\beta^T$ is symmetric in α and β iff $K_t[f]$ is symmetric. Hence from now on we require the following :

$$K_t[f] \text{ is symmetric for each } t \in F. \tag{11}$$

Then for part (i) of 10.4.1 to be satisfied it is sufficient that $g_t(\alpha) = \alpha A_t \alpha^T$, where A_t is an upper triangular matrix for which

$$A_t + A_t^T = K_t[f]. \tag{12}$$

And part (ii) of 10.4.1 is equivalent to

$$K_t - K_u \text{ is nonsingular for } t \neq u. \tag{13}$$

We say $B \in M_2(F)$ is *definite* provided $\alpha B \alpha^T = 0$ implies $\alpha = 0$ (for $\alpha \in F^2$).

If $B = \begin{pmatrix} a & b \\ c & d \end{pmatrix}$, then B is definite iff the polynomial $ax^2+(b+c)x+d$ is irreducible over F. In case q is odd, B is definite precisely when $(b+c)^2-4ad$ is a nonsquare in F. *Hence if* B *is symmetric and* q *is odd, then* B *is definite iff* $-\det B$ *is a nonsquare in* F. In either case (q odd or even) B is definite iff cB is definite ($0 \neq c \in F$) iff PBP^T is definite (P nonsingular).

It is easy to check that (iii) of 10.4.1 is equivalent to

$$A_t-A_u \text{ is definite for } t \neq u . \tag{14}$$

Now $\gamma^{\delta^{-1}(t,v)} = \gamma(K_t-K_v)^{-1}$, and (iv) of 10.4.1 is equivalent to

$$(K_t-K_v)^{-1}(A_t-A_v)((K_t-K_v)^{-1})^T+(K_v-K_u)^{-1}(A_v-A_u)((K_v-K_u)^{-1})^T \text{ is definite.} \tag{15}$$

When q is odd, $B \in M_2(F)$ is definite iff $B+B^T$ is definite. And if B is the matrix displayed in (15), then $M = B+B^T = [f]((K_t-K_v)^{-1}+(K_v-K_u)^{-1})^T = [f]((K_t-K_v)^{-1}(K_t-K_u)(K_v-K_u)^{-1})^T$. This completes the proof of the following theorem.

10.5.1. *The family* J *(as given in* 10.4 *but using* $[f]$, δ_t, *etc., as given in this section) is a* 4*-gonal family for* G *provided the following conditions* (i),...,(v) *hold :*

(i) $K_t[f]$ *is symmetric for each* $t \in F$.

(ii) $g_t(\alpha) = \alpha A_t \alpha^T$, *where* A_t *is an upper triangular matrix for which* $A_t+A_t^T = K_t[f]$, *for* $t \in F$.

(iii) K_t-K_u *is nonsingular for* $t,u \in F$, $t \neq u$.

(iv) A_t-A_u *is definite for* $t,u \in F$, $t \neq u$.

(v) $(K_t-K_v)^{-1}(A_t-A_v)((K_t-K_v)^{-1})^T+(K_v-K_u)^{-1}(A_v-A_u)((K_v-K_u)^{-1})^T$ *is definite for distinct* $t,u,v \in F$.

Moreover, if q is odd, then (v),(vi) *and* (vii) *are equivalent.*

(vi) $[f]((K_t-K_v)^{-1}+(K_v-K_u)^{-1})^T = [f]((K_t-K_v)^{-1}(K_t-K_u)(K_v-K_u)^{-1})^T$ *is definite for distinct* t,u,v *in* F.

(vii) $-\det[f]\det(K_t-K_v)\det(K_t-K_u)\det(K_v-K_u)$ *is a nonsquare in* F.

Define $\theta : G \to G$ by $(\alpha,c,\beta)^\theta = (\alpha,c-g_t(\alpha),\beta-\alpha^{\delta t})$, for some fixed $t \in F$. It is routine to check that θ is an automorphism of G fixing $A(\infty)$ elementwise and mapping $A(x)$ to $\bar{A}(x) = \{(\alpha,\bar{g}_x(\alpha),\alpha^{\bar{\delta}x}) \parallel \alpha \in F\}$, where $\bar{g}_x(\alpha) = g_x(\alpha)-g_t(\alpha)$ and $\alpha^{\bar{\delta}x} = \alpha^{\delta(x,t)}$. Moreover, $\bar{g}_t(\alpha) = 0$ and $\alpha^{\bar{\delta}t} = 0$. Hence putting

$t = 0$ we may change coordinates so as to assume that $g_0(\alpha) = 0$ and $\delta_0 = 0$. From now on we assume

$$g_0(\alpha) = 0 \textit{ and } \alpha^{\delta_0} = 0 \textit{ for all } \alpha \in F^2 (\textit{and so we assume } A_0 = K_0 = 0). \quad (16)$$

10.5.2. *Let* $A,B,C,D,E,K,M \in M_2(F)$, $x \in F$. *Define* $\theta : G \to G$ *by* $(\alpha,c,\beta)^\theta = (\alpha A+\beta B, cx+\alpha C\alpha^T+\alpha D\beta^T+\beta E\beta^T, \alpha K+\beta M)$. *Then* θ *is a group homomorphism iff the following hold :*

(i) $x[f]+D^T = M[f]A^T$.

(ii) $C+C^T = K[f]A^T$.

(iii) $E+E^T = M[f]B^T$.

(iv) $D = K[f]B^T$.

Proof. $((\alpha,c,\beta).(\alpha',c',\beta'))^\theta = (\alpha,c,\beta)^\theta.(\alpha',c',\beta')^\theta$ iff $x\beta[f](\alpha')^T+\alpha C(\alpha')^T+\alpha' C\alpha^T+\alpha D(\beta')^T+\alpha' D\beta^T+\beta E(\beta')^T+\beta' E\beta^T = \alpha K[f]A^T(\alpha')^T+\alpha K[f]B^T(\beta')^T+\beta M[f]A^T(\alpha')^T+\beta M[f]B^T(\beta')^T$. This must hold for all $\alpha,\beta,\alpha',\beta' \in F^2$. Collecting terms involving the pairs $(\beta,\alpha'),(\alpha,\alpha'),(\beta,\beta'),(\alpha,\beta')$, respectively, gives conditions (i), (ii), (iii), (iv), in that order. □

In 10.5.2 put $A = M = aI$, $B = C = D = E = K = 0$, for $0 \neq a \in F$. Then an isomorphism θ_a of G which leaves invariant each member of J is defined by

$$\theta_a : (\alpha,c,\beta) \mapsto (a\alpha, a^2c, a\beta). \quad (17)$$

Clearly $\{\theta_a \parallel a \in F\}$ may be considered to be a group of whorls about the point (∞) which is isomorphic to the multiplicative group F^o of F and which fixes the point $(0,0,0)$ of $S(G,J)$. It seems likely that an appropriate definition of *kernel* of $S(G,J)$ should lead to the field F.

Suppose an automorphism θ of the type described in 10.5.2 were to interchange $A(\infty)$ and $A(0)$. Then by (16) $A = M = 0$, and C and E are skewsymmetric (with zero diagonal). Hence we may assume $C = E = 0$. For any choice of $B \in GL(2,F)$ put $D = -x[f]^T$ and $K = -x[f]^T(B^{-1})^T[f]^{-1}$, so that the conditions of 10.5.2 are satisfied. Then if $x \neq 0$, θ is an automorphism of G interchanging $A(\infty)$ and $A(0)$ and appearing as

$$(\alpha,c,\beta)^\theta = (\beta B, x(c-\alpha[f]^T\beta^T), -x\alpha[f]^T(B^{-1})^T[f]^{-1}). \quad (18)$$

Here we tacitly assumed that B and [f] are invertible, which is indeed the case in the examples to be discussed below. Of course, we would like for θ to preserve J. So suppose there is a permutation $t \mapsto t'$ of the nonzero elements of F for which $\theta : A(t) \mapsto A(t')$. Direct calculation shows that $(\alpha,\alpha A_t\alpha^T,\alpha K_t)^\theta = (\alpha K_t B, x(\alpha A_t\alpha^T-\alpha[f]^T K_t^T\alpha^T), -x\alpha[f]^T(B^{-1})^T[f]^{-1}) = (\alpha K_t B, -x\alpha A_t\alpha^T, -x\alpha[f]^T(B^{-1})^T[f]^{-1})$. Writing out what it means for this last element to be in $A(t')$ completes the proof of the following.

10.5.3. *An automorphism* θ *of* G *as described in* 10.5.2 (*with* $x \neq 0$, B *and* [f] *in* GL(2,F)) *interchanges* $A(\infty)$ *and* A(0) *and leaves* J *invariant iff there is a permutation* $t \mapsto t'$ *of the nonzero elements of* F *for which the following hold :*

(i) $K_t B K_{t'} = -x[f]^T(B^{-1})^T[f]^{-1}$ (*is independent of* t), *and*

(ii) $K_t B A_{t'} B^T K_t^T + xA_t$ *is skewsymmetric* (*with zero diagonal*).

(Note : In these calculations we use freely the observation that $\alpha A\alpha^T = \alpha A^T\alpha^T$, since these matrices are 1×1 matrices.)

Now suppose that some θ as described in 10.5.2 fixes $A(\infty)$, so B = D = 0 and we may assume E = 0 . The conditions of 10.5.2 become $x[f] = M[f]A^T$ and $C+C^T = K[f]A^T$. Then for any choice of C and nonsingular A we must put $K = (C+C^T)(A^T)^{-1}[f]^{-1}$ and $M = x[f](A^T)^{-1}[f]^{-1}$. Hence θ appears as

$$(\alpha,c,\beta)^\theta = (\alpha A, xc+\alpha C\alpha^T, \alpha(C+C^T)(A^T)^{-1}[f]^{-1}+x\beta[f](A^T)^{-1}[f]^{-1}). \qquad (19)$$

Note that θ is an automorphism of G iff $0 \neq x$.

Suppose in addition that the θ of (19) leaves J invariant, so that there is a permutation $t \mapsto t'$ of the elements of F for which $\theta : A(t) \mapsto A(t')$. Then $(\alpha,\alpha A_t\alpha^T,\alpha K_t)^\theta = (\alpha A, x\alpha A_t\alpha^T+\alpha C\alpha^T, \alpha K+\alpha K_t M) = (\alpha A, \alpha AA_{t'}A^T\alpha^T, \alpha AK_{t'})$. From this equality the following is easily deduced.

10.5.4. *An automorphism* θ *as described in* 10.5.2 (*with* $x \neq 0$, A *and* [f] *in* GL(2,F)) *fixes* $A(\infty)$ *and leaves* J *invariant iff there is a permutation* $t \mapsto t'$ *of the elements of* F *for which the following hold for all* $t \in F$*:*

(i) $K_{t'}[f] = A^{-1}(C+C^T+xK_t[f])(A^{-1})^T$, *and*

(ii) $xA_t+C-AA_{t'}A^t$ *is skewsymmetric* (*with zero diagonal*).

To close this section we seek conditions related to the regularity of the point $A^*(\infty)$ in the GQ $S(G,J)$. In particular, consider the noncollinear pair $(A^*(\infty), (\alpha,0,0))$, $\alpha \neq 0$. With the help of Fig. 10.1 it is routine to check that

$$\{A^*(\infty),(\alpha,0,0)\}^{\perp} = \{A^*(\infty).(\alpha,0,0),(0,0,0)\} \cup \{(0,-g_u(\alpha),-\alpha^{\delta}u) \parallel 0 \neq u \in F\}, \tag{20}$$

$$\{(0,0,0),A^*(\infty).(\alpha,0,0)\}^{\perp} = \{(\alpha,0,0),A^*(\infty)\} \cup \{(\alpha,g_t(\alpha),\alpha^{\delta}t) \parallel 0 \neq t \in F\}. \tag{21}$$

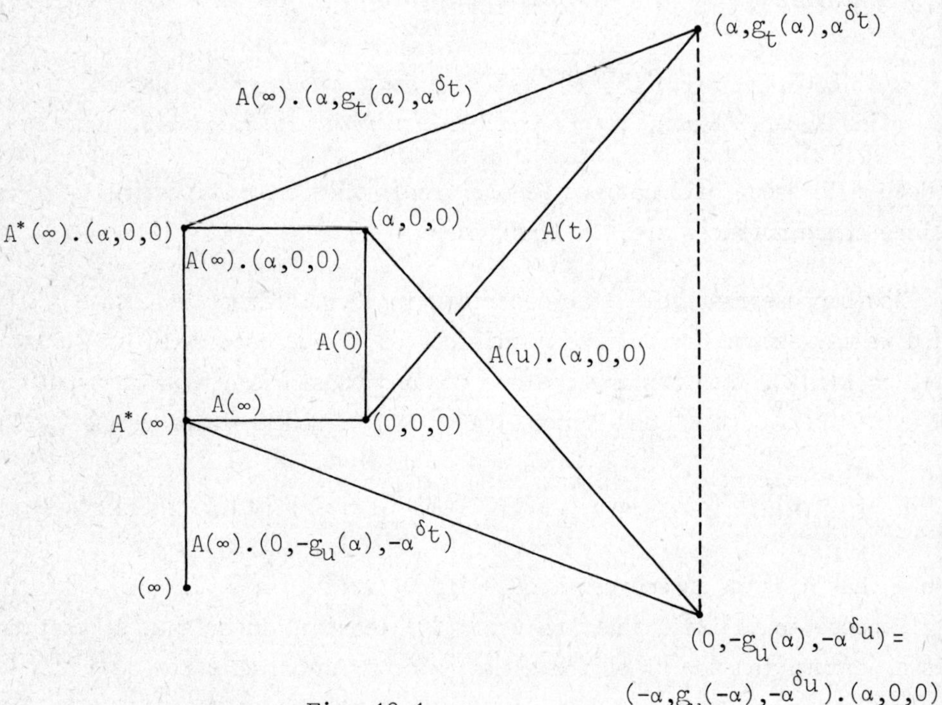

Fig. 10.1

Hence the pair $(A^*(\infty),(\alpha,0,0))$ is regular iff $(\alpha,g_t(\alpha),\alpha^{\delta}t)$ and $(0,-g_u(\alpha),-\alpha^{\delta}u)$ are collinear for all $t,u \in F^o$. This is the case precisely when there is some $v \in F$ for which $(\alpha,g_t(\alpha),\alpha^{\delta}t).(0,-g_u(\alpha),-\alpha^{\delta}u)^{-1} = (\alpha,g_t(\alpha)+g_u(\alpha),\alpha^{\delta}t+\alpha^{\delta}u) \in A(v)$. This holds iff $g_t(\alpha)+g_u(\alpha) = g_v(\alpha)$ and $\alpha^{\delta}t+\delta^{\delta}u = \alpha^{\delta}v$. This essentially completes a proof of the following.

10.5.5. *For the 4-gonal family* J *of this section, the pair* $(A^*(\infty),(\alpha,0,0))$ $(\alpha \neq 0)$ *of noncollinear points of* $S(G,J)$ *is regular iff for each choice of* $t,u \in F^o$ *there is a* $v \in F$ *for which both the following hold :*

(i) $\alpha(K_v-K_t-K_u) = 0$,

(ii) $\alpha(A_v-A_t-A_u)\alpha^T = 0$.

10.6. EXAMPLES OF STGQ WITH ORDER (q^2,q)

10.6.1. $H(3,q^2)$ *as a* STGQ (*adapted from W.M. Kantor* [88])

In the notation of the two preceding sections put $[f] = \begin{pmatrix} 2 & x_1 \\ x_1 & -2x_0 \end{pmatrix}$, where $x^2-x_1x-x_0$ is irreducible over $F = GF(q)$. It is easy to see that $[f]$ is definite iff q is odd and is nonsingular in any case. Put $K_t = tI$, so $K_t[f] = t\begin{pmatrix} 2 & x_1 \\ x_1 & -2x_0 \end{pmatrix}$ is symmetric regardless of the characteristic of F. Then $K_t-K_u = (t-u)I$ is clearly nonsingular for $t \neq u$. Put $D = \begin{pmatrix} 1 & x_1 \\ 0 & -x_0 \end{pmatrix}$, and $A_t = tD$. Then $A_t+A_t^T = K_t[f]$ and $A_t-A_u = (t-u)D$, $t \neq u$, is definite iff D is definite. When q is even, D is definite provided $x^2+x_1x+x_0$ is irreducible. When q is odd, D is definite iff $D+D^T = [f]$ is definite. Hence in either case A_t-A_u is definite for $t \neq u$. The matrix of 10.5.1 (v) is $((t-v)^{-1}+(v-u)^{-1})D$, which is definite since D is. Hence we at least have that $S = S(G,J)$ is a STGQ of order (q^2,q). Here the group of automorphisms of G leaving J invariant is doubly transitive on the elements of J (put $A = M = C = E = 0$, $D = -[f]^T$, $K = -I$, $B = I$, $x = 1$ for a θ (as in 10.5.3) that interchanges $A(\infty)$ and $A(0)$ and maps $A(t)$ to $A(-t^{-1})$; put $A = M = I$, $B = D = E = 0$, $C = A_u$, $K = uI$, $x = 1$ for a θ (as in 10.5.4) fixing $A(\theta)$ and mapping $A(t)$ to $A(t+u)$).

Now we show that all lines incident with (∞) are 3-regular. By the preceding paragraph and since S is a STGQ, it is sufficient to prove that any triple of the form $([A(\infty)],A(0),A(t)(\alpha,c,\beta))$, for $t \in GF(q)$ and $A(0) \not\sim A(t)(\alpha,c,\beta)$, is 3-regular.

The points not belonging to $(\infty)^\perp$ and incident with a line $L \in \{[A(\infty)],A(0)\}^\perp$, are of the form $(\alpha_1,f(\beta_1,\alpha_1),\beta_1)$, with $\alpha_1,\beta_1 \in F^2$. A point $(\alpha_1,f(\beta_1,\alpha_1),\beta_1)$ is incident with the line $A(t)(\alpha,c,\beta)$ if there is an $\alpha_0 \in F^2$ for which $(\alpha_0,g_t(\alpha_0),\alpha_0^\delta t)(\alpha,c,\beta) = (\alpha_1,f(\beta_1,\alpha_1),\beta_1)$, or equivalently $\alpha_0+\alpha = \alpha_1$, $g_t(\alpha_0)+c+f(\alpha_0^\delta t,\alpha) = f(\beta_1,\alpha_1)$, and $\alpha_0^\delta t+\beta = \beta_1$. Hence the point

$(\alpha_0, g_t(\alpha_0), \alpha_0^{\delta t})$ (α, c, β) is incident with a line of $\{[A(\infty)], A(0)\}^{\perp}$ iff

$$f(\beta, \alpha_0 + \alpha) + f(\alpha_0^{\delta t}, \alpha_0) = g_t(\alpha_0) + c. \tag{22}$$

These lines of $\{[A(\infty)], A(0)\}^{\perp}$ are incident with the points $(\alpha_0 + \alpha, 0, 0)$ of $A(0)$, with $\alpha_0 + \alpha$ determined by (22). Let $\alpha_0 + \alpha = (r_1, r_2)$, $\alpha = (a_1, a_2)$ and $\beta = (b_1, b_2)$, with $r_1, r_2, a_1, a_2, b_1, b_2 \in GF(q)$. Then (22) is equivalent to

$$(b_1 b_2) \begin{pmatrix} 2 & x_1 \\ x_1 & -2x_0 \end{pmatrix} \begin{pmatrix} r_1 \\ r_2 \end{pmatrix} + t\,(r_1 - a_1 \;\; r_2 - a_2) \begin{pmatrix} 2 & x_1 \\ x_1 & -2x_0 \end{pmatrix} \begin{pmatrix} r_1 - a_1 \\ r_2 - a_2 \end{pmatrix} =$$
$$t(r_1 - a_1 \;\; r_2 - a_2) \begin{pmatrix} 1 & x_1 \\ 0 & -x_0 \end{pmatrix} \begin{pmatrix} r_1 - a_1 \\ r_2 - a_2 \end{pmatrix} + c, \tag{23}$$

or

$$(2b_1 + b_2 x_1)r_1 + (b_1 x_1 - 2b_2 x_0)r_2 + t((r_1 - a_1)^2 + (r_2 - a_2)(r_1 - a_1)x_1 - x_0(r_2 - a_2)^2) = c \tag{24}$$

First, let $t \neq 0$. Then, since S has order (q^2, q), we know that (24) has exactly $q+1$ solutions (r_1, r_2). Clearly the same solutions are obtained by replacing b_1, b_2, t, c, respectively, by $\ell b_1, \ell b_2, \ell t, \ell c$, $\ell \in F^o$. Note that $A(0) \not\sim A(t)(\alpha, c, \beta)$ is equivalent to $b_1^2 + b_1 b_2 x_1 - b_2^2 x_0 + tc - t(2b_1 a_1 + x_1(a_2 b_1 + a_1 b_2) - 2x_0 a_2 b_2) \neq 0$, which clearly shows that also $A(0) \not\sim A(\ell t)(\alpha, \ell c, \ell \beta)$ for any $\ell \in F^o$. Since $\{A(0), [A(\infty)], A(\ell t)(\alpha, \ell c, \ell \beta)\}^{\perp}$ is independent of $\ell \in F^o$, the triple $(A(0), [A(\infty)], A(t)(\alpha, c, \beta))$ is 3-regular.

Now let $t = 0$. Then, since S has order (q^2, q), we know that (24) has exactly q solutions (r_1, r_2). Clearly the same solutions are obtained by replacing b_1, b_2, c, respectively by $\ell b_1, \ell b_2, \ell c$, $\ell \in F^o$. Note that $A(0) \not\sim A(0)(\alpha, c, \beta)$ is equivalent to $\beta \neq 0$, which clearly shows that also $A(0) \not\sim A(0)(\alpha, \ell c, \ell \beta)$ for any $\ell \in F^o$. Since $\{A(0), [A(\infty)], A(0)(\alpha, \ell c, \ell \beta)\}^{\perp}$ is independent of $\ell \in F^o$, the triple $(A(0), [A(\infty)], A(0)(\alpha, c, \beta))$ is 3-regular.

Hence all lines incident with (∞) are 3-regular. By 5.3.1 the GQ S is isomorphic to the dual of a $T_3(0)$. Moreover, by Step 3 of the proof of 5.3.1 all points of $(\infty)^{\perp}$ are regular. So in $T_3(0)$ all lines concurrent with some line w of type (b) are regular. Now by 3.3.3 (iii) we have $T_3(0) \cong Q(5, q)$ i.e. $S \cong H(3, q^2)$.

10.6.2. *W.M. Kantor's examples* $K^*(q)$

The description given in 3.1.6 of W.M. Kantor's examples K(q) was of a GQ of order (q,q^2). In this section we give a description of the dual GQ $K^*(q)$ as a STGQ of order (q^2,q). This construction is adapted directly from [38] (cf. [135]).

Let $q \equiv 2 \pmod 3$, q a prime power, F = GF(q). Put $[f] = \begin{pmatrix} 0 & -3 \\ 1 & 0 \end{pmatrix}$. $K_t = \begin{pmatrix} -t^2 & -2t^3 \\ 2t & 3t^2 \end{pmatrix}$, so $K_t[f] = \begin{pmatrix} -2t^3 & 3t^2 \\ 3t^2 & -6t \end{pmatrix}$ is symmetric, and put $A_t = \begin{pmatrix} -t^3 & 3t^2 \\ 0 & -3t \end{pmatrix}$. It is easy to check that $\det(K_t-K_u) = (t-u)^4 \neq 0$ for $t \neq u$, so that K_t-K_u is nonsingular.

Before checking conditions (iv) and (v) of 10.5.1 it is expedient to consider some automorphisms of G. To obtain a θ as described in 10.5.2 and 10.5.4, put B = D = E = 0, x = 1, $A = \begin{pmatrix} 1 & y \\ 0 & 1 \end{pmatrix}$, $C = \begin{pmatrix} -y^3 & -3y^2 \\ 0 & -3y \end{pmatrix}$, $M = \begin{pmatrix} 1 & 3y \\ 0 & 1 \end{pmatrix}$, $K = \begin{pmatrix} y^2 & y^3 \\ 2y & 3y^2 \end{pmatrix}$. Then θ fixes $A(\infty)$ and maps A(t) to A(t+y). To obtain a θ as described in 10.5.2 and 10.5.3 put A = C = E = M = 0, $B = \begin{pmatrix} 0 & -1 \\ 1 & 0 \end{pmatrix}$, $D = \begin{pmatrix} 0 & -1 \\ 3 & 0 \end{pmatrix}$, $K = \begin{pmatrix} 0 & -1 \\ 1 & 0 \end{pmatrix}$, x = 1. Then θ interchanges $A(\infty)$ and A(0) and maps A(t) to $A(-t^{-1})$, for $t \neq 0$. Hence the group of automorphisms of G leaving J invariant is doubly transitive on the elements of J.

From Section 10.4 (cf. (5)) and the nonsingularity of K_t-K_u for $t \neq u$ we know that condition K2 holds for the family J. And by the previous paragraph, to show that K1 holds it suffices to show that $A(\infty).A(0) \cap A(u) = 1$ if $u \neq 0$. But this is the case provided (8) holds with t = 0, which holds iff $B = A_0-A_u = \begin{pmatrix} u^3 & -3u^2 \\ 0 & 3u \end{pmatrix}$ is definite for $u \neq 0$. If q is odd, B is definite iff $-\det(B+B^T)$ is a nonsquare in F iff -3 is a nonsquare in F iff $q \equiv 2 \pmod 3$. If q is even, $B = u\begin{pmatrix} u^2 & u \\ 0 & 1 \end{pmatrix}$ is definite iff x^2+x+1 is irreducible over F iff $q \equiv 2 \pmod 3$.

Hence $K^*(q)$ really is a STGQ of order (q^2,q) for each prime power q, $q \equiv 2 \pmod 3$. The point (∞) of $K^*(q) = S(G,J)$ is regular, in fact a center of symmetry. Moreover, we claim that for $q > 2$ the point (∞) is the *unique* regular point of $K^*(q)$. Since the group of collineations of $K^*(q)$ fixing (∞)

is (doubly) transitive on the lines through (∞) and transitive on the points not collinear with (∞), we need only find some $\alpha \in F^2$, $\alpha \neq 0$, for which the pair $(A^*(\infty),(\alpha,0,0))$ is not regular. By 10.5.5 we need an α such that for some nonzero $t, u \in F$ there is no $v \in F$ for which $\alpha(K_v-K_t-K_u) = 0$ and $\alpha(A_v-A_t-A_u)\alpha^T = 0$. For q odd, put $\alpha = (0,-1)$. Then $\alpha(K_v-K_t-K_u) = (2(t+u-v), 3(t^2+u^2-v^2))$, which cannot be zero for any choice of nonzero t and u. For q even, put $\alpha = (1,0)$. Then $\alpha(K_v-K_t-K_u) = 0$ holds iff $v = t+u$, in which case $\alpha(A_v-A_t-A_u)\alpha^T = tu(t+u)$. So for $q > 2$, choose t and u to be any distinct nonzero elements of F.

Of course, when $q = 2$, $K^*(q)$ is the unique GQ of order (4,2), and hence must have all points regular. The above paragraph shows that W.M. Kantor's examples are indeed new, since the previously known examples of GQ of order (q^2,q) are just the duals of the TGQ $T_3(O)$, and $T_3(O)$ has a coregular point.

As was indicated in 3.1.6, W.M. Kantor [88] gave a geometrical description of K(q) in terms of the classical generalized hexagon H(q). To complete this section we sketch this approach to K(q).

By J. Tits [220] the generalized hexagon H(q) can be constructed as follows. Let $U_1 = \{(x,0,0,0,0,0) \parallel x \in GF(q)\},\ldots,U_6 = \{(0,0,0,0,0,x) \parallel x \in GF(q)\}$ be six groups isomorphic to the additive group of GF(q). The representative of $x \in GF(q)$ in U_i is denoted by x_i. Further, let $U_+ = U_1.U_2.U_3.U_4.U_5.U_6 = \{(x,y,z,u,v,w) \parallel x,\ldots,w \in GF(q)\}$ be defined by the commutation relations (we assume $(a,b) = a^{-1}b^{-1}ab$) :

$$(x_1,y_4) = (x_1,y_3) = (x_1,y_2) = (x_2,y_5) = (x_2,y_3) = (x_3,y_5) = (x_3,y_4) =$$
$$(x_4,y_5) = (x_3,y_6) = (x_5,y_6) = 1;$$
$$(x_1,y_5) = (-xy)_3;$$
$$(x_2,y_4) = (3xy)_3;$$
$$(x_1,y_6) = (xy)_2(-x^2y^3)_3(xy^2)_4(xy^3)_5;$$
$$(x_2,y_6) = (-3x^2y)_3(2xy)_4(3xy^2)_5;$$
$$(x_4,y_6) = (3xy)_5.$$

The generalized hexagon H(q) may now be described in terms of the group U_+.

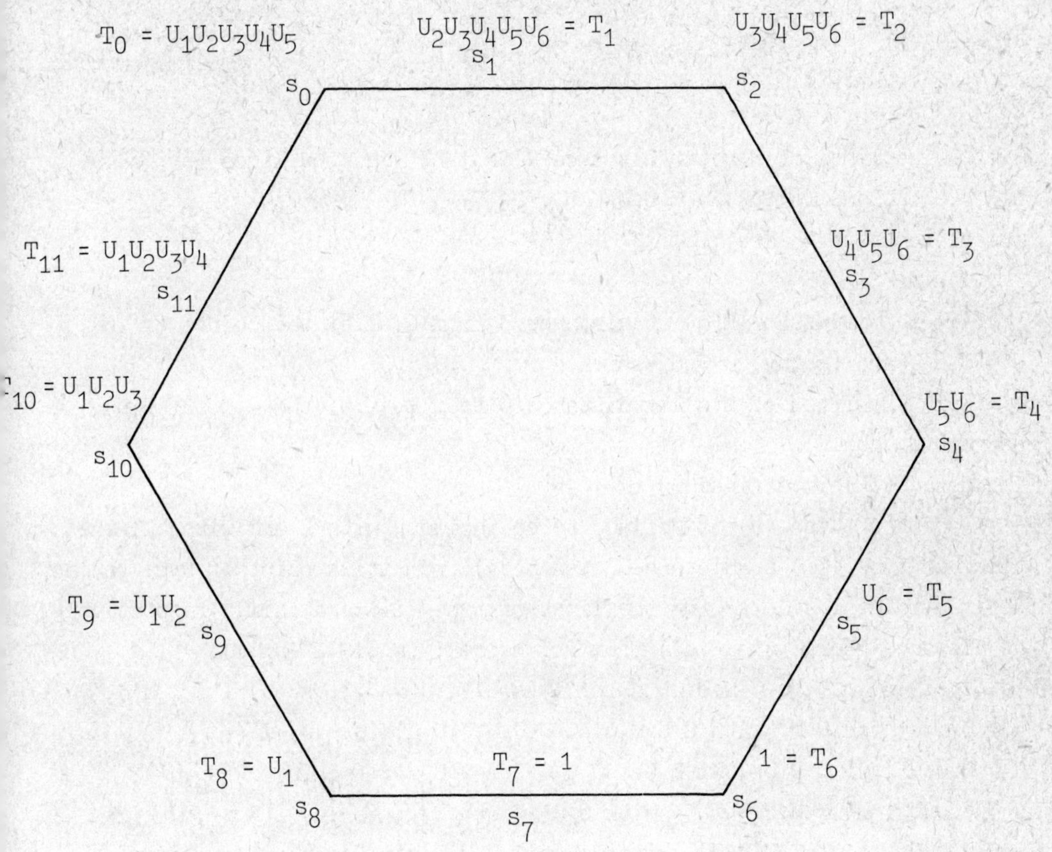

Fig. 10.2

Let T_i^*, $0 \leqslant i \leqslant 11$, be defined by

$T_i^* = T_{i+7}$ and $T_i = T_{i+7}^*$ if $1 \leqslant i \leqslant 5$ and $T_6^* = T_7^* = U_+$.
(Here subscripts are taken modulo 12.)

Points (resp., lines) of the generalized hexagon are the pairs (s_i,u) with $i \in \{0,\dots,11\}$ and i odd (resp., even), and u an element of the group T_i above s_i in Fig. 10.2. Incidence is defined as follows :
$(s_i,u)\, I\, (s_j,u') \Longleftrightarrow i \in \{j-1,j+1\}$ (mod 12) and the intersection of the cosets of T_i^* and T_j^* in U_+ containing u and u', respectively, is nonempty.

Let $q \equiv 2$ (mod 3) and define as follows the incidence structure $\mathcal{S}^* = (P^*,B^*,I^*)$, with pointset P^* and lineset B^*.

The elements of P^* are :

(a) The points of $H(q)$ on the line $L = (s_6,1)$, i.e. the points $(s_7,1)$ and (s_5,u), $u \in U_6$.

(b) The lines of $H(q)$ at distance 4 from L, i.e. the lines (s_{10},u), $u \in U_1U_2U_3$, and (s_2,u), $u \in U_3U_4U_5U_6$.

The elements of B^* are :

(i) The line $L = (s_6,1)$.

(ii) The points of $H(q)$ at distance 3 from L, i.e. the points (s_9,u), $u \in U_1U_2$, and (s_3,u), $u \in U_4U_5U_6$.

(iii) The lines of $H(q)$ at distance 6 from L, i.e. the lines (s_0,u), $u \in U_1U_2U_3U_4U_5$.

Incidence (I^*) is defined by :

A point of type (a) is defined to be incident with L and with all the lines of type (ii) at distance 2 (in $H(q)$) from it; a point of type (b) is defined to be incident with the lines of type (ii) and (iii), respectively, at distance 1 or 2 (in $H(q)$) from it. Hence : $(s_7,1)\ I^*\ (s_6,1)$; $(s_7,1)\ I^*\ (s_9,u)$, $u \in U_1U_2$; $(s_5,u)\ I^*\ (s_6,1)$, $u \in U_6$; $(s_5,u)\ I^*\ (s_3,u')$, $u \in U_6$, $u' \in U_4U_5U_6$ with $U_1U_2U_3u' \subset U_1U_2U_3U_4U_5u$; $(s_{10},u)\ I^*\ (s_9,u')$, $u \in U_1U_2U_3$, $u' \in U_1U_2$, with $U_4U_5U_6u \subset U_3U_4U_5U_6u'$; $(s_{10},u)\ I^*\ (s_0,u')$, $u \in U_1U_2U_3$, $u' \in U_1U_2U_3U_4U_5$ with $U_6u' \subset U_4U_5U_6u$; $(s_2,u)\ I^*\ (s_3,u')$, $u \in U_3U_4U_5U_6$, $u' \in U_4U_5U_6$, with $U_1U_2u \subset U_1U_2U_3u'$; $(s_2,u)\ I^*\ (s_0,u')$, $u \in U_3U_4U_5U_6$, $u' \in U_1U_2U_3U_4U_5$, with $U_6u' \cap U_1U_2u \neq \phi$.

This description of S^* may be reinterpreted as follows.

The elements of P^* are :

(a) $(s_7,1) = [\infty]$ and the cosets of $U_1U_2U_3U_4U_5$.

(b) The cosets of $U_4U_5U_6$ and U_1U_2.

The elements of B^* are :

(i) $L = (\infty)$.

(ii) The cosets of $U_3U_4U_5U_6$ and $U_1U_2U_3$.

(iii) The cosets of U_6.

Incidence (I^*) is given by :

(∞) is incident with all points of type (a); a coset of $U_3U_4U_5U_6$ is incident with $[\infty]$ and with each coset of $U_4U_5U_6$ contained in it; a coset of $U_1U_2U_3$ is incident with the coset of $U_1U_2U_3U_4U_5$ containing it and the cosets of U_1U_2 contained in it; a coset of U_6 is incident with the coset of

$U_4U_5U_6$ containing it and with the cosets of U_1U_2 having a nonempty intersection with it.

Let G be the group $U_1U_2U_3U_4U_5$ (= T_0). The elements of G are of the form (a,b,c,d,e,0) = (a,b,c,d,e). It can be shown that

(a,b,c,d,e).(a',b',c',d',e') = (a+a',b+b',c+c'+a'e-3b'd,d+d',e+e') and that

$x_6^{-1}(a,b,c,d,e)x_6 = (a,b+ax,c-3b^2x-3abx^2-a^2x^3,d+2bx+ax^2,e+3dx+3bx^2+ax^3)$,

with x_6 the representative of x in U_6.

Now let us make the following identifications :

Each coset of G in U_+ is identified with its intersection with U_6; each coset of $U_4U_5U_6$ is identified with its intersection with G; the coset $U_1U_2u_6u_1u_2u_3u_4u_5$, $u_i \in U_i$, of U_1U_2 is identified with the coset $u_6^{-1}U_1U_2u_6u_1u_2u_3u_4u_5$ of $u_6^{-1}U_1U_2u_6$ in G; each coset of $U_3U_4U_5U_6$ is identified with its intersection with G; the coset $U_1U_2U_3u_6u_1u_2u_3u_4u_5$, $u_i \in U_i$, of $U_1U_2U_3$ is identified with the coset $u_6^{-1}U_1U_2U_3u_6u_1u_2u_3u_4u_5$ of $u_6^{-1}U_1U_2U_3u_6$ in G; each coset of U_6 is identified with its intersection with G.

The description of S^* may be reinterpreted once more as follows.

The elements of P^* are :

(a) [∞] and the elements [u] with u ∈ GF(q).

(b) The cosets of $A(\infty) = U_4U_5$ in G, and the cosets of $A(u) = u_6^{-1}U_1U_2u_6$, u ∈ GF(q), in G.

The elements of B^* are :

(i) (∞).

(ii) The cosets of $A^*(\infty) = U_3U_4U_5$ in G, and the cosets of $A^*(u) = u_6^{-1}U_1U_2U_3u_6$, u ∈ GF(q), in G.

(iii) The elements of G.

Incidence (I^*) is defined by :

(∞) is incident with all points of type (a); a coset of $A^*(\infty)$ is incident with [∞] and all cosets of A(∞) contained in it ; a coset of $A^*(u)$ is incident with [u] and with the cosets of A(u) contained in it; an element of G is incident with the cosets of A(∞) and A(u) (for each u ∈ GF(q)) containing it. Since

$$A(\infty) = \{(0,0,0,d,e) \parallel d,e \in GF(q)\},$$
$$A(u) = \{(a,au+b,-a^2u^3-3abu^2-3b^2u,au^2+2bu,au^3+3bu^2) \parallel a,b \in GF(q)\},$$
$$A^*(\infty) = \{(0,0,c,d,e) \parallel c,d,e \in GF(q)\},$$

$$A^*(u) = \{(a, au+b, -a^2u^3-3abu^2-3b^2u+c, au^2+2bu, au^3+3bu^2) \parallel a,b,c \in GF(q)\},$$

or

$$A(\infty) = \{(0,0,0,d,e) \parallel d,e \in GF(q)\},$$
$$A(u) = \{(a,b,-a^2u^3+3abu^2-3b^2u, -au^2+2bu, -2au^3+3bu^2) \parallel a,b \in GF(q)\},$$
$$A^*(\infty) = \{(0,0,c,d,e) \parallel c,d,e \in GF(q)\},$$
$$A^*(u) = \{(a,b,c,-au^2+2bu,-2au^3+3bu^2) \parallel a,b,c \in GF(q)\},$$

S^* clearly is isomorphic to the dual $K(q)$ of $K^*(q)$.

Thus we have a purely geometrical description of $K(q)$ in terms of the generalized hexagon $H(q)$. It was this description which was given in 3.1.6 as the definition of $K(q)$.

10.7. 4-GONAL BASES : SPAN-SYMMETRIC GQ

Let $S = (P,B,I)$ be a GQ of order (s,t) with a regular pair (L_0,L_1) of non-concurrent lines, so that $1 < s \leq t \leq s^2$. Put $\{L_0,L_1\}^\perp = \{M_0,\ldots,M_s\}$, $\{L_0,L_1\}^{\perp\perp} = \{L_0,\ldots,L_s\}$. Let S_i be the group of symmetries about L_i, $0 \leq i \leq s$. Suppose that at least two (and hence all) of the S_i's have order s (if $\theta \in S_0$ sends L_1 onto L_j, $j \neq 0$, then $\theta^{-1}S_1\theta = S_j$). It follows that each L_i is an axis of symmetry, and we say that S is *span-symmetric with base span* $\{L_0,L_1\}^{\perp\perp}$. The general problem which this section just begins to attack is the determination of all span-symmetric GQ. However, in this section we begin to consider the general problem, with special emphasis on the case $s = t$. In that case S may be described as a kind of group coset geometry.

Put $G = \langle S_0,\ldots,S_s\rangle = \langle S_i,S_j\rangle$, $0 \leq i,j \leq s$, $i \neq j$. Using 2.4 and 2.2.2 it is routine to verify the following result.

10.7.1. *If* $\mathrm{id} \neq \theta \in G$, *then the substructure* $S_\theta = (P_\theta,B_\theta,I_\theta)$ *of elements fixed by* θ *must be given by one of the following :*

(i) $P_\theta = \phi$ *and* B_θ *is a partial spread containing* $\{L_0,L_1\}^\perp$.

(ii) *There is a line* $L \in \{L_0,L_1\}^{\perp\perp}$ *for which* P_θ *is the set of points incident with* L, *and* $M \sim L$ *for each* $M \in B_\theta$ ($\{L_0,L_1\}^\perp \subset B_\theta$).

(iii) B_θ *consists of* $\{L_0,L_1\}^\perp$ *together with a subset* B' *of* $\{L_0,L_1\}^{\perp\perp}$; P_θ *consists of those points incident with lines of* B'.

(iv) S_θ *is a subquadrangle of order* (s,t') *with* $s \leq t' < t$. *This forces* $t' = s$ *and* $t = s^2$.

10.7.2. *If* $t < s^2$ *then* G *acts regularly on the set* Ω *of* $s(s+1)(t-1)$ *points of* S *not incident with any line of* $\{L_0,L_1\}^{\perp\perp}$.

Proof. That G acts semiregularly on Ω is immediate from 10.7.1. That G is transitive on Ω follows from 9.4.1, as we now show. For suppose G is not transitive on Ω. Put $\mathcal{O} = P-\Omega$. It is easy to check that conditions (i) through (v) of 9.4.1 are satisfied, with $|\mathcal{O}| = (1+s)^2$, $\bar{b} = 2s$. Hence $s^2 \leqslant t$, contradicting the hypothesis of 10.7.2. □

Note that $S_0,\ldots,S_s$ form a complete conjugacy class of subgroups of order s in the group G. Put $S_i^* = N_G(S_i)$, $0 \leqslant i \leqslant s$. It is routine to complete a proof of the following, *assuming that* $t < s^2$.

10.7.3. (i) $|G| = s(s+1)(t-1)$.

(ii) $S_i^* = G_{L_i}$, $0 \leqslant i \leqslant s$.

(iii) $|S_i^*| = s(t-1)$, $0 \leqslant i \leqslant s$.

(iv) $|S_i^* \cap S_j^*| = t-1$, *if* $i \neq j$, $0 \leqslant i, j \leqslant s$.

(v) $|S_i^* \cap S_j| = 1$, *if* $i \neq j$, $0 \leqslant i, j \leqslant s$.

(vi) $|S_iS_j \cap S_k| = 1$, *if* $0 \leqslant i,j,k \leqslant s$ *with* i,j,k *distinct*.

Put $\Sigma = \{L_0,\ldots,L_s\}$. Then G is doubly transitive on Σ, and $S^* = \bigcap_i S_i^*$ is the kernel of this action. W.M. Kantor [89] has used results of C. Hering - W.M. Kantor - G. Seitz - E.E. Shult on doubly transitive permutation groups to give a group-theoretical proof of the following.

10.7.4. (*W.M. Kantor* [89]). *If* $s < t < s^2$, *then no span-symmetric* GQ *of order* (s,t) *exists.*

We would like to see a more elementary proof of this result. And in the case $t = s^2$ we have seen no proof that $S \cong Q(5,s)$ even using heavy group theory.

For the remainder of this section we assume that $s = t$.

Thus G is a group of order s^3-s, $s \geqslant 2$, having a collection $T = \{S_0,\ldots,S_s\}$ of 1+s subgroups, each of order s . T is a complete conjugacy class in G; $S_i \cap N_G(S_j) = 1$ if $i \neq j$, $0 \leqslant i, j \leqslant s$; and $S_iS_j \cap S_k = 1$ for distinct i,j,k, $0 \leqslant i,j,k \leqslant s$. Under these conditions we say that T is a *4-gonal basis for* G. Conversely, our main goal in this section is to show how to recover a span-symmetric GQ from a 4-gonal basis T . First we offer a simple lemma.

10.7.5. *Let S be a GQ of order* (s,s) *with a fixed regular pair* $\{L_0,L_1\}$ *of nonconcurrent lines. If each line of* $\{L_0,L_1\}^{\perp\perp}$ *is regular, then each line of* $\{L_0,L_1\}^{\perp}$ *is regular.*

Proof. We use the same notation as above : $\{L_0,L_1\}^{\perp\perp} = \{L_0,\dots,L_s\}$, $\{L_0,L_1\}^{\perp} = \{M_0,\dots,M_s\}$. Let M be any line not concurrent with M_i for some fixed i, $0 \leq i \leq s$. Then M must be incident with a point $x = L_j \cap M_k$ for some j and k (s = t implies each line meets some M_k) . Since L_j is regular, it follows that the pair (M_i,M) is regular, and hence M_i must be regular. □

Return now to the case where *S* is span-symmetric with G, S_i, Ω , etc., as above. Let x_0 be a fixed point of Ω. For each $y \in \Omega$ there is a unique element $g \in G$ for which $x_0^g = y$. In this way each point of Ω is identified with a unique element of G. Let N_i be the line through x_0 meeting L_i. Points of N_i in Ω correspond to elements of S_i. Let z_i be the point of L_i on N_i, $0 \leq i \leq s$. For $i \neq j$, $S_i^* \cap S_j^*$ acts regularly on the points of $\{z_i,z_j\}^{\perp} \cap \Omega$. It follows that $S_i^* = S_i(S_i^* \cap S_j^*) = (S_i^* \cap S_j^*)S_i$ $((S_i^* \cap S_j^*)S_i \subset S_i^*$ and $|(S_i^* \cap S_j^*)S_i| = |S_i^*| = s(t-1))$ acts regularly on the points of $z_i^{\perp} \cap \Omega$, so that the elements of a given coset $gS_i = S_ig$ of S_i in S_i^* correspond to the points of a fixed line through z_i. Hence we may identify S_i^* with z_i. Suppose that lines of $\{L_0,L_1\}^{\perp}$ are labeled so that $S_i^* = z_i$ is a point of M_i. Let $g \in G$ map x_0 to a point collinear with $L_j \cap M_i$ (keep in mind that g fixes M_i). Then each point of $x_0^{S_i^*g}$ is collinear with $L_j \cap M_i$, so we may identify $L_j \cap M_i$ with S_i^*g. In this way the points of M_i are identified with the right cosets of S_i^* in G, and a line through S_i^*g (not in $\{L_0,L_1\}^{\perp} \cup \{L_0,L_1\}^{\perp\perp}$) is a coset of S_i contained in S_i^*g. Hence the points of L_i consist of one coset of S_j^* for each $j = 0,1,\dots,s$. If $S_i^* \cap S_j^*g$ (with $S_i^* \neq S_j^*g$) contains a point $y = x_0^h$, then S_ih is the line joining S_i^* and y, and S_jh is the line joining S_j^*g and y. Hence if S_j^*g is a point of L_i (and hence collinear with S_i^*), $i \neq j$, it must be that $S_i^* \cap S_j^*g = \phi$. We show later that for each j, $j \neq i$, S_i^* is disjoint from a unique right coset of S_j^*, so that the points of L_i are uniquely determined as S_i^* and the unique right coset of S_j^* disjoint from S_i^* for $j = 0, 1,\dots,s$, $j \neq i$.

Conversely, now let G be an abstract group of order s^3-s with 4-gonal basis $T = \{S_0,\dots,S_s\}$. Put $S_i^* = N_G(S_i)$. Clearly $s+1 = (G:S_i^*)$, so $|S_i^*| = s(s-1)$. Since $S_i \cap S_j^* = 1$ for $i \neq j$, S_i acts regularly (by conjugation) on $T-\{S_i\}$, and hence S_i^* acts transitively on $T-\{S_i\}$. Since any inner automorphism of G moving S_j to S_k also moves S_j^* to S_k^*, S_i^* also acts transitively on $\{S_0^*,\dots,S_s^*\}-\{S_i^*\}$, and $\{S_0^*,\dots,S_s^*\}$ is a complete conjugacy class in G. As the number of conjugates of S_i^* in G is $1+s = (G:N_G(S_i^*))$, and $1+s = (G:S_i^*)$, it follows that $S_i^* = N_G(S_i^*)$. As S_i^* acts transitively on $T-\{S_i\}$, the subgroup of S_i^* fixing S_j, $i \neq j$, has order $|S_i^*|/s = s-1$, i.e. $|S_i^* \cap S_j^*| = s-1$, and S_i^* is a semidirect product of S_i and $S_i^* \cap S_j^*$.

10.7.6. *Let* $S_i^*g_i$ *and* $S_j^*g_j$ *be arbitrary cosets of* S_i^* *and* S_j^*, $i \neq j$. *Then* $S_i^*g_i \cap S_j^*g_j = \phi$ *iff* $g_jg_i^{-1}$ *sends* S_j^* *to* S_i^* *under conjugation. Moreover, if* $S_i^*g_i \cap S_j^*g_j \neq \phi$, *then* $|S_i^*g_i \cap S_j^*g_j| = s-1$.

Proof. If $x \in S_i^* \cap S_j^*g$, a standard argument shows that $S_i^* \cap S_j^*g = \{tx \parallel t \in S_i^* \cap S_j^*\}$, so $|S_i^* \cap S_j^*g| = s-1$. Since $|S_i^*| = s(s-1)$, S_i^* meets s cosets of S_j^* and is disjoint from the one remaining. Two elements $x,y \in G$ send S_j^* to the same S_k^* iff they belong to the same right coset of $N_G(S_j^*) = S_j^*$, i.e. iff $xy^{-1} \in S_j^*$. Suppose g maps S_j^* to S_i^* : $S_i^* = g^{-1}S_j^*g$, $i \neq j$. Then $g \notin S_j^*$, so $\phi = g^{-1}S_j^* \cap S_j^*$, implying $\phi = g^{-1}S_j^*g \cap S_j^*g = S_i^* \cap S_j^*g$. Hence $S_i^* \cap S_j^*g = \phi$ for all g in that coset of S_j^* mapping S_j^* to S_i^*. Translating by g_i, we have $S_i^*g_i \cap S_j^*gg_i = \phi$ iff $(gg_i)g_i^{-1} = g$ maps S_j^* to S_i^*. □

10.7.7. *Let* i,j,k *be distinct, and* $S_i^*g_i, S_j^*g_j, S_k^*g_k$ *be any three cosets of* S_i^*, S_j^*, S_k^*. *If* $S_i^*g_i \cap S_j^*g_j = \phi$ *and* $S_i^*g_i \cap S_k^*g_k = \phi$, *then* $S_j^*g_j \cap S_k^*g_k = \phi$.

Proof. If $S_i^*g_i \cap S_j^*g_j = \phi$ and $S_k^*g_k \cap S_i^*g_i = \phi$, then $g_jg_i^{-1}$ maps S_j^* to S_i^* and $g_ig_k^{-1}$ maps S_i^* to S_k^*. Hence $(g_jg_i^{-1})(g_ig_k^{-1}) = g_jg_k^{-1}$ maps S_j^* to S_k^*, implying $S_j^*g_j \cap S_k^*g_k = \phi$. □

We are now ready for the following main result.

10.7.8. (*S.E. Payne* [136]). *A span-symmetric* GQ *of order* (s,s) *with given base span* $\{L_0,L_1\}^{\perp\perp}$ *is canonically equivalent to a group* G *of order* s^3-s *with a 4-gonal basis* T.

Proof. We show that for each group G with 4-gonal basis T there is a span-symmetric GQ of order (s,s), denoted $S(G,T)$. However, we leave to the reader the details of showing that starting with a span-symmetric GQ S of order (s,s), with base span $\{L_0,L_1\}^{\perp\perp}$, deriving the 4-gonal basis T of the group G generated by symmetries about lines in $\{L_0,L_1\}^{\perp\perp}$, and then constructing $S(G,T)$ insures that S and $S(G,T)$ are isomorphic.

So suppose G and T are given, $|G| = s^3-s$. Then $S(G,T) = (P_T,B_T,I_T)$ is constructed as follows.

P_T consists of two kinds of points :

(a) Elements of G (s^3-s of these).

(b) Right cosets of the S_i^*'s ($(s+1)^2$ of these).

B_T consists of three kinds of lines :

(i) Right cosets of S_i, $0 \leq i \leq s$ ($(s+1)(s^2-1)$ of these).

(ii) Sets $M_i = \{S_i^*g \parallel g \in G\}$, $0 \leq i \leq s$ (s+1 of these).

(iii) Sets $L_i = \{S_j^*g \parallel S_i^* \cap S_j^*g = \phi,\ 0 \leq j \leq s,\ j \neq i\} \cup \{S_i^*\}$, $0 \leq i \leq s$ (1+s of these).

I_T is the natural incidence relation : a line S_ig of type (i) is incident with the s points of type (a) contained in it, together with that point S_i^*g of type (b) containing it. The lines of types (ii) and (iii) are already described as sets of those points with which they are to be incident. By 10.7.7 two cosets of distinct S_j^*'s are collinear (on a line of type (iii)) only if they are disjoint. In such a way there arise $(s+1)^2s/2$ pairs of disjoint cosets of distinct S_j^*'s. Since for a given coset S_j^*g and given k, $k \neq j$, there is just one coset S_k^*h disjoint from S_j^*g, the total number of pairs of disjoint cosets of distinct S_j^*'s also equals $(s+1)^2s/2$. Hence two cosets of distinct S_j^*'s are collinear iff they are disjoint.

It is now relatively straightforward to check that $S(G,T)$ is a tactical configuration with 1+s points on a line, 1+s lines through each point, at most one common line incident with two given points, $1+s+s^2+s^3$ points and also that many lines, and having no triangles. Hence $S(G,T)$ is a GQ of order (s,s) (having $\{L_i,L_j\}^\perp = \{M_0,\ldots,M_s\}$, $i \neq j$).

In the construction just given, G acts on $S(G,T)$ by right multiplication (leaving all lines M_i invariant) so that S_i is the full group of symmetries about L_i, $0 \leq i \leq s$, and S_i^* is the stabilizer of L_i in G. This can be seen as follows. For $x \in G$, let $\grave{x}$ denote the collineation determined by right multiplication by x. Clearly $\grave{x}$ fixes L_i provided $S_j^*g \cap S_i^* = \phi$ implies $S_j^*gx \cap S_i^* = \phi$, which occurs iff $S_i^*x = S_i^*$ iff $x \in S_i^*$. Moreover, if $x \in S_i^*$, then $\grave{x}$ fixes each point of L_i. Let L be some line of type (i) meeting L_i at, say S_j^*g for some $j \neq i$, where $g^{-1}S_j^*g = S_i^*$ (implying $g^{-1}S_jg = S_i$). Then L is some coset of S_j contained in S_j^*g, say $L = S_jt_jg$ where $t_j \in S_j^*$. And $\grave{x} : L \mapsto L$ iff $S_jt_jgx = S_jt_jg$ iff $g^{-1}(t_j^{-1}S_jt_j)gx = g^{-1}(t_j^{-1}S_jt_j)g$ iff $g^{-1}S_jgx = g^{-1}S_jg$ iff $S_ix = S_i$. Hence S_i is the set of all $g \in G$ for which $\grave{g}$ fixes each line of $S(G,T)$ meeting L_i. Now it is immediate that $S(G,T)$ is span-symmetric with base span $\{L_0,L_1\}^{\perp\perp}$. □

Any automorphism of G leaving $\mathcal{T}$ invariant must induce a collineation of $\mathcal{S}(G,\mathcal{T})$. In particular, for each $g \in G$, conjugation by g yields a collineation $\hat{g}$ of $\mathcal{S}(G,\mathcal{T})$. But conjugation by g followed by right multiplication by g^{-1} yields a collineation $\acute{g}$ given by left multiplication by g^{-1}. Then $g \mapsto \acute{g}$ is a representation of G as a group of collineations of $\mathcal{S}(G,\mathcal{T})$ in which S_i is a full group of symmetries about M_i, and S_i^* is the stabilizer of M_i. This is easily checked, so that we have proved the following.

10.7.9. *If* $\mathcal{S}$ *is a span-symmetric* GQ *of order* (s,s) *with base span* $\{L_0,L_1\}^{\perp\perp}$, *then each line of* $\{L_0,L_1\}^{\perp}$ *is also an axis of symmetry.*

It is natural to conjecture that a span-symmetric GQ of order (s,s) is isomorphic to Q(4,s) and $G \cong SL(2,s)$.

We bring this section to a close with the observation that the unique GQ of order (4,4) (cf. 6.3) has an easy description as a span-symmetric GQ $\mathcal{S} = \mathcal{S}(G,\mathcal{T})$, where $G = SL(2,4) \cong A_5$, the alternating group on {1,2,3,4,5}. Let S_i be the Klein 4-group on the symbols {1,2,3,4,5}-{i}, $1 \leqslant i \leqslant 5$. For example, S_1 = {e,(23)(45),(24)(35),(25)(34)}. Then $S_i^* = N_G(S_i)$ is the alternating group on the symbols {1,2,3,4,5}-{i}. It follows that $\mathcal{T} = \{S_1,\ldots,S_5\}$ is a 4-gonal basis for A_5.

11 Coordinatization of generalized quadrangles with $s = t$

11.1. ONE AXIS OF SYMMETRY

The modern theory of projective planes depends to a very great extent upon the theory of planar ternary rings, either as introduced by M. Hall, Jr. (cf. [69]) or as modified in some relatively modest way (e.g., compare the system used by D.R. Hughes and F. Piper [86]). An analogous general coordinatization theory for GQ has yet to be worked out, and indeed seems likely to be too complicated to be useful. In this chapter a preliminary version of such a theory is worked out for a special class of GQ of order (s,s), starting with those having an axis of symmetry. Throughout this chapter we assume $s > 1$.

Let $S = (P,B,I)$ be a GQ of order (s,s) having a line L_∞ that is an axis of symmetry. Then L_∞ is regular, so by 1.3.1 there is a projective plane based at L_∞ whose dual is denoted by π_∞. The lines of π_∞ are the lines of S in $L_\infty^\perp$, and the points of π_∞ are the spans of the form $\{M,N\}^{\perp\perp}$ where M,N are distinct lines in $L_\infty^\perp$. Clearly the points of the form $\{M,N\}^{\perp\perp}$ with M and N concurrent in $L_\infty^\perp$ may be identified with the points of S incident with L_∞. The coordinatization of S begins with a coordinatization of π_∞.

To begin, choose L_∞ and some three lines of S meeting L_∞ at distinct points and not lying in a same span as the coordinatizing quadrangle of π_∞. Then there is a planar ternary ring $R = (R,F)$ with underlying set R, $|R| = s$, and ternary operation F, so that R coordinatizes π_∞ as follows. There are two distinguished elements of R denoted 0 and 1, respectively. The ternary operation F is a function from $R \times R \times R$ into R satisfying five conditions (cf. [86]) :

$F(a,0,c) = F(0,b,c) = c$ for all $a,b,c \in R$. (1)

$F(1,a,0) = F(a,1,0) = a$ for all $a \in R$. (2)

Given $a,b,c,d \in R$ with $a \neq c$, there is a unique $x \in R$ for which $F(x,a,b) = F(x,c,d)$. (3)

Given $a,b,c \in R$, there is a unique $x \in R$ for which $F(a,b,x) = c$. (4)

For $a,b,c,d \in R$ with $a \neq c$, there is a unique pair of elements $x,y \in R$ for which $F(a,x,y) = b$ and $F(c,x,y) = d$. (5)

The line L_∞ of π_∞ is assigned the coordinate $[\infty]$ and the other three lines in the coordinatizing quadrangle have coordinates $[0]$, $[0,0]$, and $[1,1]$, respectively. More generally, π_∞ has lines $[\infty],[m],[a,b]$, for $a,b,m \in R$, and π_∞ has points (∞), (a), $((m,k))$, $a,k,m \in R$. Here $[\infty],[m],[a,b]$ are the lines of S in $L_\infty^\perp$ $(= [\infty]^\perp)$; (∞) and (a) are the points of S on $[\infty]$, and $((m,k))$ is a set of lines of the form $\{M,N\}^{\perp\perp}$ with M and N concurrent lines in $L_\infty^\perp$. Incidence in π_∞ is given by (6).

$[a,b]$ is incident with (a) and with $((m,k))$ provided $b = F(a,m,k)$.
$[m]$ is incident with $((m,k))$ and with (∞). (6)
$[\infty]$ is incident with (a) and with (∞).
This is for all $a,b,m,k \in R$.

As $[\infty]$ is an axis of symmetry as a line of S, there is an additively written (but not known to be abelian) group G of order s acting as the group of symmetries of S about $[\infty]$. If M is any line of π_∞ different from $[\infty]$, then G acts sharply transitively on the points of M (in S) not on $[\infty]$. Hence each point x of S not on $[\infty]$ will somehow be identified by means of the line of π_∞ through x and some element of G.

Let x be an arbitrary point of S on $[0]$ but not on $[\infty]$, and let y be the point on $[0,0]$ collinear with x. Give x the coordinates $(0,0)$, where the left-hand 0 is the zero element of R and the right hand 0 is the zero (i.e. identity) of G. Then for $g \in G$, give x^g the coordinates $(0,g)$. Similarly, let the point on $[m]$ collinear with y^g have coordinates (m,g). Then each point of S in $(\infty)^\perp$ has been assigned coordinates. Moreover, if $g \in G$, then $(\infty)^g = (\infty)$, $(a)^g = (a)$, and $(m,g_1)^{g_2} = (m,g_1+g_2)$.

Now let z be any point of S not collinear with (∞). On $[0]$ and $[\infty]$ there are unique points, say $(0,g)$ and (a), respectively, collinear with z. If the points (a) and z lie on the line $[a,b]$, we assign to z the coordinates (a,b,g). So (a,b,g) is the unique point on $[a,b]$ collinear with $(0,g)$.

Given a point (m,g) and a line $[a,b]$, there is a unique point z on $[a,b]$ collinear with (m,g). Then z must have coordinates of the form (a,b,g'), where $g' = U(a,b,m,g)$ for some function $U : R^3 \times G \to G$. By construction it has been arranged so that

$$U(0,0,m,g) = g = U(a,b,0,g). \quad (7)$$

It is also clear that if $a,b,m \in R$ are fixed, the map

$$U_{a,b,m} : g \mapsto U(a,b,m,g) \text{ permutes the elements of } G. \quad (8)$$

It now remains to assign coordinates to those lines of S not concurrent with $[\infty]$. Let L be such a line. Then L is incident with a unique point havin coordinates of the form (m,g). Moreover, $\{L,[\infty]\}^{\perp}$ consists of those lines through a unique point $((m,k))$ of the plane π_∞. Assign to L the coordinates $[m,g,k]$. Then g' in G acts as follows.

$$(a,b,g)^{g'} = (a,b,g+g'), \text{ and } [m,g,k]^{g'} = [m,g+g',k]. \quad (9)$$

For $a,m,k \in R$, $g \in G$, the following must hold :

$$(a, F(a,m,k), U(a,F(a,m,k),m,g)) \text{ is on } [m,g,k]. \quad (10)$$

Acting on the incident pair in (10) by a symmetry g', we have

$$(a, F(a,m,k), U(a,F(a,m,k),m,g)+g') \text{ is on } [m,g+g',k]. \quad (11)$$

But $(a, F(a,m,k), U(a,F(a,m,k),m,g+g'))$ is on $[m,g+g',k]$ and must be the onl point of $[a,F(a,m,k)]$ on $[m,g+g',k]$. Hence

$$U(a,F(a,m,k),m,g+g') = U(a,F(a,m,k),m,g)+g'. \quad (12)$$

So we may define $U_0 : R^3 \to G$ by $U_0(a,b,m) = U(a,b,m,0)$, giving

$$U(a,b,m,g) = U_0(a,b,m)+g. \quad (13)$$

Then (7) becomes

$$U_0(0,0,m) = U_0(a,b,0) = 0 \tag{14}$$

and (8) is automatically satisfied.

At this point we have established the following.

11.1.1. *Let S be a GQ of order* s *having a line that is an axis of symmetry. Then S may be realized in the following manner. There is a planar ternary ring $R = (R,F)$ with $|R| = s$. There is a group G (written additively but not shown to be commutative) with $|G| = s$. Finally, there is a function $U_0 : R^3 \to G$ satisfying (14). The points and lines of S have coordinates as follows :*

	Type I	*Type* II	*Type* III	*Type* IV	
points	(a,b,g)	(m,g)	(a)	(∞)	$k,a,b,m \in R$
lines	$[m,g,k]$	$[a,b]$	$[m]$	$[\infty]$	$g \in G$.

Incidence in S is described as follows :

(∞) *is on* $[\infty]$ *and on* $[m]$, $m \in R$.

(a) *is on* $[\infty]$ *and on* $[a,b]$, $a,b \in R$.

(m,g) *is on* $[m]$ *and on* $[m,g,k]$, $m,k \in R$, $g \in G$. (15)

(a,b,g') *is on* $[a,b]$, *and on* $[m,g,k]$ *provided* $b = F(a,m,k)$ *and* $g' = U_0(a,b,m)+g$, $a,b,m,k \in R$, $g,g' \in G$.

Conversely, given $R = (R, F)$, G and U_0, we would like to construct a GQ S with points and lines described above and satisfying the incidence relation given in (15). Using just the properties of R,F,U_0 described so far a routine check shows that S is at least an incidence structure with 1+s points on each line, 1+s lines through each point, two points on at most one common line, and allowing (possibly) only triangles each of whose sides is a line of type I. Hence S is indeed a GQ iff it has no triangles whose sides are lines of type I, and we now determine necessary and sufficient conditions on U_0 for this to be the case.

Consider a hypothetical triangle of S with sides of type I. There are just two cases : one vertex of the triangle is a point of type II and the other two are of type I, or all three vertices are of type I.

Case (i). One vertex is of type II.

In this case the triangle is as indicated in Fig. 11.1, where

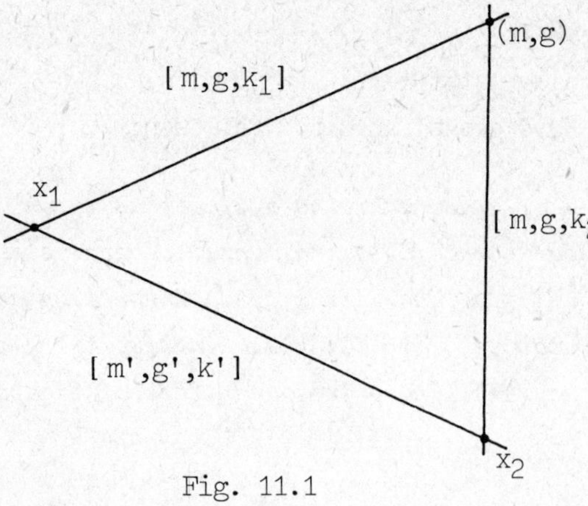

Fig. 11.1

$$x_i = (a_i, F(a_i,m,k_i), U_0(a_i,F(a_i,m,k_i),m)+g)$$
$$= (a_i, F(a_i,m',k'), U_0(a_i,F(a_i,m',k'),m')+g'), \text{ for } i = 1,2.$$

Furthermore, each $g \in G$ acts as a collineation of the resulting incidence structure (whether or not it is a GQ) in the following manner.

$$\begin{array}{l} (\infty)^g = (\infty) \quad (a)^g = (a) \quad [\infty]^g = [\infty] \quad [m]^g = [m] \\ (m,g')^g = (m,g'+g) \quad [a,b]^g = [a,b] \\ (a,b,g')^g = (a,b,g'+g) \quad [m,g',k]^g = [m,g'+g,k]. \end{array} \tag{16}$$

Hence the triangle of Case (i) may be replaced (with a slight change of notation) with the triangle of Fig. 11.2.

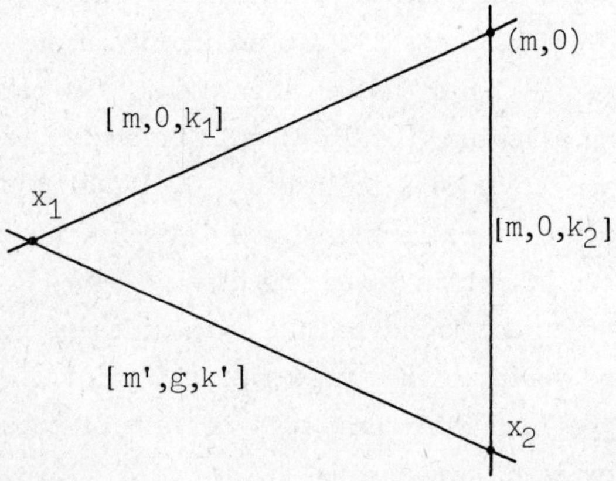

Fig. 11.2

The condition that this kind of triangle not appear is precisely the following : If $F(a_i,m,k_i) = F(a_i,m',k')$ for $i = 1,2$, and if $U_0(a_i,F(a_i,m,k_i),m) = U_0(a_i,F(a_i,m',k'),m')+g$, $i = 1,2$, then $m = m'$ or $a_1 = a_2$. Notice that $k_1 = k_2$ iff $m = m'$ or $a_1 = a_2$. We restate this as :

$$\text{If } b_i = F(a_i,m,k_i) = F(a_i,m',k') \text{ for } i = 1,2, \text{ and if}$$
$$-U_0(a_1,b_1,m')+U_0(a_1,b_1,m) = -U_0(a_2,b_2,m')+U_0(a_2,b_2,m), \tag{17}$$
$$\text{then } m = m' \text{ or } a_1 = a_2.$$

Case (ii). All three vertices are of type I.

In this case the triangle is as indicated in Fig. 11.3,

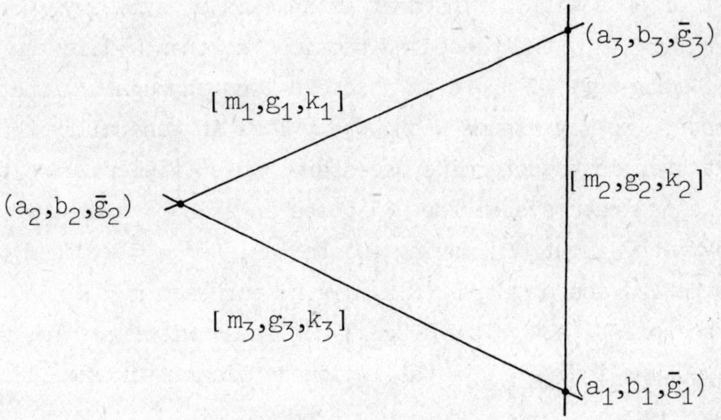

Fig. 11.3

where $b_j = F(a_j,m_i,k_i)$ and $\bar{g}_j = U_0(a_j,b_j,m_i)+g_i$, for $i \neq j$. Hence this triangle is impossible iff the following holds.

If $b_j = F(a_j,m_{j-1},k_{j-1}) = F(a_j,m_{j+1},k_{j+1})$, and if $U_0(a_j,b_j,m_{j-1})+g_{j-1} = U_0(a_j,b_j,m_{j+1})+g_{j+1}$, for $j = 1,2,3$, subscripts taken modulo 3, $(a_i,b_i,k_i,m_i \in R,\ g_i \in G)$, then it must follow that the m_i's are not distinct or the a_i's are not distinct. (18)

Let $R = (R, F)$ be a planar ternary ring as above, G a group with $|G| = |R| = s$. Let $U_0 : R^3 \to G$ be a function satisfying (14), (17) and (18).

Then U_0 is called a 4-*gonal function*, and the triple (R,G,U_0) is a 4-*gonal set up*. We have established the following theorem.

11.1.2. *The existence of a* GQ S *of order* s *with an axis of symmetry is equivalent to the existence of a* 4-*gonal set up* (R,G,U_0) *with* $|R| = s$.

It seems very difficult to study 4-gonal set ups in general. Hence in the next few sections we investigate conditions on (R,G,U_0) that correspond to the existence of additional collineations of the associated GQ, beginning with (essentially) a pair of concurrent axes of symmetry.

11.2. TWO CONCURRENT AXES OF SYMMETRY

Let S be a GQ of order s with a line that is an axis of symmetry. Moreover, let S be coordinatized as in the preceding section, so that $[\infty]$ is the hypothesized axis of symmetry. If there is a second line through (∞) that is an axis of symmetry, we may assume without any loss in generality that it is $[0]$. Our next step is to determine necessary and sufficient conditions in terms of the coordinate system for $[0]$ to be an axis of symmetry.

Let θ be a symmetry about $[0]$ moving (0) to (a), $0 \neq a \in R$. Then the point $(0,k,g)$ on $[0,k]$ and on $[m,-U_0(0,k,m)+g,k]$ for each $m \in R$ must be mapped by θ to the point (a,k,g) on $[0,g,k]$ collinear with (a). The points and lines involved are indicated in the incidence diagram of Fig. 11.4.

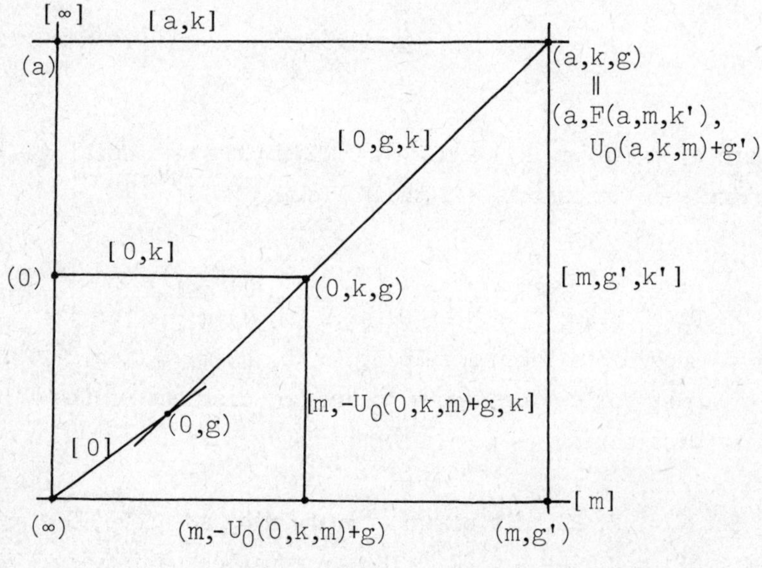

Fig. 11.4

Here $a,k,m \in R$, $0 \neq a,m$, and $g \in G$ are arbitrary. Then $g' \in G$ and $k' \in R$ are determined by $k = F(a,m,k')$ and $g' = -U_0(a,k,m)+g$. This determines the effect of θ on all points of $[m]$

$$(m,g)^{\theta} = (m,-U_0(a,k,m)+U_0(0,k,m)+g). \tag{19}$$

Hence $-U_0(a,k,m)+U_0(0,k,m)$ must be independent of k for fixed nonzero a, $m \in R$. Putting $k = 0$ yields the following

$$(m,g)^{\theta} = (m,-U_0(a,0,m)+g). \tag{20}$$

$$[m,g,k]^{\theta} = [m,-U_0(a,0,m)+g,k'], \text{ where } k = F(a,m,k'). \tag{21}$$

$$U_0(a,k,m) = U_0(0,k,m)+U_0(a,0,m), \text{ for } a,k,m \in R. \tag{22}$$

For $t,m,k \in R$, $m \neq 0$, $g \in G$, consider the incidences indicated in Fig. 11.5.

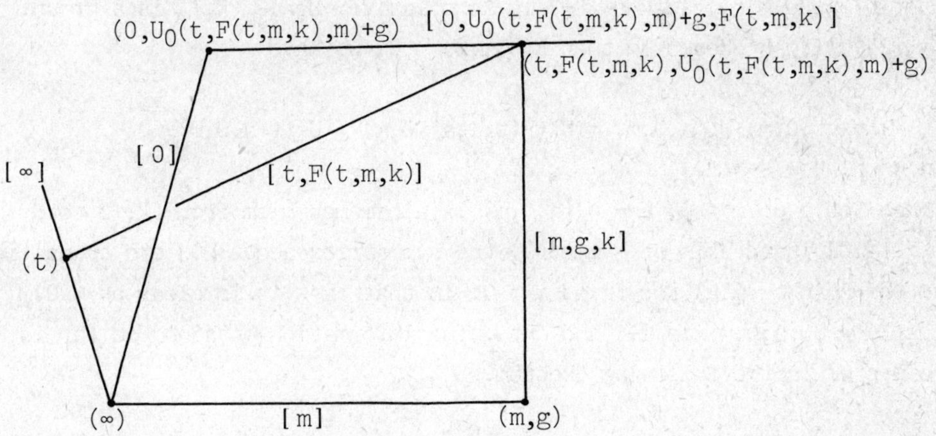

Fig. 11.5

The image of $(t,F(t,m,k),U_0(t,F(t,m,k),m)+g)$ under θ must be on $[0,U_0(t,F(t,m,k),m)+g,F(t,m,k)]$ and on $[m,-U_0(a,0,m)+g,k']$, where by (21) we have $k = F(a,m,k')$. Hence the image must be of the form

$$(\bar{t},F(t,m,k),U_0(t,F(t,m,k),m)+g), \text{ where } (\bar{t}) = (t)^{\theta} \text{ and}$$
$$F(t,m,k) = F(\bar{t},m,k'), \text{ and} \tag{23}$$

$$U_0(t,F(t,m,k),m)+g = U_0(\bar{t},F(\bar{t},m,k'),m)-U_0(a,0,m)+g. \tag{24}$$

Hence

$$F(t,m,F(a,m,k)) = F(\bar{t},m,k), \text{ where } (t)^{\theta} = (\bar{t}). \tag{25}$$

Put $m = 1$ and $k = 0$ to obtain

$$\bar{t} = F(t,1,a), \text{ so } F(t,m,F(a,m,k)) = F(F(t,1,a),m,k) \text{ for } m,t,k \in R. \tag{26}$$

For $a,b \in R$, define a binary operation "+" on R by

$$a+b = F(a,1,b). \tag{27}$$

It is easy to show that $(R,+)$ is a loop with identity 0.
By (26) with $m = 1$ we know "+" is associative. Hence $(R,+)$ is a group. Then by (24) with k chosen so that $F(t,m,k) = 0$, we have

$$U_0(t,0,m)+U_0(a,0,m) = U_0(F(t,1,a),0,m) = U_0(t+a,0,m). \tag{28}$$

Hence for each $m \neq 0$, $a \mapsto U_0(a,0,m)$ is a homomorphism from $(R,+)$ to G.

If $[0]$ is an axis of symmetry, the symmetries about $[0]$ are transitive on the points (m,g) for fixed $m \neq 0$. In that case it is clear by (20) that $a \mapsto U_0(a,0,m)$ is 1-1 and onto. The information obtained so far is collected in the following theorem.

11.2.1. *Let* $[0]$ *be an axis of symmetry (in addition to* $[\infty]$*). Then the following are true for all* $a,t,m,k \in R$

(i) $U_0(a,k,m) = U_0(0,k,m)+U_0(a,0,m)$.

(ii) $F(t,m,F(a,m,k)) = F(F(t,1,a),m,k) = F(t+a,m,k)$.

(iii) $U_0(t,0,m)+U_0(a,0,m) = U_0(F(t,1,a),0,m) = U_0(t+a,0,m)$.

(iv) *For fixed* $m \in R$, $0 \neq m$, *the map* $a \mapsto U_0(a,0,m)$ *is an isomorphism from* $(R,+)$ *onto* G .

Conversely, it is straightforward to verify that if (i), (ii) *and* (iii) *hold, then the map* θ *given below is a symmetry about* $[0]$ *moving* (0) *to* (a).

$$(\infty)^{\theta} = (\infty) \quad (t)^{\theta} = (F(t,1,a)) = (t+a)$$
$$(m,g)^{\theta} = (m,-U_0(a,0,m)+g)$$
$$(t,b,g)^{\theta} = (F(t,1,a),b,g) = (t+a,b,g) \tag{29}$$
$$[\infty]^{\theta} = [\infty] \quad [m]^{\theta} = [m]$$
$$[t,k]^{\theta} = [F(t,1,a),k] = [t+a,k]$$
$$[m,g,k]^{\theta} = [m,-U_0(a,0,m)+g,k'], \text{ where } k = F(a,m,k').$$

For the remainder of this section we assume that [0] *is an axis of symmetry.*

11.2.2. *The point* (∞) *is regular iff* $U_0(a,b,m)$ *is independent of* b. *In that case put* $U_0(a,b,m) = U_0(a,m)$. *Then* $U_0(a,m) = U_0(a,m')$ *implies either* $a = 0$ *or* $m = m'$.

Proof. Since the group generated by all symmetries about $[\infty]$ and [0] fixes (∞) and any of its orbits consisting of points not collinear with (∞) contains an element of the form (0,k,0), the point (∞) is regular iff the pair $((\infty), (0,k,0))$ is regular. But $\{(\infty),(0,0,0)\}^{\perp} = \{(0)\} \cup \{(m,0) \parallel m \in R\}$, and $\{(0),(0,0)\}^{\perp} = \{(\infty)\} \cup \{(0,k,0) \parallel k \in R\}$. Hence $((\infty),(0,k,0))$ is regular for all $k \in R$ iff $((\infty),(0,0,0))$ is regular iff (0,k,0) and (m,0) are collinear for all k and m iff $0 = U_0(0,k,m)$ for all k and m. By part (i) of 11.2.1 this is iff $U_0(a,b,m)$ is independent of b.

Now let (∞) be regular and put $U_0(a,m) = U_0(a,0,m)$. Let a,m,m' be given. Choose k_1 so that $b_1 = F(a,m,k_1) = F(a,m',0)$, and choose k_2 so that $b_2 = F(0,m,k_2) = F(0,m',0)$, i.e. $b_2 = k_2 = 0$. By (17), if $-U_0(a,m')+U_0(a,m) = 0$, then $a = 0$ or $m = m'$. □

Let τ be any nonidentity symmetry about [0] and $\theta_{g'}$ any nonidentity symmetry about $[\infty]$ as given in (16). Since *S* has no triangles, the only fixed lines of $\tau \circ \theta_{g'} = \theta$ are the lines through (∞). Then the result 1.9.1 applies to θ with $f+g = 1+s+s^2$, implying that f is odd. If $(0)^{\tau} = (a)$, the fixed points of θ must lie on lines of the form [m], $m \neq 0$, and are determined as follows :

$$(m,g)^{\theta} = (m,-U_0(a,0,m)+g+g') = (m,g) \text{ iff}$$
$$U_0(a,0,m) = g+g'-g. \tag{30}$$

The number of fixed points of $\theta = \tau \circ \theta_{g'}$ is 1 plus the number of pairs (m,g) satisfying (30). If for some m there is a g satisfying (30), then there are $|C_G(g')|$ such g (here $C_G(h)$ denotes the centralizer of h in G). So each line that has a fixed point in addition to (∞) has precisely $1+|C_G(g')|$ fixed points. If there are k such lines having fixed points other than (∞), then $f = 1+k|C_G(g')|$. If s is odd, then $|G|$ is odd, so f being odd implies k is even.

In the known examples (∞) is coregular and hence is regular when s is even and antiregular when s is odd. Under these conditions it is possible to say a bit more about the fixed points of θ.

11.2.3. *Let* $\theta = \tau_0\theta_{g'}$ *as above. Then (still under the hypothesis that both* $[\infty]$ *and* $[0]$ *are axes of symmetry) we have the following :*

(i) *If* (∞) *is regular, the fixed points of* θ *are the points of a unique line* L *through* (∞) *and the fixed lines of* θ *are precisely the lines through* (∞). *Moreover,* s *is a power of* 2 *and* G *is elementary abelian.*

(ii) *If* (∞) *is antiregular, then either* (∞) *is the unique fixed point of* θ, *or the fixed points of* θ *are precisely the points on two lines through* (∞). *The fixed lines are just the lines through* (∞).

Proof. (i) First suppose that (∞) is regular and let $\theta = \tau_0\theta_{g'}$ as above. The projective plane based at (∞) is denoted by π. Clearly θ induces a central collineation $\bar{\theta}$ on π with center (∞). Since (∞) is the only fixed point of θ on $[0]$, respectively $[\infty]$, the collineation $\bar{\theta}$ is an elation with axis some line L through (∞). If follows readily that the points of L are the fixed points of θ. We already noticed that, since S has no triangles, the only fixed lines of θ are the lines through (∞).

Since $[\infty]$ and $[0]$ are axes of symmetry in S, clearly the plane π is $((\infty),[\infty])$- and $((\infty),[0])$-transitive. By a well-known theorem [86] the group H of elations of π with center (∞) is an elementary abelian p-group of order s^2. Since the number of fixed points of θ is $f = 1+s$, which must be odd, s is even. Hence $p = 2$. As G is (isomorphic to) a subgroup of H, G is also an elementary abelian 2-group. This completes the proof of (i).

(ii) Suppose that (∞) is antiregular. By 1.5.1 s is odd. Assume that some line L through (∞) has a second point y fixed by θ. Let π_0 be the affine plane whose points are the points of $(\infty)^\perp - y^\perp$, and whose lines are the lines through (∞) different from L and sets of the form $\bar{z} = \{z,(\infty)\}^\perp - \{y\}$ for $z \in y^\perp - (\infty)^\perp$. Let π denote the projective completion of π_0. It follows that θ induces a central collineation $\bar{\theta}$ of π with center (∞). Since θ fixes no point of $P-(\infty)^{\perp}$, the only lines of π fixed by $\bar{\theta}$ must be incident with (∞). Hence $\bar{\theta}$ is an elation and must have as axis some line A of π through (∞). If A is a line of S through (∞) and distinct from L, i.e. A is not the line at infinity of π_0, then the points of A are the only fixed points of θ other than points on L. Then interchanging the roles of y and some point different from (∞) on A shows that each point of L is fixed. Hence the fixed points of θ are precisely the points of A and L. Finally, let A be the line at infinity of π_0. Then θ fixes each line through y, a contradiction since $\theta = \tau_0\theta_g$, fixes only the lines through (∞). □

11.3. THREE CONCURRENT AXES OF SYMMETRY

11.3.1. *Let* L_0,L_1,L_2 *be three distinct lines through a point* p *in a* GQ *of order* s. *Let* L_0 *be regular, and suppose that the group* H_i *of symmetries about* L_i *is nontrivial for both* i = 1 *and* i = 2. *Then the group* H_i *is elementary abelian.*

Proof. Let π be the plane based at L_0. Elements of H_i induce elations of π with center L_i and axis the set of lines through p, i = 1,2. Let $\sigma \mapsto \bar{\sigma}$ be this "induction" homomorphism. Put $\bar{H} = \langle \bar{\sigma} \parallel \sigma \in H_1H_2 \rangle$. Then $\bar{H}$ is elementary abelian by a well-known theorem [86]. Moreover, H_i is isomorphic to its image in $\bar{H}$, since the kernel of the map $\sigma \mapsto \bar{\sigma}$ has only the identity in common with H_i. Hence H_i is elementary abelian. □

We now return to the case where S is a GQ (with $[\infty]$ and $[0]$ as axes of symmetry) coordinatized by (R,G,U_0). If there is some $m \neq 0$ for which the group of symmetries about $[m]$ is nontrivial, we may suppose that m = 1. Our major goal in this section is to determine just when [1] is an axis of symmetry.

11.3.2. *Let* $0 \neq a \in R$. *If there is a symmetry about* [1] *moving* (0) *to* (a), *then*

(i) G *and* $(R,+)$ *are elementary abelian.*

(ii) $U_0(0,a+k,m) = U_0(0,a,m)+U_0(0,k,m)$, *for all* $k,m \in R$.

(iii) $F(t,m,k)+a = F(t,m,k+a)$.

Proof. Let θ be a symmetry about [1] moving (0) to (a). By 11.3.1 we know G is elementary abelian, and then by part (iv) of 11.2.1 $(R,+)$ is elementary abelian. The incidence indicated in Fig. 11.6 must be valid, where

$$\bar{g} = -U_0(0,k,m)+U_0(0,k,1)+g, \text{ and}$$
$$g' = -U_0(a,a+k,m)+U_0(a,a+k,1)+g, \text{ and} \tag{31}$$
$$a+k = F(a,m,k').$$

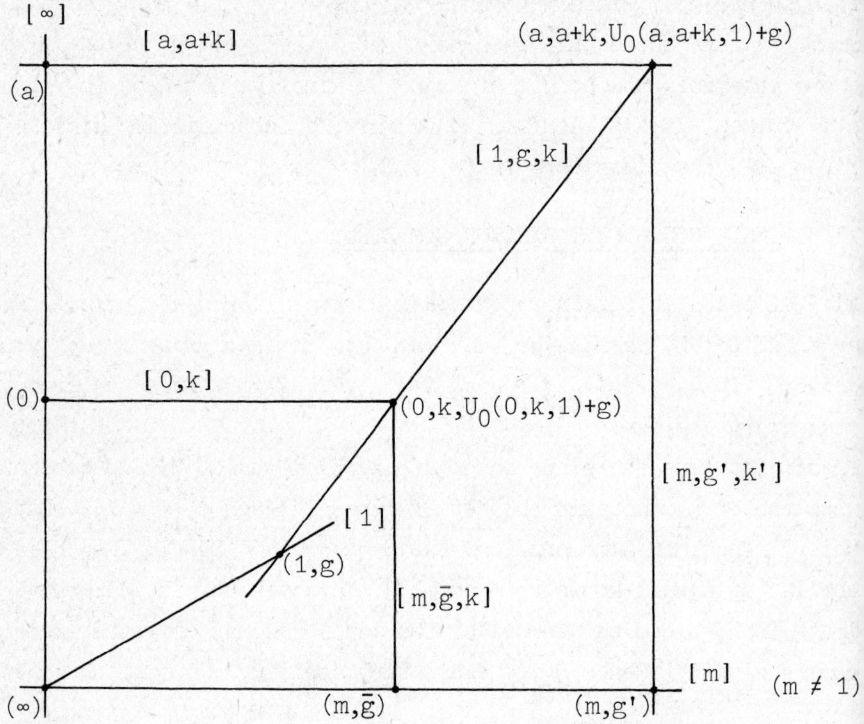

Fig. 11.6

It follows that

$$(m,\bar{g})^\theta = (m,-U_0(0,k,m)+U_0(0,k,1)+g)^\theta = \tag{32}$$
$$(m,g') = (m,-U_0(a,a+k,m)+U_0(a,a+k,1)+g).$$

For k = 0 this says

$$(m,g)^\theta = (m,-U_0(a,a,m)+U_0(a,a,1)+g). \tag{33}$$

Using (31), (32) and (33) we obtain

$$-U_0(a,a+k,m)+U_0(a,a+k,1) = -U_0(a,a,m)+U_0(a,a,1)-U_0(0,k,m)+U_0(0,k,1). \tag{34}$$

This is for fixed $a \neq 0$, all $k \in R$, and $1 \neq m \in R$. But of course it clearly holds for $m = 1$. Put $m = 0$ in (34) and use (22) to obtain

$$U_0(0,a+k,1) = U_0(0,a,1)+U_0(0,k,1). \tag{35}$$

Then use (35) in (34)

$$U_0(0,a+k,m) = U_0(0,a,m)+U_0(0,k,m). \tag{36}$$

This proves part (ii) of 11.3.2, and along with 11.2.1 (i) and (iii) and the fact that G is abelian shows the following .

For each $m \in R$, the map $(a,b) \mapsto U_0(a,b,m)$ is an additive homomorphism from $R \oplus R$ to G. (37)

Since $[m,\bar{g},k]^\theta = [m,g',k']$ in Fig. 11.6, where a is fixed, $a \neq 0$, $m \neq 1$, $a,m,k \in R$, and $g \in G$, and (31) holds, and using (37), we find that

$$[m,g,k]^\theta = [m,-U_0(a,a,m)+U_0(a,a,1)+g,k'], \tag{38}$$
for all $m,k \in R$, $g \in G$, with $a+k = F(a,m,k')$.

Now for arbitrary $t,k \in R$, $g \in G$, $x = (t,t+k,U_0(t,t+k,1)+g)$ is incident with $[1,g,k]$ and also with $[0,U_0(t,t+k,1)+g,t+k]$. Applying θ (and using (37) freely) we find that $x^\theta = (t+a,t+k+a,U_0(t+a,t+k+a,1)+g)$. It is now easy to check that θ has been completely determined as follows.

$$\begin{aligned}
&(\infty)^\theta = (\infty) \quad (t)^\theta = (t+a) \quad [\infty]^\theta = [\infty] \quad [m]^\theta = [m] \\
&(m,g)^\theta = (m,-U_0(a,a,m)+U_0(a,a,1)+g) \\
&(t,b,g)^\theta = (t+a,b+a,U_0(a,a,1)+g) \\
&[t,k]^\theta = [t+a,k+a] \\
&[m,g,k]^\theta = [m,-U_0(a,a,m)+U_0(a,a,1)+g,k'], \\
&\text{where } a+k = F(a,m,k').
\end{aligned} \tag{39}$$

But then since $(t,F(t,m,k),U_0(t,F(t,m,k),m)+g)$ is on $[m,g,k]$, it must be that $(t+a,F(t,m,k)+a,U_0(a,a,1)+U_0(t,F(t,m,k),m)+g)$ is on $[m,-U_0(a,a,m)+U_0(a,a,1)+g,k']$, where $a+k = F(a,m,k')$. This last incidence implies the following.

$$F(t,m,k)+a = F(t+a,m,k'), \text{ where } a+k = F(a,m,k'). \tag{40}$$

By (26), $F(t+a,m,k') = F(t,m,F(a,m,k')) = F(t,m,a+k)$, and the proof of 11.3.2 is complete. □

It is easy to check that the conditions of 11.3.2 are also sufficient for there to be a symmetry about $[1]$ moving (0) to (a).

11.3.3. *If* $[1]$ *is an axis of symmetry (in addition to* $[\infty]$ *and* $[0]$, *then*

(i) G *is elementary abelian.*

(ii) *For each* $m \in R$, *the map* $(a,b) \mapsto U_0(a,b,m)$ *is an additive homomorphism from* $R \oplus R$ *to* G.

(iii) *Define a multiplication "$\circ$" on* R *by* $a \circ m = F(a,m,0)$. *Then* $F(a,m,k) = (a \circ m)+k$ *for all* $a,m,k \in R$, *and* $(R,+,\circ)$ *is a right quasifield.*

(iv) *Each line* $[m], m \in R$, *is an axis of symmetry.*

Proof. Parts (i), (ii) and (iii) follow from 11.3.2 and 11.2.1. In view of part (iv) of 11.2.1 we may view $(R,+)$ as G, so that $U_0 : R^3 \to R$. Then for any $\sigma_1, \sigma_2, \sigma_3 \in R$, consider the map $\theta = \theta(\sigma_1,\sigma_2,\sigma_3)$ from S to S defined as follows.

$$\begin{aligned}
&(x,y,z)^\theta = (x+\sigma_1,y+\sigma_2,z+\sigma_3) \quad [\infty]^\theta = [\infty] \\
&(x,y)^\theta = (x,y+\sigma_3-U_0(\sigma_1,\sigma_2,x)) \quad [u]^\theta = [u] \\
&(x)^\theta = (x+\sigma_1) \quad [u,v]^\theta = [u+\sigma_1,v+\sigma_2] \\
&(\infty)^\theta = (\infty) \quad [u,v,w]^\theta = [u,v+\sigma_3-U_0(\sigma_1,\sigma_2,u),w+\sigma_2-\sigma_1 \circ u].
\end{aligned} \tag{41}$$

Using the first three parts of 11.3.3 a routine check shows that θ is a collineation of S. For each $i \in R \cup \{\infty\}$, let H_i denote the group of symmetries about $[i]$. An easy check yields the following

$$H_\infty = \{\theta(0,0,\sigma) \parallel \sigma \in R\} \tag{42}$$

$$H_m = \{\theta(\sigma,\sigma \circ m,U_0(\sigma,\sigma \circ m,m)) \parallel \sigma \in R\}, \quad m \in R. \tag{43}$$

Hence each line through (∞) is an axis of symmetry. □

11.3.4. *Let S satisfy the hypothesis of* 11.3.3. *Then*

(i) *condition* (17) *is equivalent to* (44), *and*

(ii) *condition* (18) *is equivalent to* (45).

$U_0(a,a\circ m,m) = U_0(a,a\circ m,m')$ *implies that either* $a = 0$ *or* $m = m'$. (44)

If m_0,m_1,m_2 *are distinct elements of R, and if for* $a_0,a_1,a_2 \in R$,

$0 = \sum_{i=0}^{2} a_i = \sum_{i=0}^{2} a_i\circ m_i = \sum_{i=0}^{2} U_0(a_i,a_i\circ m_i,m_i)$, *then* $a_0 = a_1 = a_2 = 0$. (45)

Proof. The proof of (i) is an easy exercise. For the proof of (ii) first note that

$$(a,b,\bar{g}) \text{ is on } [m,g,k] \text{ iff } b = a\circ m+k \text{ and } \bar{g} = U_0(a,b,m)+g. \qquad (46)$$

Then reconsider the Case (ii) of 11.1 that led to (18), i.e. assume there is a triangle whose vertices and sides are all of type I. We may assume one of the sides is $[m_0,0,k_0]$. Then the two vertices on this side are $y_i = (a_i,a_i\circ m_0+k_0,U_0(a_i,a_i\circ m_0+k_0,m_0))$, $i = 1,2$, $a_1 \neq a_2$. The other side on y_i is $L_i = [m_i,U_0(a_i,a_i\circ m_0+k_0,m_0)-U_0(a_i,a_i\circ m_0+k_0,m_i), a_i\circ m_0+k_0-a_i\circ m_i]$, $i = 1,2$. If the sides L_1 and L_2 meet at a point with first coordinate a_3, this point must be $(a_3,a_3\circ m_i+a_i\circ m_0+k_0-a_i\circ m_i, U_0(a_3,(a_3-a_i)\circ m_i+a_i\circ m_0+k_0,m_i)+U_0(a_i,a_i\circ m_0+k_0,m_0)-U_0(a_i,a_i\circ m_0+k_0,m_i)) = (a_3,(a_3-a_i)\circ m_i+a_i\circ m_0+k_0,U_0(a_3-a_i,(a_3-a_i)\circ m_i,m_i)+U_0(a_i,a_i\circ m_0+k_0,m_0))$, $i = 1,2$. Since the coordinates of this point must be the same whether $i = 1$ or $i = 2$, it follows that $(a_3-a_1)\circ m_1+(a_2-a_3)\circ m_2+(a_1-a_2)\circ m_0 = 0$, and $U_0(a_3-a_1,(a_3-a_1)\circ m_1,m_1)+U_0(a_1-a_2,(a_1-a_2)\circ m_0,m_0)+U_0(a_2-a_3,(a_2-a_3)\circ m_2,m_2) = 0$.

For the triangle to be impossible it must be that either a_1,a_2,a_3 are not distinct and/or the m_0,m_1,m_2 are not distinct. Geometrically it is clear that if the m_1,m_2,m_3 are distinct, then necessarily $a_1 = a_2 = a_3$. This is easily restated as condition (45). □

Let $R = (R,+,\circ)$ be a right quasifield with $|R| = s = p^e$, p prime. Let $U_0 : R^3 \to R$ be a function satisfying the following :

(i) $U_0(a,b,0) = 0$ for all $a,b \in R$.

(ii) The map $(a,b) \mapsto U_0(a,b,m)$ is an additive homomorphism from $R \oplus R$ to R, for each $m \in R$.

(iii) $U_0(a,a_\circ m,m) = U_0(a,a_\circ m,m')$ implies $a = 0$ or $m = m'$, for $a,m,m' \in R$.

(iv) If $0 = \sum_1^3 a_i = \sum_1^3 a_{i\circ}m_i = \sum_1^3 U_0(a_i,a_{i\circ}m_i,m_i)$, for $a_i,m_i \in R$, $i = 1,2,3,$ then either $a_1 = a_2 = a_3 = 0$ or the m_i's are not distinct.

Then the pair (R,U_0) is called a T-*set up* and U_0 is a T-*function* on R .

The following theorem summarizes the main results of this section.

11.3.5. *Let S be a* GQ *of order* s. *Then S has a point* $p_\infty = (\infty)$ *for which some three lines through* p_∞ *are axes of symmetry iff each line through* p_∞ *is an axis of symmetry iff S is coordinatized by a* T-*set up* (R,U_0) *in the following manner. Points and lines of S are as in* 11.1.1.*Then incidence in S is defined by :*

(∞) *is on* $[\infty]$ *and on* $[a]$, $a \in R$.

(m) *is on* $[\infty]$ *and on* $[m,b]$, $m,b \in R$.

(a,b) *is on* $[a]$ *and on* $[a,b,c]$, $a,b,c \in R$.

(x,y,z) *is on* $[x,y]$ *and on* $[u,v,w]$ *iff* $y = x_\circ u+w$ and $z = U_0(x,y,u)+v$, $x,y,$ $z,u,v,w \in R$.

For convenience in computing with collineations, etc., all collinearities and concurrencies are listed in the following table.

$(\infty) \sim (x)$ on $[\infty]$	$[\infty] \sim [u]$ at (∞)
$(\infty) \sim (u,v)$ on $[u]$	$[\infty] \sim [u,v]$ at (u)
$(x) \sim (y)$ on $[\infty]$	$[u] \sim [v]$ at (∞)
$(x) \sim (x,y,z)$ on $[x,y]$	$[u] \sim [u,v,w]$ at (u,v)
$(u,v) \sim (u,w)$ on $[u]$	$[u,v] \sim [u,w]$ at (u)
$(u,v) \sim (x,y,z)$ on $[u,v,y-x_\circ u]$	$[x,y] \sim [u,v,w]$ at $(x,y,v+U_0(x,y,u))$
provided $z = U_0(x,y,u)+v$	provided $y = x_\circ u+w$
$(x,y,z_1) \sim (x,y,z_2)$ on $[x,y]$	$[u,v,w_1] \sim [u,v,w_2]$ at (u,v)
$(x_1,y_1,z_1) \sim (x_2,y_2,z_2)$ on	$[u_1,v_1,w_1] \sim [u_2,v_2,w_2]$ at
$[u,z_i-U_0(x_i,y_i,u),y_i-x_{i\circ}u]$	$(x, x_\circ u_i+w_i, v_i+U_0(x,x_\circ u_i+w_i,u_i))$
$i = 1$ or 2 provided $x_1 \neq x_2$,	$i = 1$ or 2, provided $u_1 \neq u_2$,
$y_1-y_2 = (x_1-x_2)_\circ u$, and	$w_1-w_2 = -x_\circ u_1+x_\circ u_2$, and
$z_1-z_2 = U_0(x_1-x_2,y_1-y_2,u)$	$v_1-v_2 = -U_0(x,x_\circ u_1+w_1,u_1)+U_0(x,x_\circ u_2+w_2,u$

Note that a GQ S of order s coordinatized by a T-set up as above is a

TGQ with base point (∞) and the group of s^3 collineations given in (41) is the group of all translations (elations) about (∞).

11.4. THE KERNEL OF A T-SET UP

Let $S^{(\infty)}$ be a TGQ coordinatized by a T-set up (R,U_0) as in 11.3.5. By 8.6.5 we know that the multiplicative group K^o of the kernel is isomorphic to the group H of whorls about (∞) fixing $(0,0,0)$. We now study H in terms of the coordinate system.

Let $\theta \in H$, $\theta \neq id$. Then the following points and lines are fixed by θ : (∞), $(0,0,0)$, (0), $(m,0)$, for all $m \in R$; $[\infty]$, $[m]$, $[0,0]$, $[m,0,0]$, for all $m \in R$. There must be a permutation π_1 of the elements of R fixing 0 and for which

$$(a)^{\theta} = (\pi_1(a)),\ a \in R. \qquad (47)$$

Similarly, there are functions $\pi_2 : R^3 \to R$, $\pi_3 : R^3 \to R$, and $\pi_4 : R^2 \to R$, such that θ has the following partial description

$$(x,y,z)^{\theta} = (\pi_1(x),\pi_2(x,y,z),\pi_3(x,y,z)) \qquad (48)$$

(use $(a) \sim (x,y,z)$ iff $a = x$),

$$(m,g)^{\theta} = (m,\pi_4(m,g)). \qquad (49)$$

As $(x,y,z) \sim (m,g)$ iff $z = U_0(x,y,m)+g$, it follows that $(\pi_1(x),\pi_2(x,y,U_0(x,y,m)+g),\pi_3(x,y,U_0(x,y,m)+g) \sim (m,\pi_4(m,g))$, implying

$$\pi_3(x,y,U_0(x,y,m)+g) = U_0(\pi_1(x),\pi_2(x,y,U_0(x,y,m)+g),m)+\pi_4(m,g). \qquad (50)$$

Putting $m = 0$ in (50) yields

$$\pi_3(x,y,g) = \pi_4(0,g) = \pi_3(g),\ \text{i.e. } \pi_3 \text{ is a function of one variable.} \qquad (51)$$

So (50) simplifies to

$$\pi_3(U_0(x,y,m)+g) = U_0(\pi_1(x),\pi_2(x,y,U_0(x,y,m)+g),m)+\pi_4(m,g). \qquad (52)$$

Now $(x,y_1,z_1) \sim (x,y_2,z_2)$ iff $y_1 = y_2$. Put $y = y_1 = y_2$ and apply θ to obtain $(\pi_1(x),\pi_2(x,y,z_1),\pi_3(z_1)) \sim (\pi_1(x),\pi_2(x,y,z_2),\pi_3(z_2))$, so that

$$\pi_2(x,y,z) = \pi_2(x,y), \text{ i.e. } \pi_2 \text{ is a function of its first two variables only.} \tag{53}$$

As $(0,0,0)$ is fixed, $\pi_2(0,0)$ must be 0. Putting $x = y = 0$ in (52) yields

$$\pi_3(g) = \pi_4(m,g). \tag{54}$$

Hence we drop π_4 altogether, and (52) may be rewritten as

$$\pi_3(U_0(x,y,m)+g) = U_0(\pi_1(x),\pi_2(x,y),m)+\pi_3(g). \tag{55}$$

Put $g = 0$ in (55) and note that $\pi_3(0) = 0$ since $(0,0,0)$ is fixed, to obtain

$$\pi_3(U_0(x,y,m)) = U_0(\pi_1(x),\pi_2(x,y),m). \tag{56}$$

Putting this back in (55) easily yields (using e.g. 11.2.1 (iv)) that π_3 is *additive*. Also, the line $[m,0,0]$ is fixed. It is incident with the fixed point $(m,0)$ and with the points $(x,x\circ m, U_0(x,x\circ m,m))$, which must be permuted by θ. It follows that $(\pi_1(x),\pi_2(x,x\circ m),\pi_3(U_0(x,x\circ m,m)) = (\pi_1(x),\pi_1(x)\circ m, U_0(\pi_1(x),\pi_1(x)\circ m,m))$, which implies

$$\pi_2(x,x\circ m) = \pi_1(x)\circ m \tag{57}$$

and

$$\pi_3(U_0(x,x\circ m,m)) = U_0(\pi_1(x),\pi_1(x)\circ m,m). \tag{58}$$

Now $(0,g)^\theta = (0,\pi_3(g))$ implies $[0,g,k]^\theta = [0,\pi_3(g),\pi_5(0,g,k)]$, where $\pi_5 : R^3 \to R$ is defined by $[m,g,k]^\theta = [m,\pi_3(g),\pi_5(m,g,k)]$. The line $[0,g,k]$ is incident with the points (x,k,g), $x \in R$, in addition to $(0,g)$. So $(x,k,g)^\theta = (\pi_1(x),\pi_2(x,k),\pi_3(g))$ must lie on $[0,\pi_3(g),\pi_5(0,g,k)]$, implying

$$\pi_2(x,k) = \pi_5(0,g,k), \text{ i.e. } \pi_2(x,y) = \pi_2(y). \tag{59}$$

So (57) becomes

$$\pi_2(x_\circ m) = \pi_1(x)_\circ m. \tag{60}$$

With x = 1, this is

$$\pi_2(m) = \pi_1(1)_\circ m. \tag{61}$$

Put m = 1 in (60) and use (61)

$$\pi_2(x) = \pi_1(x) = \pi_1(1)_\circ x. \tag{62}$$

At the present time we know θ has the following description as a permutation of the points

$$\begin{aligned} &(x,y,z) \overset{\theta}{\mapsto} (\pi_1(1)_\circ x, \pi_1(1)_\circ y, \pi_3(z)) \\ &(x,y) \overset{\theta}{\mapsto} (x, \pi_3(y)) \\ &(x) \overset{\theta}{\mapsto} (\pi_1(1)_\circ x) \\ &(\infty) \overset{\theta}{\mapsto} (\infty). \end{aligned} \tag{63}$$

Here we also know π_3 is additive and (56) may be written as

$$\pi_3(U_0(x,y,m)) = U_0(\pi_1(1)_\circ x, \pi_1(1)_\circ y, m). \tag{64}$$

Let $t = \pi_1(1)$, and denote θ by θ_t. The effect of θ_t on the lines of S is as follows :

$$\begin{aligned} &[m,g,k] \overset{\theta_t}{\mapsto} [m, \pi_3(g), \pi_5(m,g,k)] \\ &[a,k] \overset{\theta_t}{\mapsto} [t_\circ a, t_\circ k] \\ &[m] \overset{\theta_t}{\mapsto} [m] \\ &[\infty] \overset{\theta_t}{\mapsto} [\infty]. \end{aligned} \tag{65}$$

As $(0,k,U_0(0,k,m)+g)$ is on $[m,g,k]$, it must be that $(0,t_\circ k,\pi_3(U_0(0,k,m)+g))$ is on $[m,\pi_3(g),\pi_5(m,g,k)]$, implying

$$\pi_5(m,g,k) = t_\circ k, \text{ i.e. } [m,g,k]^{\theta_t} = [m,\pi_3(g),t_\circ k]. \tag{66}$$

Then more generally, $(a,a_\circ m+k,U_0(a,a_\circ m+k,m)+g)$ on $[m,g,k]$ implies that $(t_\circ a,t_\circ(a_\circ m+k),\pi_3(U_0(a,a_\circ m+k,m)+g))$ is on $[m,\pi_3(g),t_\circ k]$. But this proves the following :

$$t_\circ(a_\circ m+k) = (t_\circ a)_\circ m+t_\circ k. \tag{67}$$

Then (67) provides an associative and a distributive law :

$$(t_\circ a)_\circ m = t_\circ(a_\circ m) \text{ and } t_\circ(a+k) = t_\circ a+t_\circ k. \tag{68}$$

The equalities (64) and (68) essentially characterize those t for which $\theta_t \in H$.

Let K denote the set of t in R satisfying the following conditions :

(i) $t_\circ(a+b) = t_\circ a+t_\circ b$ for all $a,b \in R$.

(ii) $t_\circ(a_\circ b) = (t_\circ a)_\circ b$ for all $a,b \in R$.

(iii) If $U_0(a,b,m) = U_0(a',b',m')$, then $U_0(t_\circ a,t_\circ b,m) = U_0(t_\circ a',t_\circ b',m')$, for all $a,b,m,a',b',m' \in R$.

Then K is called the *kernel* of the T-set up. By (i) and (ii) K is a subset of the kernel of the right quasifield $(R,+,\circ)$, and hence any two elements of K commute under multiplication. If θ_t is an element of H, we have seen that $t \in K-\{0\}$. It is easy to see that distinct elements of H determine distinct elements of $K-\{0\}$. Conversely, for each $t \in K-\{0\}$ there is a $\theta_t \in H$ defined by the following :

$$\begin{array}{ll} (x,y,z) \overset{\theta_t}{\mapsto} (t_\circ x,t_\circ y,\pi_3(z)) & [\infty] \overset{\theta_t}{\mapsto} [\infty] \\ (x,y) \overset{\theta_t}{\mapsto} (x,\pi_3(y)) & [m] \overset{\theta_t}{\mapsto} [m] \\ (a) \overset{\theta_t}{\mapsto} (t_\circ a) & [a,k] \overset{\theta_t}{\mapsto} [t_\circ a,t_\circ k] \\ (\infty) \overset{\theta_t}{\mapsto} (\infty) & [m,g,k] \overset{\theta_t}{\mapsto} [m,\pi_3(g),t_\circ k]. \end{array} \tag{69}$$

Here $\pi_3 : R \to R$ is the map determined by $\pi_3(U_0(a,b,m)) = U_0(t_\circ a,t_\circ b,m)$. Note that π_3 is well-defined by the definition of K and by 11.2.1 (iv). Further, for fixed $m \neq 0$, $\pi_3(U_0(a,0,m)) = U_0(t_\circ a,0,m)$ shows that π_3 is a permutation. Also, using the properties of U_0, the definition of K, and the fact that

$a \mapsto U_0(a,0,m)$ is a permutation if $m \neq 0$, it is easy to show that π_3 is additive.

Hence $\theta_t \mapsto t$, $\theta_t \in H$, defines a bijection from H onto K-{0}. As $\theta_t\theta_{t'} \mapsto t' \circ t = t \circ t'$, with $\theta_t, \theta_{t'} \in H$, it is clear that K-{0} is a commutative (cyclic!) group under the multiplication of R . And by 11.3.3 (ii) it follows that for any t , $t' \in K$ the sum $t+t'$ satisfies the condition (iii) in the definition of K . Hence K is a subfield of the kernel of $(R,+,\circ)$. Since H is isomorphic to the multiplicative group of the kernel of the TGQ $S^{(\infty)}$, the following result has been established :

11.4.1. *The kernel of a* TGQ *of order* s *is isomorphic to the kernel of a corresponding coordinatizing* T-*set up.*

12 Generalized quadrangles as amalgamations of Desarguesian planes

12.1. ADMISSIBLE PAIRS

If p^e is an odd prime power, there is (up to duality) just one known example of a GQ of order p^e. In the case of GQ of order 2^e a quite different situation prevails. There are known at least $2(\varphi(e)-1)$ pairwise nonisomorphic GQ of order 2^e, with φ the Euler function. Each of these has a regular point x_∞ incident with a regular line L_∞. S.E. Payne [124] showed that a GQ S of order s contains a regular point x_∞ incident with a regular line L_∞ if and only if it may be constructed as an "amalgamation of a pair of compatible projective planes", which of course turn out to be the planes based at x_∞ and L_∞, respectively. Moreover, in [131] it was shown that the two planes are desarguesian iff S may be "coordinatized" by means of an "admissible" pair (α,β) of permutations of the elements of F = GF(s), and in that case x_∞ is a center of symmetry, L_∞ is an axis of symmetry, and s is a power of 2. All the known GQ of order 2^e are of this type, and in this chapter we wish to proceed directly to the construction and study of such examples.

Let α and β be permutations of the elements of F = GF(s), with $s = 2^e$ and $e \geqslant 1$. For convenience we assume throughout that

$$0^\alpha = 0 \text{ and } 1^\alpha = 1. \qquad (1)$$

Define an incidence structure $S(\alpha,\beta) = (P,B,I)$ as follows. The pointset P has the following elements :

(i) (∞),

(ii) (a), $a \in F$,

(iii) (u,v), $u,v \in F$,

(iv) (a,b,c), $a,b,c \in F$.

The lineset B has the following elements :

(a) $[\infty]$,

(b) $[u]$, $u \in F$,

(c) $[a,b]$, $a,b \in F$,

(d) $[u,v,w]$, $u,v,w \in F$.

Incidence I is defined as follows : the point (∞) is incident with $[\infty]$ and with $[u]$ for all $u \in F$; the point (a) is incident with $[\infty]$ and with $[a,b]$ for all $a,b \in F$; the point (u,v) is incident with $[u]$ and with $[u,v,w]$ for all $u,v,w \in F$; the point (a,b,c) is incident with $[a,b]$ and with $[u,v,w]$ iff $b+w = au$ and $c+v = a^{\alpha}u^{\beta}$.

It is straightforward to check that $S(\alpha,\beta)$ is a tactical configuration with $s+1$ points on each line, $s+1$ lines on each point, $1+s+s^2+s^3$ points (respectively, lines) and having two points incident with at most one line. Hence a counting argument shows that $S(\alpha,\beta)$ is a GQ of order s iff $S(\alpha,\beta)$ has no triangles.

For convenience we note the following :

$$(x,xu+w,x^{\alpha}u^{\beta}+v) \text{ is on } [u,v,w] \text{ for all } x \in F;$$
$$(x,y,z) \text{ is on } [u,x^{\alpha}u^{\beta}+z\,,\, xu+y] \text{ for all } u \in F;$$
$$(x_1,y_1,z_1) \sim (x_2,y_2,z_2) \text{ iff (i) } x_1 = x_2 \text{ and } y_1 = y_2 \text{ or} \tag{2}$$
$$\text{(ii) } x_1 \neq x_2 \text{ and } ((y_1+y_2)/(x_1+x_2))^{\beta} = (z_1+z_2)/(x_1^{\alpha}+x_2^{\alpha});$$
$$(x_1,y_1,z_1) \sim (x_2,y_2) \text{ iff } z_1 = x_1^{\alpha}x_2^{\beta}+y_2.$$

12.1.1. $S(\alpha,\beta) = (P,B,I)$ *is a* GQ *of order* $s = 2^e$ *iff the following conditions on* α *and* β *hold : For distinct* $u_i \in F$ *and distinct* $x_i \in F$, $i = 1,2,3$,

$$\sum_1^3 u_i(x_{i+1}+x_{i-1}) = 0 \quad \textit{and} \quad \sum_1^3 u_i^{\beta}(x_{i+1}^{\alpha}+x_{i-1}^{\alpha}) = 0 \tag{3}$$

(*subscripts being taken modulo* 3) *never hold simultaneously*.

Proof. The proof amounts to showing that there are no triangles precisely when (3) holds and is rather tedious. We give the details only for the main case of a hypothetical triangle in which all three vertices are points of type (iv) and all three sides are lines of type (d).

So suppose $[u_3,v_3,w_3]$ is one side of the triangle having two of the vertices $(x_1,x_1u_3+w_3,x_1^{\alpha}u_3^{\beta}+v_3)$ and $(x_2,x_2u_3+w_3,x_2^{\alpha}u_3^{\beta}+v_3)$ with $x_1 \neq x_2$, which in turn lie, respectively, on the sides $[u_2,x_1^{\alpha}u_2^{\beta}+x_1^{\alpha}u_3^{\beta}+v_3,x_1u_2+x_1u_3+w_3]$ and $[u_1,x_2^{\alpha}u_1^{\beta}+x_2^{\alpha}u_3^{\beta}+v_3,x_2u_1+x_2u_3+w_3]$ with $u_2 \neq u_3 \neq u_1$. Then the third vertex of the triangle must be $(x_3,x_3u_2+x_1u_2+x_1u_3+w_3,x_3^{\alpha}u_2^{\beta}+x_1^{\alpha}u_2^{\beta}+x_1^{\alpha}u_3^{\beta}+v_3) =$ $(x_3,x_3u_1+x_2u_1+x_2u_3+w_3,x_3^{\alpha}u_1^{\beta}+x_2^{\alpha}u_1^{\beta}+x_2^{\alpha}u_3^{\beta}+v_3)$ with $x_1 \neq x_3 \neq x_2$. Setting equal these two representations of the third vertex yields the two equations of

(3). All other triangles may be ruled out without any additional conditions being introduced. □

The pair (α,β) of permutations of the elements of $F = GF(2^e)$ is said to be *admissible* provided it satisfies both (1) and (3), in which case there arises a GQ $S(\alpha,\beta)$.

12.1.2. *Let* γ *be an automorphism of* F *and let* α *and* β *be permutations of the elements of* F. *Then* (α,β) *is admissible iff* $(\alpha\gamma,\beta\gamma)$ *is admissible iff* $(\gamma\alpha,\gamma\beta)$ *is admissible, in which case* $S(\alpha,\beta) \cong S(\alpha\gamma,\beta\gamma) \cong S(\gamma\alpha,\gamma\beta)$. *Also,* (α,β) *is admissible iff* (β,α) *is admissible, in which case* $S(\alpha,\beta)$ *is isomorphic to the dual of* $S(\beta,\alpha)$. *Finally,* (α,β) *is admissible iff* (α^{-1},β^{-1}) *is admissible, but in general it is not true that* $S(\alpha,\beta) \cong S(\alpha^{-1},\beta^{-1})$.
Proof. All parts of this result are easily checked except the last claim concerning $S(\alpha,\beta) \not\cong S(\alpha^{-1},\beta^{-1})$. However, we postpone a discussion of this until later. □

12.1.3. *If* (α,β) *is admissible, then in* $S(\alpha,\beta)$ *the point* (∞) *is a center of symmetry and the line* $[\infty]$ *is an axis of symmetry. Specifically, for* $\sigma_2, \sigma_3 \in F$ *there is a collineation* θ *of* $S(\alpha,\beta)$ *defined as follows :*

$$\begin{array}{ll} [u,v,w] \overset{\theta}{\mapsto} [u,v+\sigma_3,w+\sigma_2] & (\infty) \overset{\theta}{\mapsto} (\infty) \\ [a,b] \overset{\theta}{\mapsto} [a,b+\sigma_2] & (a) \overset{\theta}{\mapsto} (a) \\ [u] \overset{\theta}{\mapsto} [u] & (u,v) \overset{\theta}{\mapsto} (u,v+\sigma_3) \\ [\infty] \overset{\theta}{\mapsto} [\infty] & (a,b,c) \overset{\theta}{\mapsto} (a,b+\sigma_2,c+\sigma_3) \end{array} \quad (4)$$

The symmetries about (∞) *are obtained by setting* $\sigma_3 = 0$; *those about* $[\infty]$ *are obtained by setting* $\sigma_2 = 0$.
Proof. Easily checked. □

It also follows readily that the planes based at (∞) and $[\infty]$, respectively, are both desarguesian, but we will not prove that here.

12.1.4. *Let* (α,β) *be admissible. Then in* $S(\alpha,\beta)$ *the following are equivalent :*
(i) *The pair* $((a_0),(a,b,c))$ *is regular for some* $a_0,a,b,c \in F$ *with* $a_0 \neq a$.
(ii) β *is additive (i.e.* $(x+y)^\beta = x^\beta+y^\beta$ *for all* $x,y \in F$).
(iii) (a) *is regular for all* $a \in F$.
(iv) (a) *is a center of symmetry for all* $a \in F$.

Dually, $([u_0],[u,v,w])$ *is regular for some* $u_0, u,v,w \in F$ *with* $u_0 \neq u$ *iff* α *is additive iff* $[u]$ *is regular for all* $u \in F$ *iff* $[u]$ *is an axis of symmetry for each* $u \in F$.

Proof. Because (∞) is regular, each point (a_0) of $[\infty]$ forms a regular pair with each point (u,v) collinear with (∞). So suppose some point (a_0) forms a regular pair with some point (a,b,c) not collinear with (∞), and with $a_0 \neq a$ so that (a_0) and (a,b,c) are not collinear. Using a collineation of the type given by (4), we see this is equivalent to saying that $((a_0),(a,0,0))$ is regular, $a \neq a_0$.

$\{(a),(a_0,0,0)\}^{\perp} = \{(a_0)\} \cup \{(a,x(a+a_0),x^{\beta}(a^{\alpha}+a_0^{\alpha})) \| \ x \in F\}$,

$\{(a_0),(a,0,0)\}^{\perp} = \{(a)\} \cup \{a_0,y(a+a_0), y^{\beta}(a^{\alpha}+a_0^{\alpha})) \| \ y \in F\}$.

Hence $((a_0),(a,0,0))$ is regular iff $(a_0,y(a+a_0),y^{\beta}(a^{\alpha}+a_0^{\alpha})) \sim (a,x(a+a_0), x^{\beta}(a^{\alpha}+a_0^{\alpha}))$ for all $x,y \in F$. This is iff (cf. (2)) $((x+y)(a+a_0)/(a+a_0))^{\beta} = ((x^{\beta}+y^{\beta})(a^{\alpha}+a_0^{\alpha}))/(a^{\alpha}+a_0^{\alpha})$, which holds iff $(x+y)^{\beta} = x^{\beta}+y^{\beta}$ for all $x,y \in F$.

At this point we clearly have (i) $\Leftrightarrow$ (ii) $\Leftrightarrow$ (iii) $\Leftarrow$ (iv). Hence to complete the proof we assume β is additive and exhibit 2^e symmetries about (t), $t \in F$. Rather, for $t, \sigma \in F$, we let the reader check that the map φ given below is a symmetry about the point (t).

$$
\begin{array}{ll}
(a,b,c) \overset{\varphi}{\mapsto} (a,\sigma(t+a)+b,\sigma^{\beta}(t^{\alpha}+a^{\alpha})+c) & (\infty) \overset{\varphi}{\mapsto} (\infty) \\
(u,v) \overset{\varphi}{\mapsto} (u+\sigma,\sigma^{\beta}t^{\alpha}+v) & (a) \overset{\varphi}{\mapsto} (a) \\
[u,v,w] \overset{\varphi}{\mapsto} [u+\sigma,\sigma^{\beta}t^{\alpha}+v,\sigma t+w] & [\infty] \overset{\varphi}{\mapsto} [\infty] \\
[a,b] \overset{\varphi}{\mapsto} [a,\sigma(t+a)+b] & [u] \overset{\varphi}{\mapsto} [u+\sigma]
\end{array}
\qquad (5)
$$

This completes the proof of the first half of the result. The dual result follows similarly. □

Note : If (α,β) is admissible and $S = S(\alpha,\beta)$, then α (respectively, β), is additive iff $S^{(\infty)}$ (respectively, $S^{[\infty]}$) is a TGQ.

12.2. ADMISSIBLE PAIRS OF ADDITIVE PERMUTATIONS

The goal of this section is to determine all admissible pairs (α,β) in which both α and β are additive. For elements $a_0,\dots,a_{e-1}$ of $F = GF(2^e)$ define the $e \times e$ matrix $[a_0,\dots,a_{e-1}] = (a_{ij})$, where $a_{ij} = a_{[j-i]}^{2^{i-1}}$, $1 \leqslant i, j \leqslant e$, where in $a_{[k]}$, $[k]$ indicates that k is to be reduced modulo e to one of $0, 1,\dots,e-1$. Put

$$D = \det([a_0,\dots,a_{e-1}]).$$

12.2.1. (*B. Segre and U. Bartocci* [163]). $D^2 = D$, *so that* $D = 0$ *or* $D = 1$. *Moreover, if* α *is the additive map defined by* $x^\alpha = \sum_{i=0}^{e-1} a_i x^{2^i}$, *then* α *is a permutation iff* $D = 1$.

Proof. Since $x \mapsto x^2$ is an automorphism of F, it follows that for any square matrix (b_{ij}), $(\det(b_{ij}))^2 = \det(b_{ij}^2)$. Hence D^2 is the determinant of a matrix whose rows are obtained by permuting cyclically the rows and columns of the matrix $[a_0,\dots,a_{e-1}]$. It follows that $D^2 = D$, implying $D = 0$ or 1.

Suppose that α is not bijective, so that for some $x \neq 0$, $0 = \sum_{i=0}^{e-1} a_i x^{2^i}$. Hence the following equalities hold :

$$0 = a_0 x + a_1 x^2 + \dots + a_{e-1} x^{2^{e-1}}$$

$$0 = a_{e-1}^2 x + a_0^2 x^2 + \dots + a_{e-2}^2 x^{2^{e-1}}$$

$$\vdots$$

$$0 = a_1^{2^{e-1}} x + a_2^{2^{e-1}} x^2 + \dots + a_0^{2^{e-1}} x^{2^{e-1}}.$$

It follows that the matrix $[a_0,\dots,a_{e-1}]$ has the characteristic vector $(x, x^2, \dots, x^{2^{e-1}})^T$ associated with the characteristic root 0, i.e. $D = 0$.

Conversely, suppose $D = 0$. If suffices to show that α is not onto. Let y be an arbitrary image under α, say

$$y = a_0 x + a_1 x^2 + \dots + a_{e-1} x^{2^{e-1}}. \text{ Hence}$$

$$y^2 = a_{e-1}^2 x + a_0^2 x^2 + \dots + a_{e-2}^2 x^{2^{e-1}},$$

$$\vdots$$

$$y^{2^{e-1}} = a_1^{2^{e-1}} x + a_2^{2^{e-1}} x^2 + \dots + a_0^{2^{e-1}} x^{2^{e-1}}.$$

Since $D = 0$, there are scalars $\lambda_0,\dots,\lambda_{e-1}$, at least one of which is nonzero, such that

$$(0,\ldots,0) = (\lambda_0,\ldots,\lambda_{e-1})\begin{pmatrix} a_0 & a_1 & \cdots & a_{e-1} \\ a_{e-1}^2 & a_0^2 & \cdots & a_{e-2}^2 \\ \vdots & \vdots & & \vdots \\ a_1^{2^{e-1}} & a_2^{2^{e-1}} & \cdots & a_0^{2^{e-1}} \end{pmatrix}.$$

Hence

$$(0,\ldots,0) = (\lambda_0,\ldots,\lambda_{e-1})[a_0,\ldots,a_{e-1}]\begin{pmatrix} x \\ \vdots \\ x^{2^{e-1}} \end{pmatrix} =$$

$$(\lambda_0,\ldots,\lambda_{e-1})\begin{pmatrix} y \\ y^2 \\ \vdots \\ y^{2^{e-1}} \end{pmatrix}.$$

This says that the homomorphism $T : y \mapsto \sum_{i=0}^{e-1} \lambda_i y^{2^i}$ of the additive group of F (which is not the zero map) must have all elements y of the form $y = x^\alpha$ in its kernel. Hence α is not onto . □

12.2.2. (*S.E. Payne* [131]). *Let* α *and* β *be additive permutations of the elements of* $F = GF(2^e)$ *that fix* 1 *and for which* $x \mapsto x^\alpha/x^\beta$ *permutes the nonzero elements of* F. *Then* $\alpha^{-1}\beta$ *is an automorphism of* F *of maximal order* e.

Proof. For $e \in \{1,2\}$ the theorem is easy to check. So assume that $e \geq 3$. Since α and β are additive maps on F there must be scalars a_i, $b_i \in F$, $0 \leq i \leq e-1$, for which $\alpha : x \mapsto \sum_{i=0}^{e-1} a_i x^{2^i}$ and $\beta : x \mapsto \sum_{i=0}^{e-1} b_i x^{2^i}$ [80]. Let $A = [a_0,\ldots,a_{e-1}] = (a_{ij})$, so $a_{ij} = a_{[j-i]}^{2^{i-1}}$, and $B = [b_0,\ldots,b_{e-1}] = (b_{ij})$, so $b_{ij} = b_{[j-i]}^{2^{i-1}}$, $1 \leq i, j \leq e$. Since α and β are permutations, by 12.2.1 both A and B are nonsingular. Since $x \mapsto x^\alpha/x^\beta$ is a permutation of the elements of $F^o = F-\{0\}$, for each $\lambda \in F^o$ there must be a unique nonzero solution x to $x^\alpha + \lambda x^\beta = 0$. Hence $\sum_{i=0}^{e-1} (a_i+\lambda b_i)x^{2^i} = 0$ has a unique nonzero solution x for each $\lambda \in F^o$. By 12.2.1 the matrix $C_\lambda = ((a_{[j-i]}+\lambda b_{[j-i]})^{2^{i-1}})$, $1 \leqslant i,j \leqslant e$, has zero determinant for each $\lambda \in F^o$. And $0 \neq \det A.\det B$, so $\det A = \det B = 1$.

It follows that $\det C_\lambda$ is a polynomial in λ of degree 2^e-1 with constant term 1, leading coefficient 1, and having each nonzero element of F as a root. This implies

$$\det C_\lambda = \lambda^{2^e-1}+1. \tag{6}$$

For $1 \leqslant t \leqslant 2^e-2$ we now calculate the coefficient of λ^t in $\det C_\lambda$ and set it equal to zero. Let $t_{i_1}, t_{i_2}, \ldots, t_{i_r}$ be the nonzero coefficients in the binary expansion $\sum_{i=0}^{e-1} t_i 2^i$ of t. Then the coefficient of λ^t in $\det C_\lambda$ is easily seen to be the determinant of the matrix obtained by replacing rows $t_{i_1}, \ldots, t_{i_r}$ of A with rows $t_{i_1}, \ldots, t_{i_r}$ of B. Hence we know the following : the rows of A are independent, the rows of B are independent, and any set of rows formed by taking some r rows of A and the complementary e-r rows of B is a linearly dependent set, $1 \leqslant r \leqslant e-1$. In particular, the first row of B is a linear combination of rows 2,3,...,e of A. Let β_i be the ith row of B, so $\beta_i = (b^{2^{i-1}}_{[1-i]}, \ldots, b^{2^{i-1}}_{[e-i]})$. Then there are scalars $d_1, \ldots, d_{e-1}$ (at least one of which is nonzero) such that $\beta_1 = (0, d_1, \ldots, d_{e-1})A$. Apply the automorphism $x \mapsto x^2$ to this latter identity to obtain

$$(b_0^2, \ldots, b_{e-1}^2) = (0, d_1^2, \ldots, d_{e-1}^2)(a^{2^k}_{[j-k]})_{1 \leqslant k,j \leqslant e}\,. \tag{7}$$

On the left hand side of (7) permute the columns cyclically, moving column j to position j+1, j = 1,...,e-1, and column e to position 1. There arises

$$\beta_2 = (d_{e-1}^2, 0, d_1^2, \ldots, d_{e-2}^2)A. \tag{8}$$

Doing this i times, $i \leqslant e-1$, we obtain

$$\beta_{i+1} = (d^{2^i}_{e-i}, \ldots, d^{2^i}_{e-1}, 0, d^{2^i}_1, \ldots, d^{2^i}_{e-i-1})A, \tag{9}$$

where the d_j's are unique.

Let α_i denote the ith row of A. For some λ_1, λ_2, not both zero, we have

$$\lambda_1\beta_1+\lambda_2\beta_2 = \sum_{j=3}^{e} f_j\alpha_j = (\lambda_2 d_{e-1}^2, \lambda_1 d_1, \lambda_1 d_2+\lambda_2 d_1^2, \ldots, \lambda_1 d_j+\lambda_2 d_{j-1}^2, \ldots)A. \quad (10)$$

Hence, as the rows of A are independent, $\lambda_2 d_{e-1}^2 = 0 = \lambda_1 d_1$.
If $\lambda_1 \neq 0$, then $d_1 = 0$. If $\lambda_2 \neq 0$, then $d_{e-1} = 0$.

Now suppose that

$$d_1 = d_2 = \ldots = d_{j-1} = 0 \text{ and } d_{e-1} = d_{e-2} = \ldots = d_{e-(k-j)} = 0 \quad (11)$$

with $k \in \{2,\ldots,e-2\}$ and $j \in \{1,\ldots,k\}$ (notice that $j-1 < e-(k-j)$ and that (11) holds for $k = 2$ and some $j \in \{1,2\}$ by $d_1 d_{e-1} = 0$). We wish to show that $d_j d_{e-(k-j+1)} = 0$, i.e. we wish to show that (11) holds for k replaced by k+1 and j replaced by at least one of j , j+1.

So assume that $d_j d_{e-(k-j+1)} \neq 0$. We have the following :

$$\beta_1 = (0,\ldots,0,d_j,\ldots,d_{e-(k-j+1)},0,\ldots,0)A,$$

$$\beta_2 = (0,\ldots,0,\underset{\substack{\nwarrow \\ \text{position } j+2}}{d_j^2},\ldots,d_{e-(k-j+1)}^2,0,\ldots,0)A \text{ if } j \neq k, \quad (12)$$

$$\beta_2 = (d_{e-1}^2,0,\ldots,0,\underset{\substack{\nearrow \\ \text{position } j+1}}{d_j^2},\ldots,d_{e-2}^2)A \text{ if } j = k,$$

etc.

Since $0 < k+1 < e$, there are scalars $\lambda_1,\ldots,\lambda_{k+1}$, at least one of which is not zero, for which $\sum_{r=1}^{k+1} \lambda_r\beta_r$ is some linear combination of $\alpha_{k+2},\ldots,\alpha_e$.

Use (12) to calculate the coefficients of $\alpha_1,\ldots,\alpha_{k+1}$ (which must be zero) in $\sum_{r=1}^{k+1} \lambda_r\beta_r$. The coefficient of α_j is $\lambda_{k+1} d_{e-(k-j+1)}^{2^k}$. Hence $\lambda_{k+1} = 0$. If $j > 1$, the coefficient of α_{j-1} is $\lambda_k d_{e-(k-j+1)}^{2^{k-1}}$, implying $\lambda_k = 0$. Continuing, we obtain $\lambda_{k+1} = \lambda_k = \ldots = \lambda_{k-j+2} = 0$. The coefficient of α_{j+1} is $\lambda_1 d_j$. Hence $\lambda_1 = 0$. The coefficient of α_{j+2} is $\lambda_2 d_j^2$. Hence $\lambda_2 = 0$. Continuing, we obtain $\lambda_1 = \lambda_2 = \ldots = \lambda_{k-j+1} = 0$, so that in fact $\lambda_r = 0$ for

$1 \leqslant r \leqslant k+1$. This impossibility implies that $d_j d_{e-(k-j+1)} = 0$ as desired. Hence by induction on k (11) holds also for $k = e-1$ and some $j \in \{1,\dots,k\}$.

It follows that only one d_i can be nonzero, say $d = d_m \neq 0$, $1 \leqslant m \leqslant e-1$. This says that

$$b_j = da^{2^m}_{[j-m]}, \quad 0 \leqslant j \leqslant e-1. \tag{13}$$

Our assumption that $1 = 1^\alpha = 1^\beta$ implies that $d = 1$. So

$$b_j = a^{2^m}_{[j-m]}, \quad 0 \leqslant j \leqslant e-1. \tag{14}$$

Clearly (14) is equivalent to $x^\beta = (x^\alpha)^{2^m}$, i.e. $\beta = \alpha . 2^m$. Since $x \mapsto x^\alpha/x^\beta$ permutes the nonzero elements of F, also $x \mapsto x^\beta/x^\alpha$ permutes the nonzero elements of F, i.e. $x \mapsto (x^\alpha)^{(2^m-1)}$ permutes the nonzero elements of F. Hence $y \mapsto y^{2^m-1}$ permutes the nonzero elements of F, implying that $(m,e) = 1$. Consequently $\alpha^{-1}\beta$ is an automorphism of F of maximal order e. □

The following immediate corollary is equivalent to the determination of all translation ovals in the desarguesian plane over $F = GF(2^e)$ and was the main result of S.E. Payne [121].

12.2.3. *If* β *is an additive permutation of the elements of* $F = GF(2^e)$ *for which* $x \mapsto x/x^\beta$ *permutes the nonzero elements of* F*, then* β *has the form* $x^\beta = dx^{2^u}$ *for fixed* $d \in F^o$, $(u,e) = 1$.

The next result is the main goal of this section.

12.2.4. (*S.E. Payne* [131]). *Let* (α,β) *be a pair of additive permutations of the elements of* $F = GF(2^e)$ *fixing* 1 . *Then the following are equivalent :*

(i) *The pair* (α,β) *is admissible.*

(ii) $0 = \sum_1^2 v_i z_i = \sum_1^2 v_i^\alpha z_i^\beta$ *for distinct, nonzero* v_1, v_2 *implies* $z_1 = z_2 = 0$.

(iii) *For each* $c \in F^o$, *the map* $\mu_c : v \mapsto v^\alpha (c/v)^\beta$ *permutes the elements of* F^o.

(iv) *For each* $c \in F^o$, *the map* $\lambda_c : z \mapsto (cz)^\alpha / z^\beta$ *permutes the elements of* F^o.

(v) α *and* β *are automorphisms of* F *for which* $\alpha^{-1}\beta$ *is an automorphism of maximal order* e.

Proof. Since β is additive, in (3) $u_{i+1}+u_{i-1}$ may be replaced by z_i, so that the condition for admissibility becomes

$$\sum_1^3 x_i z_i = 0, \quad \sum_1^3 x_i^{\alpha} z_i^{\beta} = 0, \quad \sum_1^3 z_i = 0 \tag{15}$$

cannot hold for distinct $x_i \in F$ and for distinct nonzero z_i's $\in F$. Now, using the additivity of α and of β add $\sum_1^3 x_3 z_i = 0$ to the first equation in (15) and $\sum_1^3 x_3^{\alpha} z_i^{\beta} = 0$ to the second equation of (15), and replace $x_i + x_3 = v_i$, so that $v_3 = 0$, to obtain condition (ii). In (ii) put $z_i = c/v_i$ to obtain (iii). In (ii) put $v_1 = cz_2$, $v_2 = cz_1$ to obtain (iv). It follows readily that (i)-(iv) are equivalent. The crux of the proof is to show that (iv) implies (v).

So let (α,β) be a pair of additive permutations of the elements of F fixing 1 and satisfying (iv). Putting c = 1 in (iv) we see that $\alpha^{-1}\beta$ is an automorphism of F of maximal order e by 12.2.2. For $0 \neq c \in F$, let α_c denote the additive permutation $\alpha_c : x \mapsto (cx)^{\alpha}$ for all $x \in F$. Then $\beta = \alpha_c\gamma_c = \delta_c\alpha_c$ for unique additive permutations γ_c and δ_c. For λ_c as in (iv), $\lambda_c = \alpha_c.(1-\gamma_c)$, implying that $1-\gamma_c : w \mapsto w/w^{\gamma_c}$ is also a permutation of the elements of F°. By 12.2.3 it follows that $\gamma_c : x \mapsto d_c x^{\beta_c}$ for some nonzero scalar d_c and some automorphism $\beta_c : x \mapsto x^{2^{t_c}}$, $(t_c, e) = 1$, $1 \leqslant t_c \leqslant e$. As $1 = 1^{\beta} = 1^{\alpha_c\gamma_c} = (c^{\alpha})^{\gamma_c} = d_c(c^{\alpha})^{2^{t_c}}$, d_c is easily calculated, and

$$x^{\beta} = x^{\alpha_c\gamma_c} = ((cx)^{\alpha}/c^{\alpha})^{2^{t_c}} \quad \text{for } x,c \in F,\ c \neq 0. \tag{16}$$

In particular, let $t = t_1$, so (16) implies the following :

$$\beta = \alpha.2^{t}. \tag{17}$$

It is easy to check that (α,β) is an admissible pair of additive permutations iff (α^{-1},β^{-1}) is. Hence $\beta^{-1} = \alpha_c^{-1}.\delta_c^{-1}$ implies that δ_c^{-1} (and hence δ_c) has the same form as γ_c, i.e. $\delta_c : x \mapsto \bar{d}_c x^{2^{g_c}}$ for some nonzero scalar $\bar{d}_c$, and $(g_c, e) = 1$, $1 \leqslant g_c \leqslant e$. Then $1 = 1^{\beta} = 1^{\delta_c\alpha_c} = (\bar{d}_c)^{\alpha_c} = (c\bar{d}_c)^{\alpha}$ implies

$\bar{d}_c = c^{-1}$, from which it follows that $x^\beta = x^{\delta_c \alpha_c} = (c^{-1}x^{2^{g_c}})^{\alpha_c} = x^{2^{g_c}\alpha}$, i.e. $\beta\alpha^{-1} = 2^{g_c} = 2^g$ for all c.

Hence we have

$$\beta = 2^g.\alpha = \alpha.2^t, \text{ and } \alpha = 2^g\alpha 2^{-t}. \tag{18}$$

Now we have $x^\beta = ((cx)^\alpha/c^\alpha)^{2^{t_c}}$ (by (16))

$= ((c^{2^g}x^{2^g})^{\alpha 2^{-t}}/(c^{2^g})^{\alpha 2^{-t}})^{2^{t_c}}$ (by (18))

$= ((dx^{2^g})^\alpha/d^\alpha)^{2^{t_d}.2^{-t+t_c-t_d}}$ (where $d = c^{2^g}$)

$= (x^{2^g.\beta})^{2^{-t+t_c-t_d}}$ (by (16)).

This proves the following :

$$\beta = 2^g.\beta.2^{-t+t_c-t_d}. \tag{19}$$

And so by (18)

$$\alpha = \beta 2^{-t+t_c-t_d}. \tag{20}$$

From (18) and (20) it follows that $\beta^{-1}\alpha = 2^{-t} = 2^{-t+t_c-t_d}$, i.e.

$$t_c = t_d \text{ if } d = c^{2^g}. \tag{21}$$

Since $x \mapsto x^{2^g}$ is an automorphism of maximal order, it follows that if c and d are nonzero conjugates then $t_c = t_d$. Now suppose that c and d are distinct nonzero elements of F for which $t_c = t_d$. We claim $t_{c+d} = t_c$.

$$x^\beta = ((cx)^\alpha/c^\alpha)^{2^{t_c}} = ((dx)^\alpha/d^\alpha)^{2^{t_d}} \text{ with } t_c = t_d \text{ implies } (dx)^\alpha = d^\alpha(cx)^\alpha/c^\alpha. \tag{22}$$

Then $x^\beta = (((c+d)x)^\alpha/(c+d)^\alpha)^{2^{t_{c+d}}} = (((cx)^\alpha+(dx)^\alpha)/(c^\alpha+d^\alpha))^{2^{t_{c+d}}}$

$= (((cx)^\alpha+d^\alpha(cx)^\alpha/c^\alpha)/(c^\alpha+d^\alpha))^{2^{t_{c+d}}}$ (by (22))

$= ((cx)^\alpha/c^\alpha)^{2^{t_{c+d}}}$.

Since this string of equalities holds for all $x \in F$, we have (using (16))

$$t_{c+d} = t_c. \tag{23}$$

By the Normal Basis Theorem for cyclic extensions (cf. [92]) there is an element $c \in F$ for which the conjugates of c (i.e. $c, c^2, c^4, \ldots$) form a linear basis over the prime subfield $\{0,1\}$. As $t_c = t_d$ for d any conjugate of c and then for d equal to any nonzero sum of conjugates, it follows that there is only one t : $t = t_c$ for all $c \in F^o$. Put $c = 1$ in (22) to see that α preserves multiplication. Hence α is an automorphism of F. By (18) also β is an automorphism of F. This completes the proof that (iv) implies (v). The converse is easy. □

12.3. COLLINEATIONS

Let (α,β) be an admissible pair giving rise to the GQ $S(\alpha,\beta)$ of order 2^e.

12.3.1. (*S.E. Payne* [131]). *Let* G *denote the full collineation group of* $S = S(\alpha,\beta)$. *Then at least one of the following must occur :*

(i) *All points and lines of* S *are regular and* $S \cong Q(4,2^e)$.

(ii) *Each element of* G *fixes* (∞).

(iii) *Each element of* G *fixes* $[\infty]$.

Proof. Suppose that neither (ii) nor (iii) holds. Let θ be a collineation moving (∞). First suppose that $(\infty)^\theta \not\sim (\infty)$. As $(\infty)^\theta$ is regular, by 12.1.4 it follows that $S^{[\infty]}$ is a TGQ, so that G is transitive on the set of lines not meeting $[\infty]$. In this case $[\infty]^\theta \neq [\infty]$. If $[\infty]^\theta \not\sim [\infty]$, then every line not meeting $[\infty]$ is regular, so all lines are regular and $S \cong Q(4,2^e)$. So suppose $[\infty]^\theta$ meets $[\infty]$ at (m), where $(m) \neq (\infty)$ since $(\infty)^\theta \not\sim (\infty)$. As $S^{[\infty]\theta}$ must also be a TGQ, in particular $(\infty)^\theta$ is a center of symmetry, so G must be transitive on the lines through (m) but different from $[\infty]^\theta$. It follows that each point collinear with (m) is regular, implying that each point of S is regular (by 1.3.6 (iv)). Hence if $(\infty)^\theta \not\sim (\infty)$, then $S \cong Q(4,2^e)$. Dually, if $[\infty]^\theta \not\sim [\infty]$, then $S \cong Q(4,2^e)$.

Now suppose that $(\infty)^\theta$ is a point different from (∞) on a line $[a]$, $a \in F$. Then we may suppose $[\infty]^\theta = [a]$, in which case $S^{(\infty)}$ is a TGQ. It follows that each line through $(\infty)^\theta$ is regular. But as $S^{(\infty)}$ is a TGQ, G is transitive on lines meeting $[a]$ at points different from (∞). This implies all lines meeting [a] and hence all lines of S are regular.

Finally, suppose each $\theta \in G$ maps (∞) to a point of $[\infty]$, and dually, each $\theta \in G$ maps $[\infty]$ to a line through (∞). It follows that each θ moving (∞) fixes $[\infty]$, and vice versa. But by hypothesis there is a θ moving (∞) and a ϕ moving $[\infty]$. Then $\theta\phi$ must move both (∞) and $[\infty]$, completing the proof. □

Let π denote the projective plane based at (∞), and let f denote the isomorphism from π to $PG(2,2^e)$ with homogeneous coordinates as follows :

$$\begin{array}{ll} (\infty) \overset{f}{\mapsto} (0,1,0) & [\infty] \overset{f}{\mapsto} [0,0,1]^T \\ (a) \overset{f}{\mapsto} (1,a^\alpha,0) & [m] \overset{f}{\mapsto} [1,0,m^\beta]^T \\ (m,v) \overset{f}{\mapsto} (m^\beta,v,1) & \{(a),(0,b)\}^{\perp\perp} \overset{f}{\mapsto} [a^\alpha,1,b]^T. \end{array} \tag{24}$$

Here (x,y,z) is incident in $PG(2,2^e)$ with $[u,v,w]^T$ iff $xu+yv+zw = 0$. Let θ be a collineation of S fixing (∞), so that θ induces a collineation $\bar{\theta}$ of π. Then $f^{-1}\bar{\theta}f$ must be a collineation of $PG(2,2^e)$ and hence given by a semi-linear map. This means there must be a 3×3 nonsingular matrix B over F and an automorphism δ of F for which $f^{-1}\bar{\theta}f$ is defined by

$$f^{-1}\bar{\theta}f : (x,y,z) \mapsto (x^\delta,y^\delta,z^\delta)B$$
and
$$f^{-1}\bar{\theta}f : [u,v,w]^T \mapsto B^{-1}[u^\delta,v^\delta,w^\delta]^T. \tag{25}$$

As $\bar{\theta}$ fixes (∞), we may assume that

$$B = \begin{pmatrix} b_{11} & b_{12} & b_{13} \\ 0 & 1 & 0 \\ b_{31} & b_{32} & b_{33} \end{pmatrix}. \tag{26}$$

Dually, let π' denote the projective plane based at $[\infty]$, and let g denote the isomorphism from π' to $PG(2,2^e)$ defined as follows :

$$\begin{array}{ll} [\infty] \overset{g}{\mapsto} (0,1,0) & (\infty) \overset{g}{\mapsto} [0,0,1]^T \\ [m] \overset{g}{\mapsto} (1,m,0) & (a) \overset{g}{\mapsto} [1,0,a]^T \\ [a,b] \overset{g}{\mapsto} (a,b,1) & \{[m],[0,b]\}^{\perp\perp} \overset{g}{\mapsto} [m,1,b]^T. \end{array} \tag{27}$$

Now let θ be a collineation of S fixing $[\infty]$, so that θ induces a collineation $\hat{\theta}$ of π'. Then $g^{-1}\hat{\theta}g$ must be a collineation of $PG(2,2^e)$ and hence given by a semi-linear map. This means there must be a 3×3 nonsingular matrix A over F and an automorphism γ of F for which $g^{-1}\hat{\theta}g$ is defined by

$$\begin{aligned} g^{-1}\hat{\theta}g &: (x,y,z) \mapsto (x^{\gamma},y^{\gamma},z^{\gamma})A \\ g^{-1}\hat{\theta}g &: [u,v,w]^T \mapsto A^{-1}[u^{\gamma},v^{\gamma},w^{\gamma}]^T. \end{aligned} \tag{28}$$

As $\hat{\theta}$ fixes $[\infty]$, we may assume that

$$A = \begin{pmatrix} a_{11} & a_{12} & a_{13} \\ 0 & 1 & 0 \\ a_{31} & a_{32} & a_{33} \end{pmatrix}. \tag{29}$$

For the remainder of this section we assume that θ is a collineation fixing both (∞) and $[\infty]$, so that it simultaneously induces $\bar{\theta}$ and $\hat{\theta}$ as described above. Using the fact that $\bar{\theta}$ fixes $[\infty]$ and $\hat{\theta}$ fixes (∞), we find that

$$a_{13} = 0 \neq a_{11}a_{33}; \qquad b_{13} = 0 \neq b_{11}b_{33}. \tag{30}$$

$$A^{-1} = \begin{pmatrix} \frac{1}{a_{11}} & \frac{a_{12}}{a_{11}} & 0 \\ 0 & 1 & 0 \\ \frac{a_{31}}{a_{11}a_{33}} & \frac{a_{12}a_{31}}{a_{11}a_{33}} + \frac{a_{32}}{a_{33}} & \frac{1}{a_{33}} \end{pmatrix},$$

$$B^{-1} = \begin{pmatrix} \frac{1}{b_{11}} & \frac{b_{12}}{b_{11}} & 0 \\ 0 & 1 & 0 \\ \frac{b_{31}}{b_{11}b_{33}} & \frac{b_{12}b_{31}}{b_{11}b_{33}} + \frac{b_{32}}{b_{33}} & \frac{1}{b_{33}} \end{pmatrix}. \tag{31}$$

Then calculate as follows :

$$(a)^{\theta} = (a)^{f(f^{-1}\bar{\theta}f)f^{-1}} = \left(\left(\frac{b_{12}+a^{\alpha\delta}}{b_{11}}\right)^{\alpha^{-1}}\right) \text{ and}$$
$$(a)^{\theta} = (a)^{g(g^{-1}\hat{\theta}g)g^{-1}} = \left(\frac{a_{31}+a_{11}a^{\gamma}}{a_{33}}\right) \text{ for all } a \in F. \tag{32}$$

Also

$$[m]^{\theta} = [m]^{f(f^{-1}\bar{\theta}f)f^{-1}} = \left[\left(\frac{b_{31}+b_{11}m^{\beta\delta}}{b_{33}}\right)^{\beta^{-1}}\right] \text{ and}$$
$$[m]^{\theta} = [m]^{g(g^{-1}\hat{\theta}g)g^{-1}} = \left[\frac{a_{12}+m^{\gamma}}{a_{11}}\right] \text{ for all } m \in F. \tag{33}$$

Hence we have the following necessary conditions for θ to be well defined.

$$((a_{31}+a_{11}a^{\gamma})/a_{33})^{\alpha} = (b_{12}+a^{\alpha\delta})/b_{11} \qquad \text{for all } a \in F. \tag{34}$$

$$((a_{12}+m^{\gamma})/a_{11})^{\beta} = (b_{31}+b_{11}m^{\beta\delta})/b_{33} \qquad \text{for all } m \in F. \tag{35}$$

Conversely, if (34) and (35) hold it can be shown that θ is well defined and is a collineation. We shall not need this general a θ, however, and content ourselves with the following special case.

12.3.2. *Every possible whorl of* $S(\alpha,\beta)$ *about* (∞) *fixing* $(0,0,0)$ *exists iff* α *is multiplicative. Dually, every possible whorl of* $S(\alpha,\beta)$ *about* $[\infty]$ *fixing* $[0,0,0]$ *exists iff* β *is multiplicative.*

Proof. Let θ be a whorl about (∞) fixing $(0,0,0)$, so θ fixes each $[m]$, $m \in F$. With $m = 0$ in (33), we find $b_{31} = a_{12} = 0$. Then $m = 1$ yields $a_{11} = 1$ and $b_{11} = b_{33}$, so that $m = m^{\gamma} = m^{\beta\delta\beta^{-1}}$ for all $m \in F$ implies $\gamma = \delta = \mathrm{id}$. As the point $(0,0,0)$ is fixed, so is the line $[0,0]$. But $[0,0]^{\theta} = [0,0]^{g(g^{-1}\hat{\theta}g)g^{-1}} = (a_{31},a_{32},a_{33})^{g^{-1}} = (a_{31}/a_{33}, a_{32}/a_{33},1)^{g^{-1}} = [a_{31}/a_{33},a_{32}/a_{33}]$. Hence $a_{31} = a_{32} = 0$. Since $(0)^{\theta} = (0)$ we have $b_{12} = 0$ by (32). Since $(m,0)^{\theta} = (m,0)$, we have $b_{32} = 0$. It is easily checked that (35) is now satisfied and that (34) says $(a/a_{33})^{\alpha} = a^{\alpha}/b_{11}$ for all $a \in F$. Putting $a = 1$, we obtain $(1/a_{33})^{\alpha} = 1/b_{11}$, and $(a/a_{33})^{\alpha} = a^{\alpha}(1/a_{33})^{\alpha}$. It follows that the whorl θ exists for each nonzero a_{33} iff α is multiplicative. Moreover, in that case a complete description of θ is easily worked out to be as follows, where $t = a_{33}^{-1}$.

$$\begin{array}{ll} (\infty) \overset{\theta}{\mapsto} (\infty) & [\infty] \overset{\theta}{\mapsto} [\infty] \\ (a) \overset{\theta}{\mapsto} (ta) & [m] \overset{\theta}{\mapsto} [m] \\ (a,b) \overset{\theta}{\mapsto} (a,t^{\alpha}b) & [m,v] \overset{\theta}{\mapsto} [tm,tv] \\ (a,b,c) \overset{\theta}{\mapsto} (ta,tb,t^{\alpha}c) & [m,v,w] \overset{\theta}{\mapsto} [m,t^{\alpha}v,tw]. \end{array} \qquad (36)$$

The dual result for multiplicative β is proved analogously. □

We conjecture that when α and β are both multiplicative, then α and β must be automorphisms. This has been verified for $2^e \leqslant 128$ with the aid of a computer (cf. [141]), but nothing else seems to have been done on the problem.

12.4. GENERALIZED QUADRANGLES $T_2(O)$

In this section we assume that $S^{(\infty)}$ is a TGQ whose kernel has maximal order 2^e, where $S = S(\alpha,\beta)$. Hence $S(\alpha,\beta)$ is a $T_2(O)$ of J. Tits (cf. 8.7.1). From 12.3.2, 12.1.4 and 8.6.5 this is equivalent to assuming that α is an automorphism, in which case $S(\alpha,\beta) \cong S(1,\beta\alpha^{-1})$. Hence throughout this section we assume that $\alpha = 1$ and denote $S(1,\beta)$ by S_β when it is necessary to indicate a specific β.

By (3) the set $O = \{(0,0,1)\} \cup \{(1,x,x^\beta) \parallel x \in F\}$ is an oval of $PG(2,2^e)$. Embed $PG(2,2^e)$ as the plane $x_0 = 0$ in $PG(3,2^e)$ and consider the GQ $T_2(O)$. Then we have the following isomorphism of S_β onto $T_2(O)$.

$(\infty) \mapsto (\infty)$,
$(a) \mapsto$ plane of $PG(3,2^e)$ which is tangent to O at $(0,0,0,1)$ and which contains the point $(1,a,0,0)$,
$(u,v) \mapsto$ plane of $PG(3,2^e)$ which is tangent to O at $(0,1,u,u^\beta)$ and which contains the point $(1,0,0,v)$,
$(a,b,c) \mapsto$ point $(1,a,b,c)$ of type (i) of $T_2(O)$, (37)
$[\infty] \mapsto (0,0,0,1) \in O$,
$[u] \mapsto (0,1,u,u^\beta) \in O$,
$[a,b] \mapsto$ line of type (a) of $T_2(O)$ consisting of the points

$(1,a,b,c)$, $c \in F$, and $(0,0,0,1)$ of $PG(3,2^e)$,

$[u,v,w] \to$ line of type (a) of $T_2(O)$ consisting of the points $(1,a,b,c)$, $b+w = au$ and $c+v = au^\beta$, and $(0,1,u,u^\beta)$ of $PG(3,2^e)$.

Then for each triple $(\sigma_1,\sigma_2,\sigma_3)$ of elements of F there is a translation $\tau(\sigma_1,\sigma_2,\sigma_3)$ about (∞) given by the following, where $\tau = \tau(\sigma_1,\sigma_2,\sigma_3)$:

$$\begin{array}{ll} (x,y,z) \overset{\tau}{\mapsto} (x+\sigma_1,y+\sigma_2,z+\sigma_3) & (\infty) \overset{\tau}{\mapsto} (\infty) \\ (x,y) \overset{\tau}{\mapsto} (x,y+\sigma_1x^\beta+\sigma_3) & (x) \overset{\tau}{\mapsto} (x+\sigma_1) \\ [u,v,w] \overset{\tau}{\mapsto} [u,v+\sigma_1u^\beta+\sigma_3,w+\sigma_1u+\sigma_2] & [\infty] \overset{\tau}{\mapsto} [\infty] \\ [u,v] \overset{\tau}{\mapsto} [u+\sigma_1,v+\sigma_2] & [u] \overset{\tau}{\mapsto} [u]. \end{array} \tag{38}$$

For each $t \in F$, $t \neq 0$, there is a whorl about (∞) fixing $(0,0,0)$ given as follows :

$$\begin{array}{ll} (x,y,z) \overset{\theta_t}{\mapsto} (tx,ty,tz) & (\infty) \overset{\theta_t}{\mapsto} (\infty) \\ (x,y) \overset{\theta_t}{\mapsto} (x,ty) & (x) \overset{\theta_t}{\mapsto} (tx) \\ [u,v,w] \overset{\theta_t}{\mapsto} [u,tv,tw] & [\infty] \overset{\theta_t}{\mapsto} [\infty] \\ [u,v] \overset{\theta_t}{\mapsto} [tu,tv] & [u] \overset{\theta_t}{\mapsto} [u]. \end{array} \tag{39}$$

If θ is an arbitrary collineation of S fixing (∞) and $[\infty]$, so that (24) through (35) are valid , we may follow θ by a suitable translation about (∞) and then a whorl about (∞) fixing $(0,0,0)$ so as to obtain a collineation fixing $(0,0,0)$ and (1). So we assume θ is a collineation of S fixing (∞), (1), $[\infty]$ and $(0,0,0)$. Then the corresponding matrices A and B are determined as follows :

$$A = \begin{pmatrix} a_{11} & a_{12} & 0 \\ 0 & 1 & 0 \\ 0 & 0 & a_{33} \end{pmatrix}, \quad B = \begin{pmatrix} 1 & 0 & 0 \\ 0 & 1 & 0 \\ b_{31} & 0 & b_{33} \end{pmatrix}. \tag{40}$$

In this case (34) is equivalent to

$$\gamma = \delta \text{ and } a_{11} = a_{33}. \tag{41}$$

In (35) put m = 0 to obtain

$$b_{31} = b_{33}(a_{12}/a_{11})^{\beta}. \tag{42}$$

In (35) put m = 1 to obtain

$$b_{33}^{-1} = ((a_{12}+1)/a_{11})^{\beta}+(a_{12}/a_{11})^{\beta}. \tag{43}$$

Then using (41) - (43), (35) may be rewritten as follows :

$$((a_{12}+m^{\gamma})/a_{11})^{\beta} = (a_{12}/a_{11})^{\beta}+(((a_{12}+1)/a_{11})^{\beta}+(a_{12}/a_{11})^{\beta})m^{\beta\gamma}, m \in F. \tag{44}$$

It follows that for each choice of a_{12}, a_{11}, γ, where $\gamma \in \text{Aut}(F)$, a_{11}, $a_{12} \in F$, $a_{11} \neq 0$, there is a collineation θ determined uniquely in 12.3 if and only if (44) holds. Put $d = b_{33}^{-1}$ and $\sigma = b_{31}/b_{33} = (a_{12}/a_{11})^{\beta}$. It is now possible to work out the effect of θ on points and lines.

$$\begin{aligned}
&(x,y,z) \overset{\theta}{\mapsto} (x^{\gamma}, (\sigma+dy^{\beta\gamma})^{\beta^{-1}}+(1+x^{\gamma})\sigma^{\beta^{-1}}, \sigma x^{\gamma}+dz^{\gamma})\\
&(x,y) \overset{\theta}{\mapsto} ((\sigma+dx^{\beta\gamma})^{\beta^{-1}}, dy^{\gamma})\\
&(x) \overset{\theta}{\mapsto} (x^{\gamma}) \qquad [\infty] \overset{\theta}{\mapsto} [\infty]\\
&(\infty) \overset{\theta}{\mapsto} (\infty) \qquad [u] \overset{\theta}{\mapsto} [(\sigma+du^{\beta\gamma})^{\beta^{-1}}]\\
&[u,v] \overset{\theta}{\mapsto} [u^{\gamma}, (\sigma+dv^{\beta\gamma})^{\beta^{-1}}+(1+u^{\gamma})\sigma^{\beta^{-1}}]\\
&[u,v,w] \overset{\theta}{\mapsto} [(\sigma+du^{\beta\gamma})^{\beta^{-1}}, dv^{\gamma}, (\sigma+dw^{\beta\gamma})^{\beta^{-1}}+\sigma^{\beta^{-1}}] .
\end{aligned} \tag{45}$$

For ease of reference, the collineation θ described by (45) will be denoted $\pi(\sigma,d,\gamma)$, where (44) is satisfied by a_{11}, a_{12}, γ, and $d = b_{33}^{-1}$ is defined by (43), and $\sigma = (a_{12}/a_{11})^{\beta}$.

12.4.1. *If* β *is multiplicative, there is a collineation* $\pi(0,d,\gamma)$ *of S for each* $d \in F^{\circ}$ *and each* $\gamma \in \text{Aut}(F)$. *If* β *is multiplicative, there is a colli-*

neation $\pi(\sigma,d,\gamma)$ *for some* $\sigma \neq 0$ *iff* β *is an automorphism iff* $\pi(\sigma,d,\gamma)$ *is a collineation for each choice of* $\sigma \in F$, $d \in F^o$, $\gamma \in \mathrm{Aut}(F)$.

Proof. Since $\sigma = (a_{12}/a_{11})^\beta$, $\sigma = 0$ iff $a_{12} = 0$. And it is easy to check that (44) holds if $\sigma = 0$ and β is multiplicative. We note that since β is multiplicative there is an integer i, $1 \leq i \leq 2^e-1$, with $(i,2^e-1) = 1$, for which $\beta : x \mapsto x^i$ for all $x \in F$, so that β and γ commute. Now suppose that there is a collineation $\pi(\sigma,d,\gamma)$ for some $\sigma \neq 0$ and that β is multiplicative. Hence (44) holds for some $a_{12},a_{11} \in F^o$ and $\gamma \in \mathrm{Aut}(F)$. Using the multiplicativity of β, multiply through by a_{11}^β in (44) to obtain

$$(a_{12}+m^\gamma)^\beta = a_{12}^\beta+((a_{12}+1)^\beta+a_{12}^\beta)m^{\beta\gamma} \text{ for all } m \in F. \tag{46}$$

Putting $m = a_{12}^{\gamma^{-1}}$ we obtain $(a_{12}+1)^\beta+a_{12}^\beta = 1$. So (46) becomes

$$(a_{12}+m^\gamma)^\beta = a_{12}^\beta+m^{\beta\gamma}. \tag{47}$$

Write $m^\gamma = a_{12}x$ and use the multiplicativity of β to rewrite (47) as

$$(1+x)^\beta = 1+x^\beta \text{ , for all } x \in F. \tag{48}$$

It now follows readily that β is also additive and hence an automorphism.

Conversely, if β is an automorphism (and hence an automorphism of order e), it is easy to check that (44) is satisfied for all a_{12}, $m \in F$, $a_{11} \in F^o$, $\gamma \in \mathrm{Aut}(F)$. □

12.4.2. (i) $S = S(1,\beta)$ *has a collineation moving* (∞) *iff* $\beta = 2$ *and* $S \cong Q(4,2^e)$.

(ii) *Let* β *be multiplicative and fix* $z \in P-(\infty)^\perp$. *Let* $G_{((\infty),[\infty])}$ *be the group of collineations of* S *fixing* (∞) *and* $[\infty]$, *and let* $\bar{G}_z$ *be the stabilizer of* z *in* $G_{((\infty),[\infty])}$. *Then* $\bar{G}_z$ *is transitive on the lines of* $B-[\infty]^\perp$ *through* z *if and only if* β *is an automorphism.*

Proof. If S has a collineation moving (∞) then by 4.3.3 (i) $S \cong Q(4,2^e)$. Hence O is a conic and $\beta = 2$. Suppose β is multiplicative. As $G_{((\infty),[\infty])}$ is transitive on $P-(\infty)^\perp$, we may assume $z = (0,0,0)$. If β is an automorphism, $\pi(\sigma,1,id)$ maps $[u,0,0]$ to $[u+\sigma^{\beta^{-1}},0,0]$ for each $\sigma \in F$. On the other hand, let θ be any collineation in $\bar{G}_z$ with $z = (0,0,0)$. There is a θ_t as in (39) for which $\theta.\theta_t^{-1} = \pi(\sigma,d,\gamma)$ for some choice of $\sigma \in F$, $d \in F^{\circ}$, $\gamma \in \mathrm{Aut}(F)$. Then $[u,0,0]^{\theta} = [u,0,0]^{\pi(\sigma,d,\gamma).\theta_t} = [(\sigma+du^{\beta\gamma})^{\beta^{-1}},0,0]$. Since β is multiplicative, $\{\pi(0,d,\gamma) \parallel d \in F^{\circ}, \gamma \in \mathrm{Aut}(F)\}$ is transitive on the set of lines of the form $[u,0,0]$, $u \neq 0$. But $[0,0,0]$ is moved by some $\pi(\sigma,d,\gamma)$ iff $\sigma \neq 0$, so the proof of (ii) is complete by 12.4.1. □

12.4.3. *If* $c \in F$ *satisfies* $(c^{\beta})^{\beta} \neq (c^{\beta})^2$, *then* (∞) *is the unique regular point on the line* $[c]$.

Proof. Let $(c^{\beta})^{\beta} \neq (c^{\beta})^2$, so that $0 \neq c \neq 1$. As the translations about (∞) are transitive on the points of $[c]$ different from (∞), it suffices to show that the pair $((c,0),(0,0,1))$ is not regular. Use (2) to check the following.

$$\{(c,1),(0,0,0)\}^{\perp} = \{(c,0),(0,0,1)\} \cup \{(\frac{1}{m^{\beta}+c^{\beta}}, \frac{m}{m^{\beta}+c^{\beta}}, \frac{m^{\beta}}{m^{\beta}+c^{\beta}}) \parallel m \in F, m \neq c\}.$$

$$\{(c,0),(0,0,1)\}^{\perp} = \{(c,1),(0,0,0)\} \cup (\frac{1}{u^{\beta}+c^{\beta}}, \frac{u}{u^{\beta}+c^{\beta}}, \frac{c^{\beta}}{u^{\beta}+c^{\beta}}) \parallel u \in F, u \neq c\}.$$

Hence $((c,0),(0,0,1))$ is regular iff $(\frac{1}{m^{\beta}+c^{\beta}}, \frac{m}{m^{\beta}+c^{\beta}}, \frac{m^{\beta}}{m^{\beta}+c^{\beta}}) \sim$

$(\frac{1}{u^{\beta}+c^{\beta}}, \frac{u}{u^{\beta}+c^{\beta}}, \frac{c^{\beta}}{u^{\beta}+c^{\beta}})$ whenever $u \neq c \neq m$. Put $m = 0$ and $u = 1$ and use (2) to obtain $(c^{\beta})^{\beta} = (c^{\beta})^2$ if $(c,0)$ is regular. □

Put $A = \{\beta \parallel (1,\beta)$ is admissible$\}$. Using (3) with $\alpha = \mathrm{id}, u_3=x_3=0, u_1=x_1, u_2=x_2$, it is easy to show that the map $x \mapsto x^{\beta}/x$ permutes the elements of F°. Since β^{-1} is a permutation of F° as well as the map $x \mapsto x^{-1}$, it follows that the map $\lambda : x \mapsto x(x^{-1})^{\beta^{-1}}$ permutes the elements of F°. Let β^* be the inverse of λ. With a little juggling it can be seen that for $x,y,z \in F^{\circ}$, the following holds

$$(y/x)^{\beta} = z/x \text{ iff } (y/z)^{\beta^*} = x/z. \tag{49}$$

12.4.4. *If* $\beta \in A$, *then* $\beta^* \in A$ *and there is an isomorphism* $\tau^* : S_{\beta} \to S_{\beta^*}$ *in which* $(\infty)_{\beta} \overset{\tau^*}{\mapsto} (\infty)_{\beta^*}$, $[\infty]_{\beta} \overset{\tau^*}{\mapsto} [0]_{\beta^*}$, $[0]_{\beta} \overset{\tau^*}{\mapsto} [\infty]_{\beta^*}$. *(Subscripts* β, β^* *are used to indicate to which structure,* $S(1,\beta)$ *or* $S(1,\beta^*)$, *the given object belongs.)*

Proof. $\beta^* \in A$ iff $S(1,\beta^*)$ is a GQ, and it suffices to exhibit an isomorphism $\tau^* : S_\beta \to S_{\beta^*}$. In fact, it suffices to exhibit τ^* as a collinearity-preserving bijective mapping on points. Then using (2) and (49) it is routine to check that the τ^* exhibited in (50) satisfies $x \sim y$ in $S(1,\beta)$ iff $x^{\tau^*} \sim y^{\tau^*}$ in $S(1,\beta^*)$.

$$\begin{aligned}
&(x,y,z)_\beta \overset{\tau^*}{\mapsto} (z,y,x)_{\beta^*}\\
&(x_0,x_1)_\beta \overset{\tau^*}{\mapsto} ((1/x_0^\beta)^{(\beta^*)^{-1}}, x_1/x_0^\beta)_{\beta^*}, \text{ if } x_0 \neq 0\\
&(0,x_1)_\beta \overset{\tau^*}{\mapsto} (x_1)_{\beta^*} \qquad\qquad (50)\\
&(x_0)_\beta \overset{\tau^*}{\mapsto} (0,x_0)_{\beta^*}\\
&(\infty)_\beta \overset{\tau^*}{\mapsto} (\infty)_{\beta^*}.
\end{aligned}$$

We leave the details to the reader. □

Put $M = \{\beta \,\|\, \beta$ is a multiplicative permutation of the elements of F for which $\beta : x \mapsto x^\beta/x$ permutes the elements of $F^o\}$.
Put $\mathcal{D} = M \cap A$. For $\beta \in \mathcal{D}$, it follows that $\beta^* = \beta/(\beta-1)$, using exponential notation, and $(\beta^*)^* = \beta$. (In fact, $(\beta^*)^* = \beta$ for all $\beta \in A$). Hence we can extend the definition of the map $*: \beta \mapsto \beta^*$ to $A \cup M$ by defining $\beta^* = \beta/(\beta-1)$ for all $\beta \in M$. It still follows that $(\beta^*)^* = \beta$. Moreover, for $\beta \in M$ it follows that $\beta \in \mathcal{D}$ iff $\beta^{-1} \in \mathcal{D}$ iff $\beta^* \in \mathcal{D}$. Hence for $\beta \in M$ each of the following elements of M is in $\mathcal{D}$ or none of them is in $\mathcal{D}$:

$$\beta, \beta^* = \beta/(\beta-1), (\beta-1)/\beta, 1-\beta, (1-\beta)^{-1}, \beta^{-1}. \qquad (51)$$

12.4.5. *Let* $\beta \in \mathcal{D}$. *Then one of the following must occur :* (i) $\beta = 2$ *and* $S_\beta \cong Q(4,2^e)$; (ii) $\beta \neq 2$ *and* S_β *has* $2^e+1 = s+1$ *collinear regular points, either on* $[\infty]_\beta$ *or* $[0]_\beta$ *according as* β *or* β^* *is an automorphism of* F; (iii) $\beta \neq 2$ *and* $(\infty)_\beta$ *is the unique regular point of* S_β.
Proof. Suppose $\beta \in \mathcal{D}$ and that some point x ($x \neq (\infty)_\beta$) is regular . If $x \in P-(\infty)_\beta^\perp$, clearly all points are regular, $S_\beta \cong Q(4,2^e)$, and $\beta = 2$. So suppose $S_\beta \not\cong Q(4,2^e)$. First consider the case where x is incident with $[\infty]_\beta$. In this case β is an automorphism of F by 12.1.4 and the group

$G_{((\infty),[\infty])_\beta}$ is transitive on the 2^e lines $[m]_\beta, m \in F$, by 12.4.1. Since $S_\beta^{(\infty)}$ is a TGQ, the group $G_{((\infty),[\infty])_\beta}$ acts transitively on the 2^e points incident with the line $[\infty]_\beta$ (resp., $[m]_\beta$, $m \in F$) and distinct from $(\infty)_\beta$. If S_β has a regular point not incident with the line $[\infty]_\beta$, then it follows readily that all points of $(\infty)_\beta^\perp$ are regular. By 1.3.6 (iv) all points of S_β are regular, a contradiction. It follows that a point is regular iff it is incident with $[\infty]_\beta$. Now suppose x is incident with $[0]_\beta$. Using the isomorphism $\tau^* : S_\beta \to S_{\beta^*}$, we see that β^* is an automorphism and $[0]_\beta$ is the unique line of regular points of S_β. Finally, suppose x is incident with some line $[c]_\beta$, $0 \neq c \in F$. Since $G_{((\infty),[\infty])_\beta}$ is transitive on the lines of the form $[c]_\beta$, $0 \neq c \in F$, it follows from 12.4.3 that $m^\beta = m^2$ for all $m \neq 0$, and hence that $\beta = 2$, i.e. $S_\beta \cong T_2(O) \cong Q(4,2^e)$, a contradiction. □

12.4.6. *For* $\beta \in A$, *if* S_β *has a regular point other than* $(\infty)_\beta$, *then* $S_\beta \cong S_\gamma$ *for some* $\gamma \in \mathrm{Aut}(F)$.

Proof. If $x \notin (\infty)_\beta^\perp$ is regular, then $S_\beta \cong Q(4,2^e)$ and $\beta = 2$. So suppose $x \in (\infty)_\beta^\perp - \{(\infty)_\beta\}$ is regular. If x I $[\infty]_\beta$, then by 12.1.4 β is additive, and so by 12.2.4 β is an automorphism. Finally, assume x I $[u]_\beta$, $u \in F$. In the plane $PG(2,2^e)$ of the oval $O = \{(0,0,1)\} \cup \{(1,x,x^\beta) \parallel x \in F\}$ a new coordinate system is chosen in such a way that the point $(1,u,u^\beta)$ is the new point (0,0,1), that the new points (1,0,0),(1,1,1) are on O, and that the nucleus of O is again the point (0,1,0). Then in the new system $O = \{(0,0,1)\} \cup \{(1,x,x^\gamma) \parallel x \in F\}$ with $\gamma \in A$. We have $S_\beta \cong T_2(O) \cong S_\gamma$. Since there is a regular point other than $(\infty)_\gamma$ and incident with $[\infty]_\gamma$, γ is an automorphism. □

12.5. ISOMORPHISMS

Let (α_1,β_1) and (α_2,β_2) be admissible pairs. We begin this section by seeking necessary and sufficient conditions for the existence of a type-preserving isomorphism θ from $S(\alpha_1,\beta_1)$ to $S(\alpha_2,\beta_2)$. Let $(\infty)_i,(a)_i,(a,b)_i,(a,b,c)_i$ denote the points of $S(\alpha_i,\beta_i)$, i = 1,2. Use analogous notation for lines. Let π_i denote the plane based at $(\infty)_i$ and π_i' the plane based at $[\infty]_i$, i = 1,2. Functions $f_i : \pi_i \to PG(2,2^e)$, i = 1,2, are defined as in (24). Similarly, functions $g_i : \pi_i' \to PG(2,2^e)$ are defined as in (27). Let $\theta : S(\alpha_1,\beta_1) \to S(\alpha_2,\beta_2)$ be an isomorphism for which $\theta : (\infty)_1 \mapsto (\infty)_2$ and $\theta : [\infty]_1 \mapsto [\infty]_2$, i.e. θ is type-preserving on points and lines. Then θ induces an isomorphism $\bar{\theta} : \pi_1 \to \pi_2$ and an isomorphism $\hat{\theta} : \pi_1' \to \pi_2'$. Just as in Section 12.3, $f_1^{-1}\bar{\theta}f_2$ is a semi-linear map of $PG(2,2^e)$ as in (25) and $g_1^{-1}\hat{\theta}g_2$ is a semi-linear map as in (28). Using symmetries about $(\infty)_2$ and about $[\infty]_2$ we may assume that the image of $(0,0)_1$ under θ is of the form $(c,0)_2$ for some $c \in F$ and that the image of $[0,0]_1$ under θ is of the form $[d,0]_2$ for some $d \in F$. Hence there are nonsingular matrices A,B and automorphisms δ,γ of F for which

$$
\begin{aligned}
f_1^{-1}\bar{\theta}f_2 &: (x,y,z) \mapsto (x^\delta,y^\delta,z^\delta)B;\ [u,v,w]^T \mapsto B^{-1}[u^\delta,v^\delta,w^\delta]^T \\
g_1^{-1}\hat{\theta}g_2 &: (x,y,z) \mapsto (x^\gamma,y^\gamma,z^\gamma)A;\ [u,v,w]^T \mapsto A^{-1}[u^\gamma,v^\gamma,w^\gamma]^T.
\end{aligned} \tag{52}
$$

The specific assumptions on θ allow us to write

$$
A = \begin{pmatrix} a_{11} & a_{12} & 0 \\ 0 & 1 & 0 \\ a_{31} & 0 & a_{33} \end{pmatrix}, \quad A^{-1} = \begin{pmatrix} \frac{1}{a_{11}} & \frac{a_{12}}{a_{11}} & 0 \\ 0 & 1 & 0 \\ \frac{a_{31}}{a_{11}a_{33}} & \frac{a_{12}a_{31}}{a_{11}a_{33}} & \frac{1}{a_{33}} \end{pmatrix},
$$

$$
B = \begin{pmatrix} b_{11} & b_{12} & 0 \\ 0 & 1 & 0 \\ b_{31} & 0 & b_{33} \end{pmatrix}, \quad B^{-1} = \begin{pmatrix} \frac{1}{b_{11}} & \frac{b_{12}}{b_{11}} & 0 \\ 0 & 1 & 0 \\ \frac{b_{31}}{b_{11}b_{33}} & \frac{b_{12}b_{31}}{b_{11}b_{33}} & \frac{1}{b_{33}} \end{pmatrix}. \tag{53}
$$

And $\theta : (a)_1 \mapsto (a)_1^{g_1(g_1^{-1}\hat{\theta}g_2)g_2^{-1}} = ([1,0,a]^T)^{(g_1^{-1}\hat{\theta}g_2)g_2^{-1}} =$

$$(A^{-1}[1,0,a^\gamma]^T)^{g_2^{-1}} = ([\frac{1}{a_{11}},0,\frac{a_{31}+a_{11}a^\gamma}{a_{11}a_{33}}]^T)^{g_2^{-1}} = ([1,0,\frac{a_{31}+a_{11}a^\gamma}{a_{33}}]^T)^{g_2^{-1}} =$$

$$= (\frac{a_{31}+a_{11}a^\gamma}{a_{33}})_2 .$$

Also $\theta : (a)_1 \mapsto (a)_1^{f_1(f_1^{-1}\bar{\theta}f_2)f_2^{-1}} = (1,a^{\alpha_1},0)^{(f_1^{-1}\bar{\theta}f_2)f_2^{-1}} =$

$= ((1,a^{\alpha_1\delta},0)B)^{f_2^{-1}} = (b_{11},b_{12}+a^{\alpha_1\delta},0)^{f_2^{-1}} =$

$= (1,\frac{b_{12}+a^{\alpha_1\delta}}{b_{11}},0)^{f_2^{-1}} = ((\frac{b_{12}+a^{\alpha_1\delta}}{b_{11}})^{\alpha_2^{-1}})_2 .$

Equating these two values of the image of $(a)_1$ under θ yields

$$\frac{b_{12}+a^{\alpha_1\delta}}{b_{11}} = (\frac{a_{31}+a_{11}a^{\gamma}}{a_{33}})^{\alpha_2} \text{ for all } a \in F. \tag{54}$$

Similarly, equating the two values for the image of $[m]_1$ under θ yields

$$(b_{31}+b_{11}m^{\beta_1\delta})/b_{33} = ((a_{12}+m^{\gamma})/a_{11})^{\beta_2} \quad \text{for all } m \in F. \tag{55}$$

This proves the following :

12.5.1. *If there is an isomorphism* $\theta : S(\alpha_1,\beta_1) \to S(\alpha_2,\beta_2)$ *with* $\theta : (\infty)_1 \mapsto (\infty)_2$ *and* $\theta : [\infty]_1 \mapsto [\infty]_2$, *then there are automorphisms* γ, δ *of* F *and scalars* $b_{12},b_{11},b_{31}, b_{33},a_{11},a_{12},a_{31},a_{33}$ *in* F *with* $a_{11}a_{33}b_{11}b_{33} \neq 0$ *for which* (54) *and* (55) *both hold.*

A converse holds, but we won't need it here. We now restrict our attention to S_β, $\beta \in A$.

Let $\delta \in \text{Aut}(F)$, and let π_δ be the permutation of points and lines of S_β obtained by replacing each coordinate by its image under δ. Hence for $\beta \in \mathcal{D}$, π_δ is just the collineation $\pi(0,1,\delta)$ of 12.4.1.

12.5.2. *Let* $\alpha,\beta \in A$ *and suppose that* θ *is an isomorphism from* S_α *to the dual of* S_β. *Then* S_β *is self-dual and is isomorphic to* S_γ *for some* $\gamma \in \text{Aut}(F)$. *Moreover, for* $\beta \in \mathcal{D}$, S_β *is self-dual iff* $\beta \in \text{Aut}(F)$ *or* $\beta^* \in \text{Aut}(F)$, *in which case* S_β *is self-polar iff* e *is odd.*

Proof. Let $\alpha,\beta \in A$ and suppose θ is an isomorphism from S_α to the dual of S_β. If $\beta = 2$, then $S_\beta \cong Q(4,2^e)$ is self-dual. Suppose $\beta \neq 2$. Suppose $\theta : (\infty)_\alpha \mapsto L$. As $\beta \neq 2$ and L is coregular, L must be incident with $(\infty)_\beta$ and could have been chosen as the line through $(\infty)_\beta$ which is an axis of symmetry with desarguesian plane at L when coordinates were set up, perhaps with a permutation γ different from β . So $S_\beta \cong S_\gamma$ with $[\infty]_\gamma$ coregular and hence γ is an automorphism by 12.1.4 and 12.2.4. We have already seen that if $\beta \in \mathcal{D}$, $\beta \neq 2$, then S_β has a line of regular points iff β or β^* is an automorphism. So we must now exhibit a duality of S_β when β is an automorphism.

Let β be an automorphism of F of order e. Then θ defined by (56) is a duality

$$\begin{array}{ll} (x,y,z) \overset{\theta}{\mapsto} [x,y^\beta,z] & [u,v,w] \overset{\theta}{\mapsto} (u^\beta,v,w^\beta) \\ (u,v) \overset{\theta}{\mapsto} [u^\beta,v] & [x,y] \overset{\theta}{\mapsto} (x,y^\beta) \\ (x) \overset{\theta}{\mapsto} [x] & [u] \overset{\theta}{\mapsto} (u^\beta) \\ (\infty) \overset{\theta}{\mapsto} [\infty] & [\infty] \overset{\theta}{\mapsto} (\infty). \end{array} \tag{56}$$

It is easy to check that θ preserves incidence and $\theta^2 = \pi_\beta$. If e is even, then S_β is not self-polar by 1.8.2. If e is odd, there is a $\sigma \in \mathrm{Aut}(F)$ for which $\beta\sigma^2 = \mathrm{id}$. Then $\gamma = \theta\pi_\sigma = \pi_\sigma\theta$ is easily seen to be a polarity. □

12.5.3. *Let* $\alpha,\beta \in \mathcal{D}$. *Then*

(i) S_α *is isomorphic to the dual of* S_β *iff* α *or* α^* *is an automorphism and* $\alpha = \beta$ *or* $\alpha = \beta^*$.

(ii) $S_\alpha \cong S_\beta$ *iff* $\alpha = \beta$ *or* $\alpha = \beta^*$.

Proof. If α or α^* is an automorphism and $\alpha = \beta$ or $\alpha = \beta^*$, then clearly S_α is isomorphic to the dual of S_β. Let $\alpha,\beta \in \mathcal{D}$ and suppose S_α is isomorphic to the dual of S_β. Then as S_α and S_β have coregular lines, by 12.4.5 either α or α^* is an automorphism and either β or β^* is an automorphism. As $S_\alpha \cong S_{\alpha^*}$ and $S_\beta \cong S_{\beta^*}$, using the duality in (56) we see that all four of these GQ and their duals are isomorphic. Hence the result will follow from (ii), which we now prove.

If $\alpha = \beta$ or $\alpha = \beta^*$, then clearly $S_\alpha \cong S_\beta$. Let $\alpha,\beta \in \mathcal{D}$ and suppose $\theta : S_\alpha \to S_\beta$ is an isomorphism. We may suppose also that $\alpha \neq 2 \neq \beta$, since otherwise the conclusion is clear. In this case it is also clear that $\theta : (\infty)_\alpha \mapsto (\infty)_\beta$.

First suppose that either α or α^* is an automorphism. If α is an automorphism, then S_α is self-dual, hence S_β is self-dual, implying that β or β^* is an automorphism; if α^* is an automorphism, then S_{α^*} is self-dual, hence S_α and S_β are self-dual, implying that β or β^* is an automorphism. By 12.1.4 and 12.4.4 each point incident with the line $[\infty]_\alpha$ or $[0]_\alpha$ (resp. $[\infty]_\beta$ or $[0]_\beta$) is regular. By 12.4.5 θ maps at least one of the lines $[\infty]_\alpha$,

$[0]_\alpha$ to at least one of the lines $[\infty]_\beta, [0]_\beta$. By means of τ^* (cf. (50)) we may replace α by α^* and/or β by β^* if necessary and assume that θ : $[\infty]_\alpha \mapsto [\infty]_\beta$ (i.e. we assume that α and β are automorphisms). Now we apply 12.5.1 with $\alpha_1 = \alpha_2 = \mathrm{id}$, $\beta_1 = \alpha$, $\beta_2 = \beta$. Then in the notation of (54) and (55), (54) becomes

$$\frac{b_{12}}{b_{11}} + \frac{a^\delta}{b_{11}} = \frac{a_{31}}{a_{33}} + \left(\frac{a_{11}}{a_{33}}\right)a^\gamma \text{ for all } a \in F. \tag{57}$$

It follows readily that $\dfrac{b_{12}}{b_{11}} = \dfrac{a_{31}}{a_{33}}$, $b_{11}a_{11} = a_{33}$, and $\delta = \gamma$.

Then (55) becomes

$$\frac{b_{31}}{b_{33}} + \left(\frac{b_{11}}{b_{33}}\right)m^{\alpha\delta} = \frac{a_{12}^\beta}{a_{11}^\beta} + \frac{m^{\delta\beta}}{a_{11}^\beta} \quad \text{for all } m \in F. \tag{58}$$

Putting $m = 0$, we obtain $\dfrac{b_{31}}{b_{33}} = \dfrac{a_{12}^\beta}{a_{11}^\beta}$, and then putting $m = 1$ we have $\dfrac{b_{11}}{b_{33}} = \dfrac{1}{a_{11}^\beta}$. Hence $m^{\alpha\delta} = m^{\delta\beta}$ for all $m \in F$. As $\delta\beta = \beta\delta$, clearly $\alpha = \beta$.

Now suppose that no one of $\alpha, \alpha^*, \beta, \beta^*$ is an automorphism. We show that θ maps the two lines $[\infty]_\alpha, [0]_\alpha$ to the two lines $[\infty]_\beta, [0]_\beta$. Since α (resp., β) is not an automorphism, each collineation of S_α (resp., S_β) fixing $(\infty)_\alpha$ and $[\infty]_\alpha$ (resp., $(\infty)_\beta$ and $[\infty]_\beta$) also fixes $[0]_\alpha$ (resp., $[0]_\beta$) by 12.4.1. Since α^* (resp., β^*) is not an automorphism, each collineation of S_{α^*} (resp., S_{β^*}) fixing $(\infty)_{\alpha^*}$ and $[\infty]_{\alpha^*}$ (resp., $(\infty)_{\beta^*}$ and $[\infty]_{\beta^*}$) also fixes $[0]_{\alpha^*}$ (resp. , $[0]_{\beta^*}$). Hence each collineation of S_α (resp., S_β) fixing $(\infty)_\alpha$ and $[0]_\alpha$ (resp., $(\infty)_\beta$ and $[0]_\beta$) also fixes $[\infty]_\alpha$ (resp., $[\infty]_\beta$). Suppose that $[\infty]_\alpha^\theta$ (resp., $[0]_\alpha^\theta$) is $[u]_\beta$, with $u \neq 0, \infty$. Then each collineation of S_β fixing $[u]_\beta$ and $(\infty)_\beta$ also fixes some line $[v]_\beta$, $u \neq v$, and each collineation of S_β fixing $[v]_\beta$ and $(\infty)_\beta$ also fixes $[u]_\beta$. Since β is multiplicative there is a collineation $\pi(0,u^{-1},2)$. Since $[u]_\beta^{\pi(0,u^{-1},2)} = [u]_\beta$, also $[v]_\beta^{\pi(0,u^{-1},2)} = [v]_\beta$, i.e. $u^{-1}v^2 = v$, or $v = 0$. It follows that a collineation of S_β fixing $(\infty)_\beta$ will fix $[\infty]_\beta$ iff it fixes $[0]_\beta$ iff it fixes $[u]_\beta$. Consequently $[u]_\beta^{\pi(0,d,\gamma)} = [u]_\beta$, i.e. $u = du^\gamma$, for all $d \in F^o$ and each $\gamma \in \mathrm{Aut}(F)$. Hence

F = GF(2), implying α and β are automorphisms contrary to hypothesis. This shows that $\{[\infty]_\alpha^\theta, [0]_\alpha^\theta\} = \{[\infty]_\beta, [0]_\beta\}$. By means of τ^* (cf. (50)) we may replace β with β^* if necessary so as to assume that $\theta : [\infty]_\alpha \mapsto [\infty]_\beta$.

Now we apply 12.5.1 with $\alpha_1 = \alpha_2 = \mathrm{id}$, $\beta_1 = \alpha$, $\beta_2 = \beta$. Then in the notation of (54) and (55), (54) becomes

$$b_{12}/b_{11}+a^\delta/b_{11} = a_{31}/a_{33}+(a_{11}/a_{33})a^\gamma \text{ for all } a \in F. \tag{59}$$

It follows readily that $b_{12}/b_{11} = a_{31}/a_{33}$, $b_{11}a_{11} = a_{33}$, and $\delta = \gamma$. Then (55) becomes

$$(b_{31}+b_{11}m^{\alpha\delta})/b_{33} = ((a_{12}+m^\delta)/a_{11})^\beta \text{ for all } m \in F. \tag{60}$$

(60) came from the fact that $\theta : [m]_\alpha \mapsto [\frac{a_{12}+m^\delta}{a_{11}}]_\beta = [(\frac{b_{31}+b_{11}m^{\alpha\delta}}{b_{33}})^{\beta^{-1}}]$.

Since $[0]_\alpha \overset{\theta}{\mapsto} [0]_\beta$, it follows that $a_{12} = b_{31} = 0$. Hence (60) says (using $\beta \in \mathcal{D}$) that $m^{\alpha\delta}(b_{11}a_{11}^\beta/b_{33}) = m^{\delta\beta}$, for all $m \in F$. Put m = 1 and use $\delta\beta = \beta\delta$ to see that $\alpha = \beta$. This completes the proof. □

12.6. <u>NONISOMORPHIC GQ</u>

For $\alpha = \mathrm{id}$, condition (3) of 12.1.1 may be rewritten to say that

$\beta \in A$ iff $y \mapsto (x^\beta+y^\beta)/(x+y)$, $y \neq x$, is an injection for each $x \in F$. (Compare with 10.3.1.) (61)

Since the determination of all $\beta \in A$ is equivalent to the determination of all ovals in PG$(2,2^e)$, it is unlikely that such a project will be completed in the near future. However, all known complete ovals, except the one in PG(2,16) not arising from a conic (cf. D. Glynn [65], M. Hall, Jr. [71] and S.E. Payne and J.E. Conklin [139]), do arise from an oval $O = \{(0,0,1)\} \cup \{(1,x,x^\beta) \parallel x \in F\}$ with $\beta \in \mathcal{D}$. Hence we consider the known examples arising from $\beta \in \mathcal{D}$.

It is an easy exercise to prove the following :

For $\beta \in M$, $\beta \in \mathcal{D}$ iff $u \mapsto (1+(1+u)^\beta)/u$ permutes the elements of F^o. (62)

For e = 1 and e = 2 there is a unique GQ of order 2^e. For e = 3 it is not too difficult to show that there are exactly two TGQ, both self-polar, given by $\beta = 2$ and $\beta = 4$ (cf. S.E. Payne [130]). For e = 4, there are exactly three $T_2(O)$'s : S_2 and S_8 are self-dual (and distinct by 12.5.3), and there is one other nonself-dual example arising from the unique nonconical complete oval in PG(2,16) (cf. [71,139]). Now let $e \geq 5$. Let $\beta_1 = 2$, $\beta_2 = 2^{-1} = 2^{e-1}$, β_3, $\beta_4 = \beta_3^{-1}, \ldots, \beta_{2t-1}$, $\beta_{2t} = \beta_{2t-1}^{-1}$ be the $2t = \varphi(e)$ automorphisms of F of order e arranged in pairs so that $\beta_1 = 2$ and $\beta_{2i} = \beta_{2i-1}^{-1}$. Then S_{β_j} is self-dual for $1 \leq j \leq \varphi(e)$. Moreover, for $t-1 \geq i \geq 1$, $\bar{\beta}_i = 1-\beta_{2i+1}$ yields an additional, nonself-dual example. This gives a total of $2(\varphi(e)-1)$ pairwise nonisomorphic GQ of order 2^e with $\varphi(e)$ of them being self-dual. If e is odd, there are some additional examples arising from ovals in PG(2,2^e) discovered by B. Segre and U. Bartocci [163], and D. Glynn [65].

12.6.1. *For* e *odd*, $6 \in \mathcal{D}$.

Proof. Since e is odd, $z \mapsto z^6$ and $z \mapsto z^5$ permute the elements of F^o. Hence we need to show that $z \mapsto (1+(1+z)^6)/z = z+z^3+z^5$ permutes the elements of F^o. So suppose $0 = (x+x^3+x^5)+(y+y^3+y^5) = (x+y)((x^2+y^2+1)^2+(x^2+y^2+1)(xy+1)+(xy+1)^2)$, with $x \neq y$. Since e is odd, $z^2+z+1 = 0$ has no solution in F. It follows that if $xy+1 = 0$, then $(x^2+y^2+1)^2 = 0$ has no solution. And if $xy \neq 1$, then for $T = (x^2+y^2+1)/(xy+1)$, $T^2+T+1 = 0$ has no solution. Hence $6 \in \mathcal{D}$. □

If e = 5, then $6^{-1} = -5$, so that $(6^*)^{-1} = (6-1)/6 = 1-6^{-1} = 1+5 = 6$. It can be shown (by hand calculations) that all the distinct S arising from $\mathcal{D}$ are the following : $S_2, S_{16}, S_4, S_8, S_{28}$ and its dual , S_6 and its dual.

Now suppose $e \geq 7$. Then for e odd, let 6^{-1} denote the multiplicative inverse of 6 modulo 2^e-1. Then S_6, $S_{6^{-1}}$, S_{-5} and their duals provide six additional examples. This proves the following.

12.6.2. *If* e *is odd*, $e \geq 7$, *there are at least* $2(\varphi(e)+2)$ *pairwise nonisomorphic* GQ *of order* 2^e.

M. Eich and S.E. Payne [56], and J.W.P. Hirschfeld [79,80], have independently verified that for e = 7 there are precisely two additional examples arising from $\mathcal{D}$: S_{20} and its dual. Also, for e = 8, it follows from computations in J.W.P. Hirschfeld [80] that the only distinct GQ arising from $\mathcal{D}$ are the $2(\varphi(8)-1) = 6$ mentioned just preceeding the statement of 12.6.1.

12.7. THE OVALS OF D. GLYNN

Let $F = GF(q)$, $q = 2^e$, e odd, $e \geq 3$. Define two automorphisms $x \mapsto x^\sigma$ and $x \mapsto x^\gamma$ of F as follows :

$$\sigma = 2^{(e+1)/2}, \tag{63}$$

$$\gamma = \begin{cases} 2^n, \text{ if } e = 4n-1 \\ 2^{3n+1}, \text{ if } e = 4n+1. \end{cases} \tag{64}$$

It follows that $\gamma^2 \equiv \sigma$ and $\gamma^4 \equiv \sigma^2 \equiv 2 \pmod{q-1}$. The goal of this section is to prove the following.

12.7.1. (*D. Glynn* [65]). (i) $\sigma+\gamma \in \mathcal{D}$; (ii) $3\sigma+4 \in \mathcal{D}$.

Before beginning the proof of this result we review certain facts about F.

Let α be an automorphism of F of maximal order e, say $\alpha : x \mapsto x^{2^t}$, $(t,e) = 1$. Define $L_\alpha : F \to F$ by $L_\alpha(\xi) = \xi^\alpha+\xi$. Then L_α is an additive homomorphism of $(F,+)$ with kernel $\{0,1\}$, so that the image of L_α is a subgroup of order 2^{e-1}. Suppose $\delta \in \mathrm{Im}(L_\alpha)$, say $\xi^\alpha = \xi+\delta$. Then a finite induction shows that $\xi^{\alpha^r} = \xi+\delta+\delta^\alpha+\delta^{\alpha^2}+\ldots+\delta^{\alpha^{r-1}}$. Since $\xi^{\alpha^e} = \xi$ and α has maximal order e, there holds $0 = \delta+\delta^\alpha+\delta^{\alpha^2}+\ldots+\delta^{\alpha^{e-1}} = \sum_{i=0}^{e-1} \delta^{2^i}$. The map $\delta \mapsto \sum_{i=0}^{e-1} \delta^{2^i}$ is an additive map of F whose kernel contains the image of L_α. It is well known that such a map is never the zero map, but of course with e odd it is clear that 1 is not in the kernel. Moreover, since $\sum_{i=0}^{e-1} \delta^{2^i}$ is invariant under the map $\lambda \mapsto \lambda^2$, its value is always 0 or 1. This completes a proof of the following lemma.

12.7.2. *The elements of* F *are partitioned into two sets, an additive subgroup* C_1 *of order* 2^{e-1} *whose elements are said to be of first category, and its coset* $C_2 = 1+C_1$ *whose elements are said to be of second category. For* $\delta \in F$, *and for any automorphism* α *of maximal order*

$$\sum_{i=0}^{e-1} \delta^{2^i} = \begin{cases} 0 \text{ iff } \delta \in C_1 \text{ iff } \delta \in \mathrm{Im}(L_\alpha), \\ 1 \text{ iff } \delta \in C_2 \text{ iff } \delta \notin \mathrm{Im}(L_\alpha). \end{cases} \tag{65}$$

Moreover $C_i^\theta = C_i$, $i = 1,2$, *for any automorphism* θ *of* F .

Of course, since the kernel of L_α has order 2, each element of first category is the image under L_α of exactly two elements of F.

12.7.3. *Let* α *be an automorphism of* F *of maximal order* e. *Then*

$$x^\alpha+ax+b = 0 \text{ has } \begin{cases} \text{one solution iff } a = 0, \\ \text{two solutions iff } a \neq 0 \text{ and } b/a^{\alpha/(\alpha-1)} \in C_1, \\ \text{no solutions iff } a \neq 0 \text{ and } b/a^{\alpha/(\alpha-1)} \in C_2. \end{cases}$$

Proof. This is an easy corollary of 12.7.2. □

For an integer k, $1 \leqslant k$, put $D(k) = \{(0,1,0),(0,0,1)\} \cup \{(1,\lambda,\lambda^k) \parallel \lambda \in F\}$. Then we know that D(k) is a (q+2)-arc of PG(2,q) iff $\rho : x \mapsto x^k$ is in $\mathcal{D}$ iff $(k,q-1) = 1$ and $y \mapsto (x^k+y^k)/(x+y)$ is a bijection from $F-\{x\}$ to $F-\{0\} = F^\circ$ (for each $x \in F$) iff $(k,q-1) = (k-1,q-1) = 1$ and $t \mapsto ((1+t)^k+1)/t$ permutes the elements of F°.

We are now ready for the proof of 12.7.1.

Proof. (i) $(\sigma+\gamma)(-\gamma^{-1}+\sigma-\gamma+1) \equiv 1 \pmod{q-1}$ (use $\gamma^2 \equiv \sigma$ and $\sigma^2 \equiv 2 \pmod{q-1}$), so that $(\sigma+\gamma,q-1) = 1$. Further, $(\sigma+\gamma-1)(\sigma\gamma+\gamma-1)3^{-1} \equiv 1 \pmod{q-1}$, so that $(\sigma+\gamma-1,q-1) = 1$. Hence it remains to show that $t \mapsto ((1+t)^{\sigma+\gamma}+1)/t = t^{\sigma+\gamma-1}+t^{\sigma-1}+t^{\gamma-1} = f(t)$ permutes the elements of F°. It $f(t) = 0$ and $t \neq 0$, then $(1+t)^{\sigma+\gamma} = 1$. Since $(\sigma+\gamma,q-1) = 1$, we have $1+t = 1$ and $t = 0$, a contradiction. It follows that $f(t) = 0$ iff $t = 0$. Hence it suffices to show that $f(s) \neq f(t)$ if $st(s+t) \neq 0$.

From now on we assume $st(s+t) \neq 0$, and put $Y = st(s+t)^{-2}$. For each non-negative integer a, put $\alpha_a = (s^a+t^a)/(s+t)$ and $\beta_a = st\alpha_a(s+t)^{-(a+1)}$. Then $f(t)+f(s) \neq 0$ iff $\alpha_{\sigma+\gamma-1}+\alpha_{\sigma-1}+\alpha_{\gamma-1} \neq 0$ iff $\beta_{\sigma+\gamma-1}+\beta_{\sigma-1}(s+t)^{-\gamma}+\beta_{\gamma-1}(s+t)^{-\sigma} \neq 0$ iff $X^\gamma\beta_{\gamma-1}+X\beta_{\sigma-1}+\beta_{\sigma+\gamma-1} \neq 0$, where $X = (s+t)^{-\gamma}$, so that $(s+t)^{-\sigma} = X^\gamma$.
Hence it suffices to show that

$$X^{\gamma}\beta_{\gamma-1}+X\beta_{\sigma-1}+\beta_{\sigma+\gamma-1} \neq 0 \tag{66}$$

(for $s,t \in F$ with $st(s+t) \neq 0$).

$$\beta_0 = st\alpha_0/(s+t) = 0; \ \beta_1 = st/(s+t)^2 = Y. \tag{67}$$

If $\theta = 2^k$, $1 \leqslant k$, then

$$\beta_{\theta} = st(s+1)^{\theta-1}/(s+t)^{\theta+1} = Y. \tag{68}$$

Now notice that if $1 \leqslant r \leqslant a$, $(s^r+t^r)(s^{a-r+1}+t^{a-r+1})+st(s^{r-1}+t^{r-1}) \times (s^{a-r}+t^{a-r}) = (s+t)(s^a+t^a)$. Multiply this equation by $st(s+t)^{-(a+3)}$ to obtain

$$\beta_a = Y^{-1}\beta_r\beta_{a-r+1}+\beta_{r-1}\beta_{a-r} \text{ for } 1 \leqslant r \leqslant a. \tag{69}$$

Thus β_a is a polynomial in Y for each nonnegative integer a. With $a = 2^{m+1}-1$ $r = 2^m$, $m \geqslant 0$, and using (68) and (69), a finite induction shows

$$\beta_{2^{m+1}-1} = \sum_{i=0}^{m} Y^{2^i}. \tag{70}$$

From (70) it follows that

$$\begin{aligned} \beta_{\gamma-1} &= Y+Y^2+Y^4+\ldots+Y^{\gamma/2} \quad \text{and} \\ \beta_{\sigma-1} &= Y+Y^2+Y^4+\ldots+Y^{\sigma/2}. \end{aligned} \tag{71}$$

Also from (68) and (69) we have

$$\beta_{\sigma+\gamma-1} = Y+\beta_{\sigma-1}\beta_{\gamma-1}. \tag{72}$$

Furthermore, since $Y = st(s+t)^{-2} = s/(s+t)+(s/(s+t))^2$ is of first category, it must be that

$$\sum_{i=0}^{e-1} Y^{2^i} = 0. \tag{73}$$

Using (71) and $\gamma^2 \equiv \sigma \pmod{q-1}$ we have

$$\beta_{\gamma-1}+(\beta_{\gamma-1})^{\gamma} = \beta_{\sigma-1}. \tag{74}$$

Put $K = \beta_{\gamma-1}$, so $\beta_{\sigma-1} = K+K^{\gamma}$ and $\beta_{\sigma+\gamma-1} = Y+K^2+K^{\gamma+1}$ (by (72)). Since $K = Y+Y^2+Y^4+\ldots+Y^{\gamma/2}$ and (73) holds, it follows that $K+K^{\gamma}+K^{\gamma^2}+K^{\gamma^3} = \sum_{i=0}^{e-1} Y^{2^i}+Y = Y$. Hence from (66) we know that $\sigma+\gamma \in \mathcal{D}$ iff $X^{\gamma}K+X(K+K^{\gamma})+K+K^2+K^{\gamma}+K^{\gamma^2}+K^{\gamma^3}+K^{\gamma+1} \neq 0$, which we write as

$$X^{\gamma}K+X(K+K^{\gamma})+K+K^2+K^{\gamma}+K^{\gamma+1}+K^{\sigma}+K^{\sigma\gamma} \neq 0 \tag{75}$$

for all $X \neq 0$ and for all $K = \beta_{\gamma-1}$ $(st(s+t) \neq 0)$.
Since $st(s+t) \neq 0$ we have both $K = \beta_{\gamma-1} \neq 0$ and $\beta_{\sigma-1} \neq 0$.
Since $\beta_{\sigma-1} \neq 0$, we have $K+K^{\gamma} = \beta_{\sigma-1} \neq 0$, and hence $K^{\gamma-1} \neq 1$. Then divide (75) by K and use 12.7.3 to obtain

$$\sigma+\gamma \in \mathcal{D} \text{ iff } W = (1+K+K^{\gamma-1}+K^{\gamma}+K^{\sigma-1}+K^{\sigma\gamma-1})/(1+K^{\gamma-1})^{g+1} \in C_2 \tag{76}$$

where $g \equiv (\gamma-1)^{-1} \equiv (\sigma+1)(\gamma+1) \pmod{q-1}$, so that $\gamma/(\gamma-1) \equiv g+1 \pmod{q-1}$.

Now put $A = 1+K^{\gamma-1}$, so $K = (A+1)^g$; $K^{\sigma-1}A^{\sigma} = K^{\sigma-1}+K^{\sigma\gamma-1}$; $K^{\sigma-1} = (A+1)^{g(\sigma-1)} = (A+1)^{g/(\sigma+1)} = (A+1)^{\gamma+1}$. Then substituting into (76) we have $W = (A+KA+K^{\sigma-1}A^{\sigma})/A^{g+1} = A^{-g}(1+(A+1)^g+(A+1)^{\gamma+1}A^{\sigma-1}) = A^{-\sigma\gamma-\sigma-\gamma-1}(1+(A^{\sigma\gamma}+1)(A^{\sigma}+1)(A^{\gamma}+1)(A+1)+(A^{\gamma}+1)(A+1)A^{\sigma-1})$. After expanding and regrouping, this becomes

$$W = (1+(A^{-1}+A^{-\gamma})+(A^{-\gamma-1}+(A^{-\gamma-1})^{\sigma})+(A^{-\sigma}+(A^{-\sigma})^{\gamma})+(A^{-\sigma-1}+(A^{-\sigma-1})^{\gamma})+ (A^{-\sigma-\gamma}+(A^{-\sigma-\gamma})^{\sigma}+(A^{-\sigma-\gamma-1}+(A^{-\sigma-\gamma-1})^{\gamma})+(A^{-\sigma\gamma-\sigma-1}+(A^{-\sigma\gamma-\sigma-1})^{\gamma}). \tag{77}$$

It follows by 12.7.2 that $W \in C_2$, completing the proof of (i).

(ii) Since $(3\sigma+3,q-1) = 1$, then by 12.4.4 we have $3\sigma+4 \in \mathcal{D}$ iff $(\sigma+2)/3 = (3\sigma+4)(3\sigma+3)^{-1} \in \mathcal{D}$. Let $h \equiv (\sigma+2)/3 \pmod{q-1}$ with h a positive integer . Our strategy is to show that each line of PG(2,q) intersects D(h) in at most two points.

a) $\ell x_0+x_1 = 0$ intersects D(h) in $\{(0,0,1),(1,\ell,\ell^h)\}$.

b) $\ell x_0+x_2 = 0$ intersects D(h) in $\{(0,1,0),(1,\ell',\ell)\}$, with $\ell'^h = \ell$.

c) If $k,\ell \in F$ with $k \neq 0$, $\ell x_0+kx_1+x_2 = 0$ intersects D(h) in $\{(1,x,x^h) \parallel \ell+kx+x^h = 0\}$. Define m and y by the substitutions $x = k^{3(\sigma+1)}y$ and $m = \ell k^{-3\sigma-4}$. Then $\ell+kx+x^h = \ell+kx+x^{(\sigma+2)/3} = \ell+k^{3\sigma+4}y^3+k^{3\sigma+4}y^{\sigma+2} = k^{3\sigma+4}(m+y^3+y^{\sigma+2})$. Hence we have

$$(\sigma+2)/3 \in \mathcal{D} \text{ iff } 0 = m+y^3+y^{\sigma+2} \text{ has at most two solutions } y. \tag{78}$$

We may suppose (78) has at least two distinct solutions $\alpha,\beta \in F$. Then $(y+\alpha)(y+\beta) = y^2+(\alpha+\beta)y+\alpha\beta = 0$ for $y \in \{\alpha,\beta\}$. Consequently $y^4 = (\alpha+\beta)^2y^2+\alpha^2\beta^2 = (\alpha+\beta)^2((\alpha+\beta)y+\alpha\beta)+\alpha^2\beta^2 = (\alpha+\beta)^3y+\alpha\beta(\alpha+\beta)^2+\alpha^2\beta^2$. Proceeding in this way we obtain

$$y^\sigma = ay+b \tag{79}$$

for some a,b (functions of α and β) if $y \in \{\alpha,\beta\}$. Hence $y^{\sigma^2} = y^2 = a^\sigma y^\sigma+b^\sigma = a^\sigma(ay+b)+b^\sigma = a^{\sigma+1}y+a^\sigma b+b^\sigma$, and then $\alpha+\beta = a^{\sigma+1}$ and $\alpha\beta = a^\sigma b+b^\sigma$. Now substitute $y^\sigma = ay+b$ and $y^2 = a^{\sigma+1}y+a^\sigma b+b^\sigma$ into (78) to obtain (after some simplification) :

$$\begin{aligned} 0 = y^2(y^\sigma+y)+m &= (a^{2\sigma+2}(a+1)+(a^\sigma b+b^\sigma)(a+1)+a^{\sigma+1}b)y+ \\ &(a^\sigma b+b^\sigma)(a^{\sigma+1}(a+1)+b)+m, \end{aligned} \tag{80}$$

for $y \in \{\alpha,\beta\}$. But since the equation in (80) is linear and has two distinct roots, it must be trivial. In particular

$$a^{2\sigma+2}(a+1) = b^\sigma(a+1)+ba^\sigma. \tag{81}$$

If $a = 0$, then $y^\sigma = ay+b$ has only one solution, an impossibility. So $a \neq 0$. Multiply (81) by $(a+1)^{\sigma+1}/a^{2\sigma+2}$ to obtain

$$(a+1)^{\sigma+2} = \left(\frac{(a+1)^{\sigma+1}b}{a^{\sigma+2}}\right) + \left(\frac{(a+1)^{\sigma+1}b}{a^{\sigma+2}}\right)^\sigma \in C_1. \tag{82}$$

Then $a^{\sigma+2} = (a^\sigma+1)(a^2+1)+a^\sigma+a^2+1 = (a+1)^{\sigma+2}+(a^{\sigma/2}+a)^2+1 \in C_1+C_1+C_2 = C_2$. Hence

$$\alpha+\beta = a^{\sigma+1} = (a^{\sigma+2})^{\sigma/2} \in C_2. \tag{83}$$

The essence of (83) is that the sum of any two roots of (78) must be in C_2. Hence if there were a third root ρ, it would follow that each of $\alpha+\beta$, $\alpha+\rho, \beta+\rho$ would be in C_2. But then their sum, which is zero, would also be in C_2, a blatant impossibility. This shows that the equation in (78) has at most two solutions and completes the proof that $3\sigma+4 \in \mathcal{D}$. □

13 Generalizations and related topics

13.1. PARTIAL GEOMETRIES, PARTIAL QUADRANGLES AND SEMI PARTIAL GEOMETRIES

A (finite) *partial geometry* is an incidence structure $S = (P,B,I)$ in which P and B are disjoint (nonempty) sets of objects called points and lines, respectively, and for which I is a symmetric point-line incidence relation satisfying the following axioms :

(i) Each point is incident with $1+t$ ($t \geq 1$) lines and two distinct points are incident with at most one line.

(ii) Each line is incident with $1+s$ ($s \geq 1$) points and two distinct lines are incident with at most one point.

(iii) If x is a point and L is a line not incident with x, then there are exactly α ($\alpha \geq 1$) points $x_1,x_2,\ldots,x_\alpha$ and α lines $L_1,L_2,\ldots,L_\alpha$ such that $x \mathrel{I} L_i \mathrel{I} x_i \mathrel{I} L$, $i = 1,2,\ldots,\alpha$.

Partial geometries were introduced by R.C. Bose [16]. Clearly the partial geometries with $\alpha = 1$ are the generalized quadrangles.

It is easy to prove that $|P| = v = (1+s)(st+\alpha)/\alpha$ and $|B| = b = (1+t)(st+\alpha)/\alpha$. Further, the following hold : $\alpha(s+t+1-\alpha) \mid st(s+1)(t+1)$ [17,77], $(t+1-2\alpha)s \leq (t+1-\alpha)^2(t-1)$ [34], and dually $(s+1-2\alpha)t \leq (s+1-\alpha)^2(s-1)$

For a survey on the subject we refer to F. De Clerck [44,45], J.A. Thas [193], and A.E. Brouwer and J.H. van Lint [22].

A (finite) *partial quadrangle* is an incidence structure $S = (P,B,I)$ of points and lines satisfying (i) and (ii) above and also :

(iii)' If x is a point and L is a line not incident with x, then there is at most one pair $(y,M) \in P \times B$ for which $x \mathrel{I} M \mathrel{I} y \mathrel{I} L$.

(iv)' If two points are not collinear, then there are exactly μ ($\mu > 0$) points collinear with both.

Partial quadrangles were introduced and studied by P.J. Cameron [31]. A partial quadrangle is a generalized quadrangle iff $\mu = t+1$.

We have $|P| = v = 1+(t+1)s(1+st/\mu)$, and $v(t+1) = b(s+1)$ with $|B| = b$ [31]. The following hold : $\mu \leq t+1$, $\mu \mid s^2t(t+1)$, and $b \geq v$ if $\mu \neq t+1$ [31]. Moreover $D = (s-1-\mu)^2+4((t+1)s-\mu)$ is a square (except in the case

$\mu = s = t = 1$, where $D = 5$ (and then S is a pentagon)) and $((t+1)s+(v-1)(s-1-\mu+\sqrt{D})/2)/\sqrt{D}$ is an integer [31].

A (finite) *semi partial geometry* is an incidence structure $S = (P,B,I)$ of points and lines satisfying (i) and (ii) above and also satisfying :

(iii)" If x is a point and L is a line not incident with x, then there are 0 or α ($\alpha \geq 1$) points which are collinear with x and incident with L.

(iv)" If two points are not collinear, then there are μ ($\mu > 0$) points collinear with both.

Semi partial geometries were introduced by I. Debroey and J.A. Thas [43]. A semi partial geometry is a partial geometry iff $\mu = (t+1)\alpha$; it is a generalized quadrangle iff $\alpha = 1$ and $\mu = t+1$.

We have $|P| = v = 1+(t+1)s(1+t(s-\alpha+1)/\mu)$, and $v(t+1) = b(s+1)$ with $|B| = b$ [43]. The following also hold : $\alpha^2 \leq \mu \leq (t+1)\alpha$, $(s+1) \mid t(t+1)(\alpha t+\alpha-\mu)$, $\mu \mid (t+1)st(s+1-\alpha)$, $\alpha \mid st(t+1)$, $\alpha \mid st(s+1)$, $\alpha \mid \mu$, $\alpha^2 \mid \mu st$, $\alpha^2 \mid t((t+1)\alpha-\mu)$, and $b \geq v$ if $\mu \neq (t+1)\alpha$ [43]. Moreover, $D = (t(\alpha-1)+s-1-\mu)^2+4((t+1)s-\mu)$ is a square (except in the case $\mu = s = t = \alpha = 1$ where $D = 5$ (here S is a pentagon)) and $((t+1)s+(v-1)(t(\alpha-1)+s-1-\mu+\sqrt{D})/2)/\sqrt{D}$ is an integer [43].

For a survey on the subject we refer to I. Debroey [41,42], and I. Debroey and J.A. Thas [43].

If we write " → " for "generalizes to", then we have the following scheme :

generalized quadrangle	→	partial geometry
↓		↓
partial quadrangle	→	semi partial geometry

13.2. PARTIAL 3-SPACES

Partial 3-spaces (involving points , lines and planes) have been defined as follows by R. Laskar and J. Dunbar [93].

A *partial 3-space* S is a system of points , lines and planes, together with an incidence relation for which the following conditions are satisfied :

(i) If a point p is incident with a line L, and L is incident with a plane π, then p is incident with π.

(ii) (a) A pair of distinct planes is incident with at most one line.

(b) A pair of distinct planes not incident with a line is incident with at most one point.

(iii) The set of points and lines incident with a plane forms a partial geometry with parameters s,t and α.

(iv) The set of lines and planes incident with a point p forms a partial geometry with parameters s^*, t and α^*, where the points and lines of the geometry are the planes and lines through p, respectively, and incidence is that of S.

(v) Given a plane π and a line L not incident with π, π and L not intersecting in a point, there exist exactly u planes through L intersecting π in a line and exactly w-u planes through L intersecting π in a point but not in a line.

(vi) Given a point p and a line L, p and L not incident with a common plane, there exist exactly u^* points on L which are collinear with p, and w^*-u^* points on L coplanar but not collinear with p.

(vii) Given a point p and a plane π not containing p, there exist exactly x planes through p intersecting π in a line.

Many properties of S are deduced in R. Laskar and J. Dunbar [93], and several examples are described in R. Laskar and J.A. Thas [94]. In J.A. Thas [202] all partial 3-spaces for which the lines are lines of PG(n,q), for which the points are all the (projective) points on these projective lines, and for which the incidence of points and lines is that of PG(n,q), are determined. Among these "embeddable" partial 3-spaces there are several examples for which the partial geometries of axiom (iii) (resp., axiom (iv)) are classical generalized quadrangles.

13.3. PARTIAL GEOMETRIC DESIGNS

A "non-linear" generalization of partial geometries is due to R.C. Bose, S.S. Shrikhande and N.M. Singhi [20].

A (finite) *partial geometric design* is an incidence structure $S = (P,B,I)$ of points and blocks for which the following properties are satisfied :

(i) Each point is incident with 1+t $(t \geq 1)$ blocks, and each block is incident with 1+s $(s \geq 1)$ points.

(ii) For each given point-block pair (x,L), x $\not{I}$ L (resp., x I L), we have $\sum_{y I L} [x,y] = \alpha$ (resp., β), where $[x,y]$ denotes the number of blocks incident with x and y.

For the structure S we also use the notation $D(s,t,\alpha,\beta)$. A $D(s,t,\alpha,s+t+1)$

is just a partial geometry; a D(s,t,1,s+t+1) is just a generalized quadrangle.

13.4. GENERALIZED POLYGONS

Let S = (P,B,I) be an arbitrary incidence structure of points and blocks. A *chain* in S is a finite sequence $X = (x_0,\dots,x_h)$ of elements in $P \cup B$ such that x_{i-1} I x_i for $i = 1,\dots,h$. The integer h is the *length* of the chain, and the chain X is said to *join* the elements x_0 and x_h of S. If S is connected, in the obvious sense that any two of its elements can be joined by some chain, then $d(x,y) = \min\{h \parallel \text{some chain of length } h \text{ joins } x \text{ and } y\}$ is a well-defined positive integer for all distinct $x,y \in P \cup B$. Put $d(x,x) = 0$ for any element x of S.

We now define a *generalized* n-*gon*, $n \geq 3$, as a connected incidence structure S = (P,B,I) satisfying the following conditions :

(i) $d(x,y) \leq n$ for all $x,y \in P \cup B$.

(ii) If $d(x,y) = h < n$, there is a unique chain of length h joining x and y.

(iii) For each $x \in P \cup B$ there is a $y \in P \cup B$ such that $d(x,y) = n$.

Generalized n-gons were introduced by J. Tits [216] in 1959, in connection with certain group theoretical problems.

A *generalized polygon* is an incidence structure which is a generalized n-gon for some integer n. Clearly any two distinct points (resp., blocks) of a generalized polygon are incident with at most one block (resp., point). From now on the blocks of a generalized n-gon will be called *lines*. A finite generalized n-gon has order (s,t), $s \geq 1$ and $t \geq 1$, if there are exactly s+1 points incident with each line and exactly t+1 lines incident with each point. A generalized polygon of order (s,t) is called *thick* if $s > 1$ and $t > 1$. Notice that the generalized n-gons of order (1,1) are just the polygons with n vertices and n sides in the usual sense.

If S is a generalized n-gon of order (s,t), then by a celebrated theorem of W. Feit and G. Higman [57,91,153] we have (s,t) = (1,1) or $n \in \{3,4,6,8,12\}$. Further, they prove that there are no thick generalized 12-gons of order (s,t) and they show that if a thick generalized n-gon of order (s,t) exists, then 2st is a square if n = 8 and st is a square if n = 6.

The thick generalized 3-gons of order (s,t) have s = t and are just the projective planes of order s. The generalized 4-gons of order (s,t) are just the generalized quadrangles of order (s,t).

In [67] W. Haemers and C. Roos prove that $s \leqslant t^3 \leqslant s^9$ for thick generalized 6-gons of order (s,t), and in [78] D.G. Higman shows that $s \leqslant t^2 \leqslant s^4$ for thick generalized 8-gons of order (s,t).

For more information about generalized polygons we refer to P. Dembowski [49], W. Feit and G. Higman [57], W. Haemers [66], M.A. Ronan [148,149,151, 152], J. Tits [216,220,221,222,224], A. Yanushka [236,237].

13.5. POLAR SPACES AND SHULT SPACES

A *polar space of rank* n, $n \geqslant 2$, is a pointset P together with a family of subsets of P called *subspaces*, satisfying :

(i) A subspace, together with the subspaces it contains, is a d-dimensional projective space with $-1 \leqslant d \leqslant n-1$ (d is called the *dimension* of the subspace).

(ii) The intersection of two subspaces is a subspace.

(iii) Given a subspace V of dimension n-1 and a point $p \in P-V$, there is a unique subspace W such that $p \in W$ and $V \cap W$ has dimension n-2; W contains all points of V that are joined to p by a line (a *line* is a subspace of dimension 1).

(iv) There exist two disjoint subspaces of dimension n-1.

This definition is due to J. Tits [219]. Notice that the polar spaces of rank 2 which are not grids or dual grids (cf. 1.1) are just the generalized quadrangles of order (s,t) with $s > 1$ and $t > 1$.

By a deep theorem due to F.D. Veldkamp [226,227,228] and J. Tits [219] all polar spaces of finite rank $\geqslant 3$ have been classified. In particular, if P is finite, then the subspaces of the polar space (of rank $\geqslant 3$) are just the totally isotropic subspaces [49] with respect to a polarity of a finite projective space, or the projective spaces on a nonsingular quadric of a finite projective space.

In [30] F. Buekenhout and E.E. Shult reformulate the polar space axioms in terms of points and lines. Let P be a pointset from which distinguished subsets of cardinality $\geqslant 2$ are called lines (we assume that the lineset is nonempty). Then P together with its lines is a *Shult space* if and only if for each line L of P and each point $p \in P-L$, the point p is collinear with either one or all points of L.

A Shult space is *nondegenerate* if no point is collinear with all other points, and is *linear* if two distinct lines have at most one common point. A *subspace* X of the Shult space is a nonempty set of pairwise collinear points such that any line meeting X in more than one point is contained in X. If there exists an integer n such that every chain of distinct subspaces $X_1 \subset X_2 \subset \ldots \subset X_\ell$ has at most n members, then S is a finite rank.

F. Buekenhout and E.E. Shult [30] prove the following fundamental theorem :

(a) *Every nondegenerate Shult space is linear.*

(b) *If* P *together with its lines is a nondegenerate Shult space of finite rank, and if all lines contain at least three points, then the Shult space together with its subspaces is a polar space.*

13.6. PSEUDO-GEOMETRIC AND GEOMETRIC GRAPHS

A *graph* consists of a finite set of vertices together with a set of edges, where each edge is a subset of order 2 of the vertex set. The two elements of an edge are called *adjacent*. A graph is *complete* if every two vertices are adjacent, and *null* if it has no edges at all. If p is a vertex of a graph Γ,the *valency* of p is the number of edges containing p, i.e. the number of vertices adjacent to p. If every vertex has the same valency, the graph is called *regular*, and the common valency is the *valency* of the graph. A *strongly regular* graph is a graph which is regular, but not complete or null, and which has the property that the number of vertices adjacent to p_1 and p_2 ($p_1 \neq p_2$) depends only on whether or not p_1 and p_2 are adjacent. Its parameters are v,k,λ,μ, where v is the number of vertices, k is the valency, and λ (resp., μ) is the number of vertices adjacent to two adjacent (resp., nonadjacent) vertices.

Let $S = (P,B,I)$ be a partial geometry (cf. 13.1) with parameters s , t and α. Then a graph is defined as follows : vertices are the points of S and two vertices are adjacent if they are collinear as points of S. This graph is called the *point graph* of the partial geometry. Clearly, for $\alpha \neq s+1$, this point graph is strongly regular with parameters $v = |P| = (s+1)(st+\alpha)/\alpha$, $k = s(t+1)$, $\lambda = s-1+(\alpha-1)t$ and $\mu = (t+1)\alpha$. For a generalized quadrangle, $v = (s+1)(st+1)$, $k = s(t+1)$, $\lambda = s-1$, and $\mu = t+1$. The point graph of a partial geometry is called a *geometric graph*, and a strongly regular graph which has the parameters of a geometric graph is called a *pseudo-geometric graph* [17]. An interesting but difficult problem is the following : for which values of s,t,α are pseudo-geometric graphs always geometric ?

In this context we mention the following important theorem due to P.J. Cameron, J.-M. Goethals and J.J. Seidel [34] (see also W. Haemers [66], p. 61).

Every pseudo-geometric graph with parameters $v = (q+1)(q^3+1)$, $k = q(q^2+1)$, $\lambda = q-1$ *and* $\mu = q^2+1$, *is geometric, i.e. it is the point graph of a generalized quadrangle of order* (q,q^2).

Bibliography

1 R.W. Ahrens and G. Szekeres, On a combinatorial generalization of 27 lines associated with a cubic surface, *J. Austral. Math. Soc.* 10 (1969) 485-492.

2 R. Artzy, A Pascal theorem applied to Minkowski geometry, *J. Geom.* 3 (1973) 93-102.

3 R. Artzy, Addendum to "A Pascal theorem applied to Minkowski geometry", *J. Geom.* 3 (1973) 103-105.

4 H.F. Baker, Principles of geometry, 6 vol.'s, Cambridge Univ. Press 1921-1934.

5 A. Barlotti, Un'estensione del teorema di Segre-Kustaanheimo, *Boll. Un. Mat. Ital.* (3) 10 (1955) 498-506.

6 A. Barlotti, Some topics in finite geometrical structures, Univ. of North Carolina at Chapel Hill, Institute of Stat. Mimeo Series no. 439, 1965.

7 L.M. Batten, An introduction to synthetic geometry, Monograph, preprint.

8 L.M. Batten and F. Buekenhout, Quadriques partielles d'indice deux, *J. Geom.* 16 (1981) 93-102.

9 C.T. Benson, Minimal regular graphs of girths eight and twelve, *Canad. J. Math.* 18 (1966) 1091-1094.

10 C.T. Benson, On the structure of generalized quadrangles, *J. Algebra* 15 (1970) 443-454.

11 W. Benz, Vorlesungen über Geometrie der Algebren, Springer Verlag 1973.

12 A. Bichara, Caratterizzazione dei sistemi rigati immersi in $A_{3,q}$, *Riv. Mat. Univ. Parma* (4) 4 (1978) 277-290.

13 A. Bichara, Sistemi rigati immersi in uno spazio combinatorio, *Sem. Geom. Comb. Univ. Roma* 6 (1978).

14 A. Bichara, F. Mazzocca and C. Somma, Sulla classificazione dei sistemi rigati immersi in AG(3,q), *Sem. Geom. Comb. Univ. Roma* 9 (1978).

15 A. Bichara, F. Mazzocca and C. Somma, On the classification of generalized quadrangles in a finite affine space $AG(3,2^h)$, *Boll. Un. Mat. Ital.* (5) 17B (1980) 298-307.

16 R.C. Bose, Strongly regular graphs, partial geometries and partially balanced designs, *Pacific J. Math.* 13 (1963) 389-419.

17 R.C. Bose, Graphs and designs, in : *Finite geometric structures and their applications* (ed. A. Barlotti), Ed. Cremonese Roma (1973) 1-104.

18 R.C. Bose, W.G. Bridges and M.S. Shrikhande, A characterization of partial geometric designs, *Discrete Math.* (1976) 1-7.

19 R.C. Bose and S.S. Shrikhande, Geometric and pseudo-geometric graphs $(q^2+1,q+1,1)$ *J. Geom.* 2/1 (1972) 75-94.

20 R.C. Bose, S.S. Shrikhande and N.M. Singhi, Edge regular multigraphs and partial geometric designs with an application to the embedding of quasi-residual designs, in : *Teorie combinatorie* (ed. B. Segre), Roma Accad. Naz. Lincei (1976) 49-81.

21 A.E. Brouwer, Private communication 1981.

22 A.E. Brouwer and J.H. van Lint, Strongly regular graphs and partial geometries, *Proc. "Silver Jubilee" Conf.*, Waterloo 1982, to appear.

23 A.E. Brouwer and H.A. Wilbrink, The structure of near polygons with quads, *Geom. Dedicata* 14 (1983) 145- 176.

24 A.A. Bruen and J.W.P. Hirschfeld, Applications of line geometry over finite fields. II. The Hermitian surface, *Geom. Dedicata* 7 (1978) 333-353.

25 A.A. Bruen and J.A. Thas, Partial spreads, packings and Hermitian manifolds, *Math. Z.* 151 (1976) 207-214.

26 F. Buekenhout, Une caractérisation des espaces affins basée sur la notion de droite, *Math. Z.* 111 (1969) 367-371.

27 F. Buekenhout, Characterizations of semi quadrics : A survey, in : *Teorie combinatorie* (ed. B. Segre), Roma Accad. Naz. Lincei (1976) 393-421.

28 F. Buekenhout and X. Hubaut, Locally polar spaces and related rank 3 groups, *J. Algebra* 45 (1977) 393-434.

29 F. Buekenhout and C. Lefèvre, Generalized quadrangles in projective spaces, *Arch. Math.* 25 (1974) 540-552.

30 F. Buekenhout and E.E. Shult, On the foundations of polar geometry, *Geom. Dedicata* 3 (1974) 155-170.

31 P.J. Cameron, Partial quadrangles, *Quart. J. Math. Oxford* (3) 25 (1974) 1-13.

32 P.J. Cameron, Dual polar spaces, *Geom. Dedicata* 12 (1982) 75-85.

33 P.J. Cameron, P. Delsarte and J.-M. Goethals, Hemisystems, orthogonal configurations, and dissipative conference matrices, *Philips J. Res.* 34 (1979) 147-162.

34 P.J. Cameron, J.-M. Goethals and J.J. Seidel, Strongly regular graphs having strongly regular subconstituents, *J. Algebra* 55 (1978) 257-280.

35 P.J. Cameron and W.M. Kantor, 2-transitive and antiflag transitive collineation groups of finite projective spaces, *J. Algebra* 60 (1979) 384-422.

36 P.J. Cameron, J.A. Thas and S.E. Payne, Polarities of generalized hexagons and perfect codes, *Geom. Dedicata* 5 (1976) 525-528.

37 L.R.A. Casse, J.A. Thas and P.R. Wild, (q^n+1)-sets of PG(3n-1,q), generalized quadrangles, and Laguerre planes, *Simon Stevin*, to appear.

38 Y. Chen, A characterization of some geometries of chains, *Canad. J. Math.* 26 (1974) 257-272.

39 Y. Chen and G. Kaerlein, Eine Bemerkung über endliche Laguerre- und Minkowski-Ebenen, *Geom. Dedicata* 2 (1973) 193-194.

40 B. Cooperstein and M. Walker, The non-existence of cousins for Kantor's quadrangle and a characterization of the Frobenius automorphism of $GF(2^n)$, preprint.

41 I. Debroey, Semi partiële meetkunden, Dissertation Ph.D., Univ. of Ghent 1978.

42 I. Debroey, Semi partial geometries, *Bull. Soc. Math. Belg.* (B) 31 (1979) 183-190.

43 I. Debroey and J.A. Thas, On semipartial geometries, *J. Combin. Theory* (A) 25 (1978) 242-250.

44 F. De Clerck, Een kombinatorische studie van de eindige partiële meetkunden, Dissertation Ph.D., Univ. of Ghent 1978.

45 F. De Clerck, Partial geometries - a combinatorial survey, *Bull. Soc. Math. Belg.* (B) 31 (1979) 135-145.

46 F. De Clerck and J.A. Thas, Partial geometries in finite projective spaces, *Arch. Math.* 30 (1978) 537-540.

47 P. Dembowski, Inversive planes of even order, *Bull. Amer. Math. Soc.* 69 (1963) 850-854.

48 P. Dembowski, Möbiusebenen gerader Ordnung, *Math. Ann.* 157 (1964) 179-205.

49 P. Dembowski, Finite geometries, Springer Verlag 1968.

50 M. De Soete and J.A. Thas, A characterization theorem for the generalized quadrangle $T_2^*(O)$ of order (s,s+2), *Ars Combin.*, to appear.
51 K.J. Dienst, Verallgemeinerte Vierecke in Pappusschen projektiven Räumen, *Geom. Dedicata* 9 (1980) 199-206.
52 K.J. Dienst, Verallgemeinerte Vierecke in projektiven Räumen, *Arch. Math* 35 (1980) 177-186.
53 S. Dixmier and F. Zara, Etude d'un quadrangle généralisé autour de deux de ses points non liés, preprint 1976.
54 S. Dixmier and F. Zara, Essai d'une méthode d'étude de certains graphes liés aux groupes classiques, *C.R. Acad. Sci. Paris Sér.* A 282 (1976) 259-262.
55 C.E. Ealy, Jr., Generalized quadrangles and odd transpositions, Dissertation Ph. D. , The University of Chicago (1977).
56 M.M. Eich and S.E. Payne, Nonisomorphic symmetric block designs derived from generalized quadrangles, *Rend. Accad. Naz. Lincei* 52 (1972) 893-902.
57 W. Feit and G. Higman, The nonexistence of certain generalized polygons, *J. Algebra* 1 (1964) 114-131.
58 J.C. Fisher and J.A. Thas, Flocks in PG(3,q), *Math. Z.* 169 (1979) 1-11.
59 P. Fong and G. Seitz, Groups with a (B,N)-pair of rank 2, I, *Invent. Math* 21 (1973) 1-57.
60 P. Fong and G. Seitz, Groups with a (B,N)-pair of rank 2, II, *Invent. Math.* 24 (1974) 191-239.
61 M. Forst, Topologische 4-gone, Dissertation Ph.D. , Christian-Albrechts-Universität Kiel (1979).
62 H. Freudenthal, Une interprétation géométrique des automorphismes extérieurs du groupe symétrique, *Rend. Sem. Mat. Fis. Milano* 42 (1972) 47-56.
63 H. Freudenthal, Une étude de quelques quadrangles généralisés, *Ann. Mat . Pura Appl.* (4) 102 (1975) 109-133.
64 D.G. Glynn, Finite projective planes and related combinatorial systems, Dissertation Ph. D. , University of Adelaide 1978.
65 D.G. Glynn, Two new sequences of ovals in finite Desarguesian planes, *Proc. 10th Conf. ACM,* Adelaide 1982, to appear.
66 W. Haemers, Eigenvalue techniques in design and graph theory, Dissertation Ph. D. , Technological University of Eindhoven 1979.

67 W. Haemers and C. Roos, An inequality for generalized hexagons, *Geom. Dedicata* 10 (1981) 219-222.

68 H.R. Halder and W. Heise, Einführung in die Kombinatorik, Carl Hanser Verlag 1976.

69 M. Hall Jr., Projective planes, *Trans. Amer. Math. Soc.* 54 (1943) 229-277.

70 M. Hall Jr., Affine generalized quadrilaterals, *Studies in Pure Math.* (ed. L. Mirsky), Academic Press (1971) 113-116.

71 M. Hall Jr., Ovals in the Desarguesian plane of order 16, *Ann. Mat. Pura Appl.* (4) 102 (1975) 159-176.

72 W. Heise, Minkowski-Ebenen gerader Ordnung, *J. Geom.* 5 (1974) 83.

73 W. Heise and H. Karzel, Symmetrische Minkowski-Ebenen, *J. Geom.* 3 (1973) 5-20.

74 C. Hering, W.M. Kantor and G. Seitz, Finite groups with a split BN-pair of rank 1, I , *J. Algebra* 20 (1972) 435-475.

75 C. Hering, W.M. Kantor and G. Seitz, Finite groups with a split BN-pair of rank 1, II, *J. Algebra* 20 (1972) 476-494.

76 M.D. Hestenes and D.G. Higman, Rank 3 permutation groups and strongly regular graphs, in : *Computers in algebra and number theory, SIAM-AMS Proc.*, Vol. 4, Providence, R.I. (1971).

77 D.G. Higman, Partial geometries, generalized quadrangles and strongly regular graphs, in : *Atti convegno di geometria e sue applicazioni* (ed. A. Barlotti), Perugia (1971) 263-293.

78 D.G. Higman, Invariant relations, coherent configurations and generalized polygons, in : *Combinatorics, part 3* (eds. M. Hall Jr. and J.H. van Lint), Math. Centre Tracts 57 (1974) 27-43.

79 J.W.P. Hirschfeld, Ovals in Desarguesian planes of even order, *Ann. Mat. Pura Appl.* 102 (1975) 79-89.

80 J.W.P. Hirschfeld, Projective geometries over finite fields, Clarendon Press Oxford 1979.

81 J.W.P. Hirschfeld and X. Hubaut, Sets of even type in PG(3,4), alias the binary (85,24) projective geometry code, *J. Combin. Theory* (A) 29 (1980) 101-112.

82 J.W.P. Hirschfeld and J.A. Thas, Sets of type (1,n,q+1) in PG(d,q), *Proc. London Math. Soc.* 41 (1980) 254-278.

83 J.W.P. Hirschfeld and J.A. Thas, The characterization of projections of quadrics over finite fields of even order, *J. London Math. Soc.* 22 (1980) 226-238.

84 J.W.P. Hirschfeld and J.A. Thas, The generalized hexagon H(q) and the associated generalized quadrangle K(q), *Geom. Dedicata,* preprint.

85 S.G. Hoggar, A complex polytope as generalized quadrangle, preprint.

86 D.R. Hughes and F.C. Piper, Projective planes, Springer Verlag 1973.

87 N. Jacobson, Basic algebra II, W.H. Freeman and Company 1980.

88 W.M. Kantor, Generalized quadrangles associated with $G_2(q)$, *J. Combin. Theory* (A) 29 (1980) 212-219.

89 W.M. Kantor, Span-symmetric generalized quadrangles, Private communication.

90 W.M. Kantor, Ovoids and translation planes, *Canad. J. Math.* 34 (1982) 1195-1207.

91 R. Kilmoyer and L. Solomon, On the theorem of Feit-Higman, *J. Combin. Theory* (A) 15 (1973) 310-322.

92 S. Lang, Algebra, Addison-Wesley Publishing Company 1965.

93 R. Laskar and J. Dunbar, Partial geometry of dimension three, *J. Combin. Theory* (A) 24 (1978) 187-200.

94 R. Laskar and J.A. Thas, On some generalizations of partial geometry, in : *Graph theory and related topics* (eds. J.A. Bondy and U.S.R. Murty), Academic Press (1979) 277-287.

95 C. Lefèvre, Semi-quadriques et sous-ensembles des espaces projectifs, Dissertation Ph.D., University of Brussels 1976.

96 C. Lefèvre-Percsy, Polar spaces embedded in a projective space, in : *Finite geometries and designs* (eds. P.J. Cameron, J.W.P. Hirschfeld, D.R. Hughes), Cambridge Univ. Press, London Math. Soc. Lecture Note Ser. 49 (1980) 216-220.

97 C. Lefèvre-Percsy, Espaces polaires faiblement plongés dans un espace projectif, *J. Geom.* 16 (1982) 126-137.

98 C. Lefèvre-Percsy, Quadrilatères généralisés faiblement plongés dans PG(3,q), *European J. Combin.* 2 (1981) 249-255.

99 H. Lenz, Zur Begründing der analytischen Geometrie, *Sitzungsber. Bayer. Akad. Wiss. Math.-Natur. Kl.* (1954) 17-72.

100 H. Lüneburg, Die Suzukigruppen und ihre Geometrien, Springer Verlag, Lecture Notes in Math. no. 10 (1965).

101 M. Masuyama, A note on partial geometric association scheme, *Rep. Statist. Appl. Res. Un. Japan Sci. Engrs.* 17 (1970) 53-56 and 44.

102 F. Mazzocca, Sistemi grafici rigati di seconda specie, *Ist. Mat. Univ. Napoli Rel.* n.28 (1973) 3-21.

103 F. Mazzocca, Caratterizzazione dei sistemi rigati isomorfi ad una quadrica ellittica dello $S_{5,q}$, con q dispari, *Rend. Accad. Naz. Lincei* 57 (1974) 360-368.

104 F. Mazzocca, Immergibilità in $S_{4,q}$ di certi sistemi rigati di seconda specie, *Rend. Accad. Naz. Lincei* 56 (1974) 189-196.

105 F. Mazzocca, Sistemi rigati di seconda specie U-regolari, *Ist. Mat. Univ. Napoli Rel.* n.32 (1974).

106 F. Mazzocca and D. Olanda, Alcune caratterizzazioni dei sistemi rigati di prima specie, *Ricerche Mat.* 28 (1979) 101-108.

107 F. Mazzocca and D. Olanda, Sistemi rigati in spazi combinatori, *Rend. Mat.* (2) 12 (1979) 221-229.

108 U. Melchior, Homologien von verallgemeinerten Vierecken, in : *Beiträge zur Geometrischen Algebra*, Birkhäuser, Lehrbücher Monograph. Geb. Exakten Wissensch., Math. Reihe 21 (1977) 265-268.

109 D. Olanda, Quadragoni di Tits e sistemi rigati, *Rend. Accad. Sci. Fis. Mat. Napoli* (4) 39 (1972) 81-87.

110 D. Olanda, Sistemi rigati immersi in uno spazio proiettivo, *Ist. Mat. Univ. Napoli Rel.* n. 26 (1973) 1-21.

111 D. Olanda, Sistemi rigati immersi in uno spazio proiettivo, *Rend. Accad. Naz. Lincei* 62 (1977) 489-499.

112 U. Ott, Some remarks on representation theory in finite geometry, in : *Geometries and groups* (eds. M. Aigner and D. Jungnickel), Springer Verlag, Lecture Notes in Math. n. 893 (1981) 68-110.

113 U. Ott, Eine Bemerkung über Polaritäten eines verallgemeinerten Hexagons, *Geom. Dedicata* 11 (1981) 341-345.

114 U. Ott, Private communication 1982.

115 A. Pasini, Nonexistence of proper epimorphisms of finite thick generalized polygons, preprint.

116 S.E. Payne, Symmetric representations of nondegenerate generalized n-gons, *Proc. Amer. Math. Soc.* 19 (1968) 1321-1326.

117 S.E. Payne, Collineations of affinely represented generalized quadrangles, *J. Algebra* 16 (1970) 496-508.

118 S.E. Payne, Affine representations of generalized quadrangles, *J. Algebra* 16 (1970) 473-485.

119 S.E. Payne, Nonisomorphic generalized quadrangles, *J. Algebra* 18 (1971) 201-212.

120 S.E. Payne, The equivalence of certain generalized quadrangles, *J. Combin. Theory* 10 (1971) 284-289.

121 S.E. Payne, A complete determination of translation ovoids in finite Desarguesian planes, *Rend. Accad. Naz. Lincei* 51 (1971) 328-331.

122 S.E. Payne, On the non-existence of a class of configurations which are nearly generalized n-gons, *J. Combin. Theory* (A) 12 (1972) 268-282.

123 S.E. Payne, Quadrangles of order (s-1,s+1), *J. Algebra* 22 (1972) 97-119.

124 S.E. Payne, Generalized quadrangles as amalgamations of projective planes, *J. Algebra* 22 (1972) 120-136.

125 S.E. Payne, Finite generalized quadrangles : a survey, *Proc. Intern. Conf. Proj. Planes,* Washington State Univ. (1973) 219-261.

126 S.E. Payne, A restriction on the parameters of a subquadrangle, *Bull. Amer. Math. Soc.* 79 (1973) 747-748.

127 S.E. Payne, Generalized quadrangles of even order, *J. Algebra* 31 (1974) 367-391.

128 S.E. Payne, Skew-translation generalized quadrangles, *Congress. Numer.* XIV, *Proc. 6th S.E. Conf. Comb., Graph Th. and Comp.*(1975) 485-504.

129 S.E. Payne, All generalized quadrangles of order 3 are known, *J. Combin. Theory* 18 (1975) 203-206.

130 S.E. Payne, Translation generalized quadrangles of order eight, *Congress. Numer.* XVII, *Proc. 7th S.E. Conf. Comb., Graph Th. and Comp.* (1976) 469-474.

131 S.E. Payne, Generalized quadrangles with symmetry II, *Simon Stevin* 50 (1977) 209-245.

132 S.E. Payne, Generalized quadrangles of order 4, I, *J. Combin. Theory* 22 (1977) 267-279.

133 S.E. Payne, Generalized quadrangles of order 4, II, *J. Combin. Theory* 22 (1977) 280-288.

134 S.E. Payne, An inequality for generalized quadrangles, *Proc. Amer. Math. Soc.* 71 (1978) 147-152.

135 S.E. Payne, Generalized quadrangles as group coset geometries, *Congress. Numer.* XXIX, *Proc. 12th S.E. Conf. Comb., Graph Th. and Comp.* (1980) 717-734.

136 S.E. Payne, Span-symmetric generalized quadrangles, in : *The Geometric Vein* (eds. C. Davis, B. Grünbaum, P. Sherk), Springer Verlag (1981) 231-242.

137 S.E. Payne, On the structure of translation generalized quadrangles, *Ann. Discrete Math.* 18 (1983) 661-666.

138 S.E. Payne, Collineations of finite generalized quadrangles, in : *Finite geometries* (eds. N.L. Johnson, M.J. Kallaher, C.T. Long), Marcel Dekker AG Verlag, Lecture Notes in Pure and Appl. Math. 82 (1983) 361-390.

139 S.E. Payne and J.E. Conklin, An unusual generalized quadrangle of order sixteen, *J. Combin. Theory* (A) 24 (1978) 50-74.

140 S.E. Payne and R.B. Killgrove, Generalized quadrangles of order sixteen, *Congress. Numer.* XXI, *Proc. 9th S.E. Conf. Comb., Graph Th. and Comp.* (1978) 555-565.

141 S.E. Payne, R.B. Killgrove and D.I. Kiel, Generalized quadrangles as amalgamations of desarguesian planes : the multiplicative case, *Congress. Numer.* XXIV , *Proc. 10th S.E. Conf. Comb., Graph Th. and Comp.* II (1979) 787-794.

142 S.E. Payne and J.A. Thas, Generalized quadrangles with symmetry, Part I, *Simon Stevin* 49 (1975) 3-32.

143 S.E. Payne and J.A. Thas, Generalized quadrangles with symmetry, Part II, *Simon Stevin* 49 (1976) 81-103.

144 S.E. Payne and J.A. Thas, Moufang conditions for finite generalized quadrangles, in : *Finite geometries and designs* (eds. P.J. Cameron, J.W.P. Hirschfeld and D.R. Hughes), Cambridge Univ. Press, London Math. Soc. Lecture Note Ser. 49 (1980) 275-303.

145 S.E. Payne and M.F. Tinsley, On $v_1 \times v_2$ (n,s,t)-configurations, *J. Combin. Theory* 7 (1969) 1-14.

146 N. Percsy, A characterization of classical Minkowski planes over a perfect field of characteristic 2, *J. Geom.* 5 (1974) 191-204.

147 O. Prohaska and M. Walker, A note on the Hering type of inversive planes of even order, *Arch. Math.* 28 (1977) 431-432.

148 M.A. Ronan, Generalized hexagons, Dissertation Ph.D. , University of Oregon 1978.

149 M.A. Ronan, A geometric characterization of Moufang hexagons, *Invent. Math.* 57 (1980) 227-262.

150 M.A. Ronan, Semiregular graph automorphisms and generalized quadrangles, *J. Combin. Theory* (A) 29 (1980) 319-328.

151 M.A. Ronan, A combinatorial characterization of the dual Moufang hexagons, *Geom. Dedicata* 11 (1981) 61-67.

152 M.A. Ronan, A note on the $^3D_4(q)$ generalized hexagons, *J. Combin. Theory* 29 (1980) 249-250.

153 C. Roos, An alternative proof of the Feit-Higman theorem on generalized polygons, *Delft Progr. Rep. Series* F : *Math. Eng.* 5 (1980) 67-77.

154 A. Russo, Calotte hermitiane di un $S_{r,4}$, *Ricerche Mat.* 20 (1971) 297-307.

155 N.S.N. Sastry, Codes, partial geometries and generalized n-gons, preprint.

156 G.L. Schellekens, On a hexagonic structure, Part I, *Indag. Math.* 24 (1962) 201-217.

157 G.L. Schellekens, On a hexagonic structure, Part II, *Indag. Math.* 24 (1962) 218-234.

158 B. Segre, Sulle ovali nei piani lineari finiti, *Rend. Accad. Naz. Lincei* 17 (1954) 141-142.

159 B. Segre, Lectures on modern geometry, Ed. Cremonese Roma 1961.

160 B. Segre, Ovali e curve σ nei piani di Galois di caratteristica due, *Rend. Accad. Naz. Lincei* 32 (1962) 785-790.

161 B. Segre, Forme e geometrie hermitiane, con particolare riguardo al caso finito, *Ann. Mat. Pura Appl.* (4) 70 (1965) 1-202.

162 B. Segre, Introduction to Galois geometries, *Mem. Accad. Naz. Lincei* 8 (1967) 137-236.

163 B. Segre and U. Bartocci, Ovali ed altre curve nei piani di Galois di caratteristica due, *Acta Arith.* 18 (1971) 423-449.

164 J.J. Seidel, Strongly regular graphs with (-1,1,0) adjacency matrix having eigenvalue 3, *Linear Algebra Appl.* 1 (1968) 281-298.

165 S. Shad and E.E. Shult , The near n-gon geometries, preprint.

166 E.E. Shult, Characterizations of certain classes of graphs, *J. Combin. Theory* (B) 13 (1972) 142-167.

167 E.E. Shult and A. Yanushka, Near n-gons and line systems, part I, *Geom. Dedicata* 9 (1980) 1-72.

168 R.R. Singleton, Minimal regular graphs of maximal even girth, *J. Combin. Theory* 1 (1966) 306-332.

169 C. Somma, Generalized quadrangles with parallelism, *Ann. Discrete Math.* 14 (1982) 265-282.

170 A.P. Sprague, A characterization of 3-nets, *J. Combin. Theory* (A) 27 (1979) 223-253.

171 J.J. Sylvester, 1844, Collected Math. Papers 1, Cambridge Univ. Press 1904.

172 J.J. Sylvester, Note sur l'involution de six lignes dans l'espace, *Comptes Rendus Hebdomadaires des Séances* 52 (1861) 815-817; Reprinted in *Collected Math. Papers* 2, Cambridge Univ. Press 1908.

173 G. Tallini, Sistemi grafici rigati, *Ist. Mat. Univ. Napoli Rel.* n. 8 (1971).

174 G. Tallini, Strutture di incidenza dotate di polarità, *Rend. Sem. Mat. Fis. Milano* 41 (1971) 3-42.

175 G. Tallini, Ruled graphic systems, in : *Atti convegno di geometria e sue applicazioni* (ed. A. Barlotti), Perugia (1971) 385-393.

176 G. Tallini, Problemi e risultati sulle geometrie di Galois, *Ist. Mat. Univ. Napoli Rel.* n. 30 (1973).

177 G. Tallini, Graphic characterization of algebraic varieties in a Galois space, in : *Teorie combinatorie* (ed. B. Segre), Roma Accad. Naz. Lincei (1976) 153-165.

178 M. Tallini Scafati, Caratterizzazione grafica delle forme hermitiane di un $S_{r,q}$, *Rend. Mat.* 26 (1967) 273-303.

179 J.A. Thas, The m-dimensional projective space $S_m(M_n(GF(q)))$ over the total matrix algebra $M_n(GF(q))$ of the $n \times n$-matrices with elements in the Galois field GF(q), *Rend. Mat.* (6) 4 (1971) 459-532.

180 J.A. Thas, 4-gonal subconfigurations of a given 4-gonal configuration, *Rend. Accad. Naz. Lincei* 53 (1972) 520-530.

181 J.A. Thas, Ovoidal translation planes, *Arch. Math.* 23 (1972) 110-112.

182 J.A. Thas, Flocks of finite egglike inversive planes, in : *Finite geometric structures and their applications* (ed. A. Barlotti), Ed. Cremonese Roma (1973) 189-191.
183 J.A. Thas, A combinatorial problem, *Geom. Dedicata* 1 (1973) 236-240.
184 J.A. Thas, On 4-gonal configurations, *Geom. Dedicata* 2 (1973) 317-326.
185 J.A. Thas, 4-gonal configurations, in : *Finite geometric structures and their applications* (ed. A. Barlotti), Ed. Cremonese Roma (1973) 251-26
186 J.A. Thas, On semi ovals and semi ovoids, *Geom. Dedicata* 3 (1974) 229-231.
187 J.A. Thas, Geometric characterization of the [n-1]-ovaloids of the projective space PG(4n-1,q), *Simon Stevin* 47 (1974) 97-106.
188 J.A. Thas, On 4-gonal configurations with parameters $r = q^2+1$ and $k = q+1$, part I, *Geom. Dedicata* 3 (1974) 365-375.
189 J.A. Thas, Translation 4-gonal configurations, *Rend. Accad. Naz. Lincei* 56 (1974) 303-314.
190 J.A. Thas, A remark concerning the restriction on the parameters of a 4-gonal subconfiguration, *Simon Stevin* 48 (1974-75) 65-68.
191 J.A. Thas, Flocks of non-singular ruled quadrics in PG(3,q), *Rend. Accad. Naz. Lincei* 59 (1975) 83-85.
192 J.A. Thas, 4-gonal configurations with parameters $r = q^2+1$ and $k = q+1$, part II, *Geom. Dedicata* 4 (1975) 51-59.
193 J.A. Thas, Combinatorics of partial geometries and generalized quadrangles, in : *Higher combinatorics* (ed. M. Aigner), Reidel Publ. Comp. NATO adv. Study Inst. Ser., Ser. C : Math. Phys. Sci. (1976) 183-199.
194 J.A. Thas, On generalized quadrangles with parameters $s = q^2$ and $t = q$ *Geom. Dedicata* 5 (1976) 485-496.
195 J.A. Thas, Combinatorial characterizations of the classical generalized quadrangles, *Geom. Dedicata* 6 (1977) 339-351.
196 J.A. Thas, Partial geometries in finite affine spaces, *Math. Z.* 158 (1978) 1-13.
197 J.A. Thas, Combinatorial characterizations of generalized quadrangles with parameters $s = q$ and $t = q^2$, *Geom. Dedicata* 7 (1978) 223-232.
198 J.A. Thas, Geometries in finite projective and affine spaces, in : *Surveys in combinatorics* (ed. B. Bollobàs), Cambridge Univ. Press, London Math. Soc. Lecture Note Ser. 38 (1979) 181-211.

199 J.A. Thas, Generalized quadrangles satisfying at least one of the Moufang conditions, *Simon Stevin* 53 (1979) 151-162.

200 J.A. Thas, Polar spaces, generalized hexagons and perfect codes, *J. Combin. Theory* (A) 29 (1980) 87-93.

201 J.A. Thas, A remark on the theorem of Yanushka and Ronan characterizing the "Generalized Hexagon" H(q) arising from the group $G_2(q)$, *J. Combin. Theory* (A) 29 (1980) 361-362.

202 J.A. Thas, Partial three-spaces in finite projective spaces, *Discrete Math.* 32 (1980) 299-322.

203 J.A. Thas, Ovoids and spreads of finite classical polar spaces, *Geom. Dedicata* 10 (1981) 135-144.

204 J.A. Thas, New combinatorial characterizations of generalized quadrangles, *European J. Combin.* 2 (1981) 299-303.

205 J.A. Thas, Combinatorics of finite generalized quadrangles : a survey, *Ann. Discrete Math.* 14 (1982) 57-76.

206 J.A. Thas, Semi partial geometries and spreads of classical polar spaces, *J. Combin. Theory* 35 (1983) 58-66.

207 J.A. Thas, Ovoids and spreads of generalized quadrangles, postgraduate course, Univ. of Ghent 1981.

208 J.A. Thas, Extensions of finite generalized quadrangles, in : *Symp. Math.*, Academic Press, to appear.

209 J.A. Thas, 3-regularity in generalized quadrangles of order (s,s^2), preprint.

210 J.A. Thas, Characterization of generalized quadrangles by generalized homologies, *J. Combin. Theory*, to appear.

211 J.A. Thas, The theorems of Dembowski and Heise-Percsy from the point of view of generalized quadrangles, preprint.

212 J.A. Thas and P. De Winne, Generalized quadrangles in finite projective spaces, *J. Geom.* 10 (1977) 126-137.

213 J.A. Thas and S.E. Payne, Classical finite generalized quadrangles : a combinatorial study, *Ars Combin.* 2 (1976) 57-110.

214 J.A. Thas and S.E. Payne, Generalized quadrangles and the Higman-Sims technique, *European J. Combin.* 2 (1981) 79-89.

215 J.A. Thas and S.E. Payne, Moufang conditions for finite generalized quadrangles, *Ann. Discrete Math.* 18 (1983) 745-752.

216 J. Tits, Sur la trialité et certains groupes qui s'en déduisent, *Inst. Hautes Etudes Sci. Publ. Math.* 2 (1959) 14-60.

217 J. Tits, Ovoides et groupes de Suzuki, *Arch. Math.* 13 (1962) 187-198.

218 J. Tits, Géométries polyédriques finies, *Rend. Mat.* 23 (1964) 156-165.

219 J. Tits, Buildings and BN-pairs of spherical type, Springer Verlag, Lecture Notes in Math. n. 386 (1974).

220 J. Tits, Classification of buildings of spherical type and Moufang polygons : a survey, in : *Teorie combinatorie* (ed. B. Segre), Roma Accad. Naz. Lincei (1976) 229-246.

221 J. Tits, Non-existence de certains polygons généralisés I, *Invent. Math.* 36 (1976) 275-284.

222 J. Tits, Non-existence de certains polygons généralisés II, *Invent. Math.* 51 (1979) 267-269.

223 J. Tits, Quadrangles de Moufang, I, preprint, 1976.

224 J. Tits, Moufang octagons and Ree groups of type $^2F_4$, *Invent. Math.*, to appear.

225 O. Veblen and J.W. Young, Projective geometry, 2 vol's, Ginn & Co, Boston 1916.

226 F.D. Veldkamp, Polar geometry, Dissertation Ph.D., University of Utrecht 1959.

227 F.D. Veldkamp, Polar geometry I-IV, *Indag. Math.* 21 (1959) 512-551.

228 F.D. Veldkamp, Polar geometry V, *Indag. Math.* 22 (1960) 207-212.

229 M. Walker, On the structure of finite collineation groups containing symmetries of generalized quadrangles, *Invent. Math.* 40 (1977) 245-265.

230 M. Walker, On central root automorphisms of finite generalized hexagons, *J. Algebra* 78 (1982) 303-340.

231 M. Walker, On generalized hexagons which admit a fundamental system of root automorphisms, *Proc. London Math. Soc.* (3) 45 (1982) 281-299.

232 M. Walker, On central root automorphisms of finite generalized octagons *European J. Combin.* 4 (1983) 65-86.

233 R. Weiss, The non-existence of certain Moufang polygons, *Invent. Math.* 51 (1979) 261-266.

234 D.J.A. Welsh, Matroid theory, Academic Press, London Math. Soc. Monographs no. 8 (1976).

235 H.A. Wilbrink, A characterization of the classical unitals, in : *Finite geometries* (eds. N.L. Johnson, M.J. Kallaher, C.T. Long),Marcel Dekker AG Verlag, Lecture Notes in Pure and Appl. Math. 82 (1983).

236 A. Yanushka, Generalized hexagons of order (t,t), *Israel J. Math.* 23 (1976) 309-324.

237 A. Yanushka, On order in generalized polygons, *Geom. Dedicata* 10 (1981) 451-458.

Index of names

* It seemed appropriate not to list citations of works whose authors formed a subset of {S.E. Payne, J.A. Thas}.

General index